AF615213

Lecture Notes in Physics

Springer
Berlin
Heidelberg
New York
Barcelona
Budapest
Hong Kong
London
Milan
Paris
Santa Clara
Singapore
Tokyo

The Editorial Policy for Proceedings

The series Lecture Notes in Physics reports new developments in physical research and teaching – quickly, informally, and at a high level. The proceedings to be considered for publication in this series should be limited to only a few areas of research, and these should be closely related to each other. The contributions should be of a high standard and should avoid lengthy redraftings of papers already published or about to be published elsewhere. As a whole, the proceedings should aim for a balanced presentation of the theme of the conference including a description of the techniques used and enough motivation for a broad readership. It should not be assumed that the published proceedings must reflect the conference in its entirety. (A listing or abstracts of papers presented at the meeting but not included in the proceedings could be added as an appendix.)
When applying for publication in the series Lecture Notes in Physics the volume's editor(s) should submit sufficient material to enable the series editors and their referees to make a fairly accurate evaluation (e.g. a complete list of speakers and titles of papers to be presented and abstracts). If, based on this information, the proceedings are (tentatively) accepted, the volume's editor(s), whose name(s) will appear on the title pages, should select the papers suitable for publication and have them refereed (as for a journal) when appropriate. As a rule discussions will not be accepted. The series editors and Springer-Verlag will normally not interfere with the detailed editing except in fairly obvious cases or on technical matters.
Final acceptance is expressed by the series editor in charge, in consultation with Springer-Verlag only after receiving the complete manuscript. It might help to send a copy of the authors' manuscripts in advance to the editor in charge to discuss possible revisions with him. As a general rule, the series editor will confirm his tentative acceptance if the final manuscript corresponds to the original concept discussed, if the quality of the contribution meets the requirements of the series, and if the final size of the manuscript does not greatly exceed the number of pages originally agreed upon. The manuscript should be forwarded to Springer-Verlag shortly after the meeting. In cases of extreme delay (more than six months after the conference) the series editors will check once more the timeliness of the papers. Therefore, the volume's editor(s) should establish strict deadlines, or collect the articles during the conference and have them revised on the spot. If a delay is unavoidable, one should encourage the authors to update their contributions if appropriate. The editors of proceedings are strongly advised to inform contributors about these points at an early stage.
The final manuscript should contain a table of contents and an informative introduction accessible also to readers not particularly familiar with the topic of the conference. The contributions should be in English. The volume's editor(s) should check the contributions for the correct use of language. At Springer-Verlag only the prefaces will be checked by a copy-editor for language and style. Grave linguistic or technical shortcomings may lead to the rejection of contributions by the series editors. A conference report should not exceed a total of 500 pages. Keeping the size within this bound should be achieved by a stricter selection of articles and not by imposing an upper limit to the length of the individual papers. Editors receive jointly 30 complimentary copies of their book. They are entitled to purchase further copies of their book at a reduced rate. As a rule no reprints of individual contributions can be supplied. No royalty is paid on Lecture Notes in Physics volumes. Commitment to publish is made by letter of interest rather than by signing a formal contract. Springer-Verlag secures the copyright for each volume.

The Production Process

The books are hardbound, and the publisher will select quality paper appropriate to the needs of the author(s). Publication time is about ten weeks. More than twenty years of experience guarantee authors the best possible service. To reach the goal of rapid publication at a low price the technique of photographic reproduction from a camera-ready manuscript was chosen. This process shifts the main responsibility for the technical quality considerably from the publisher to the authors. We therefore urge all authors and editors of proceedings to observe very carefully the essentials for the preparation of camera-ready manuscripts, which we will supply on request. This applies especially to the quality of figures and halftones submitted for publication. In addition, it might be useful to look at some of the volumes already published. As a special service, we offer free of charge LaTeX and TeX macro packages to format the text according to Springer-Verlag's quality requirements. We strongly recommend that you make use of this offer, since the result will be a book of considerably improved technical quality. To avoid mistakes and time-consuming correspondence during the production period the conference editors should request special instructions from the publisher well before the beginning of the conference. Manuscripts not meeting the technical standard of the series will have to be returned for improvement.

For further information please contact Springer-Verlag, Physics Editorial Department II, Tiergartenstrasse 17, D-69121 Heidelberg, Germany

Hans Hippelein Klaus Meisenheimer
Hermann-Josef Röser (Eds.)

Galaxies in the Young Universe

Proceedings of a Workshop
Held at Ringberg Castle,
Tegernsee, Germany, 22–28 September 1994

Springer

Editors

Hans Hippelein
Klaus Meisenheimer
Hermann-Josef Röser
Max-Planck-Institut für Astronomie
Königstuhl, D-69117 Heidelberg, Germany

Cataloging-in-Publication Data applied for.

Die Deutsche Bibliothek - CIP-Einheitsaufnahme

Galaxies in the young universe : proceedings of a workshop held at Ringberg Castle, Tegernsee, Germany, 22 - 28 September 1994 / Hans Hippelein ... (ed.). - Berlin ; Heidelberg ; New York ; Barcelona ; Budapest ; Hong Kong ; London ; Milan ; Paris ; Santa Clara ; Singapore ; Tokyo : Springer, 1995
(Lecture notes in physics ; Vol. 463)
ISBN 3-540-60586-X
NE: Hippelein, Hans [Hrsg.]; GT

ISBN 3-540-60586-X Springer-Verlag Berlin Heidelberg New York

Printed in Germany

Typesetting: Camera-ready by the authors
SPIN: 10515235 55/3142-543210 - Printed on acid-free paper

Surroundings of the weak radio galaxy 53W002 at $z = 2.390$. The color plate shows a composite of 12 orbits in both the V_{606} and I_{814} filters taken with the HST Wide Field Planetary Camera 2. See Windhorst *et al.*, p.265.

(Color image from Driver *et al.* 1995, *Astrophys. J.* **449**, L23).

Preface

The problem of the formation and early evolution of galaxies is certainly one of the most exciting topics in astronomy today. With the recent developments in observational capabilities – a new generation of large telescopes and an increased access to a wider range of the electromagnetic spectrum, efficient detectors, Hubble Space Telescope, *etc.* – the direct study of galaxies in their first stages of evolution seems to come into reach. But also the theoretical background has evolved rapidly due to model simulations on fast computers and constraints set by important observational facts, such as the smoothness of the cosmic microwave background, deep galaxy counts and the detection of quasars at very high redshift.

Despite strong observational efforts no genuine galaxy has been detected in its formation phase so far. One problem is that we do not exactly know what to look for and it seems that only extensive and deep surveys can bring progress in the problem of galaxy formation and early evolution. This led us to the idea to organize a workshop on the subject of galaxies in the early universe, with the aim to bring together a selection of leading scientists, theorists and observers in order to learn from each other and to optimize the methods of these surveys. The remote location of the Ringberg Castle in Bavaria provided an ideal atmosphere for such a workshop.

The meeting was organized by H. Hippelein (Max-Planck-Institut für Astronomie), R. Kron (Yerkes Observatory), K. Meisenheimer and H.-J. Röser (both Max-Planck-Institut für Astronomie), and J. Silk (Univ. of California, Berkeley).

A generous grant from the Max-Planck Gesellschaft made it possible to support invited speakers and several other participants. The organizers would like to thank Mr. Hörmann and the staff at the castle for their hospitality and excellent support which made the stay at Ringberg castle so enjoyable.

Heidelberg
September 1995

Hans Hippelein
Klaus Meisenheimer
Hermann-Josef Röser

Contents

VI Survey Activities

List of participants

Name	Institution	Address	Country
Beckwith Steven V.W.	MPI für Astronomie	Königstuhl 17 69117 Heidelberg	Germany
Belloni Paola	Universitäts-Sternwarte	Scheinerstr. 1 89679 München	Germany
Bender Ralf	Universitäts-Sternwarte	Scheinerstr. 1 81679 München	Germany
Bershady Matthew A.	Penn State University Astronomy Dept.	525 Davey Lab. PA 16802	USA
Börner Gerhard	MPI für Astrophysik	Schwarzschild-Str. 1 85748 Garching	Germany
Broadhurst Tom	John Hopkins Univ.	3400N Charles St. Baltimore, MD 21218	USA
Bruzual Gustavo	C.I.D.A.	Apartado Postal 264 Mérida	Venezuela
Burkert Andreas	MPI für Astrophysik	Schwarzschild-Str. 1 85748 Garching	Germany
Carlberg Ray G.	Dept. of Astronomy University of Toronto	Toronto M5S 1A7	Canada
Charlot Stephane	Astronomy Department University of California	Berkeley, CA 94720	USA
Condon James J.	NRAO Charlottesville	520 Edgemont Road VA 22903	USA
Dickinson Mark	STScI	San Martin Drive Baltimore, MD 21218	USA
Dunlop James S.	Institute of Astronomy Univ. of Edinburgh	Blackford Hill Edinburgh EH9 3HJ	UK
Efstathiou George	Department of Physics University of Oxford	Keble Rd. Oxford OX1 3RH	UK
Ellis George F.R.	Applied Mathem. Dept. Cape Town University	Rondebosch 7700	South- Africa
Gehren Thomas	Universitäts-Sternwarte	Scheinerstr. 1 81679 München	Germany
Hähnelt Martin	MPI für Astrophysik	Schwarzschild-Str. 1 85748 Garching	Germany
Hamilton Donald	MPI für Astronomie	Königstuhl 17 69117 Heidelberg	Germany
Hammer Francois	DAEC Observatoire de Meudon	92195 Meudon Principal Cedex	France
Henkel Christian	MPI Radioastronomie	Auf dem Huegel 69 53121 Bonn	Germany
Hensler Gerhard	Inst. für Astrophysik Universität Kiel	Olshausenstr. 40 24098 Kiel	Germany

Hippelein Hans	MPI für Astronomie	Königstuhl 17 69117 Heidelberg	Germany
Hunstead Richard	School of Physics University of Sydney	Sydney, NSW 2006	Australia
Kauffmann Guinevere	MPI für Extraterr. Physik	85748 Garching	Germany
Kron Richard G.	Fermilab	PO Box 500 Batavia, IL 60510	USA
Meisenheimer Klaus	MPI für Astronomie	Königstuhl 17 69117 Heidelberg	Germany
Möller Palle	Space Tel. Science Institute	San Martin Drive Baltimore, MD 21218	USA
Neeser Mark	MPI für Astronomie	Königstuhl 17 69117 Heidelberg	Germany
Peacock John A.	Royal Observatory	Blackfors Hill Edinburgh EH9 3HJ	Scotland
Rawlings Steve	Astrophysics Dept Univers. of Oxford	Keble Road Oxford OX1 3RH	UK
Rocca- Volmerange Brigitte	Institut d'Astrophys. de Paris	98bis Boulev. Arago 75014 Paris	France
Röser Hermann-J.	MPI für Astronomie	Königstuhl 17 69117 Heidelberg	Germany
Röttgering Huub	Cavendish Laboratory MRAO Cambridge	Madingley Road Cambridge CB3 0HE	UK
Schneider Peter	MPI für Astrophysik	Schwarzschild-Str. 1 85748 Garching	Germany
Silk Joseph	Univ. of California	Berkeley CA 94720	USA
Spinrad Hyron	Univ. of Califonlia	Berkeley CA 94720	USA
Steinmetz Matthias	MPI für Astrophysik	Schwarzschild-Str. 1 85748 Garching	Germany
Theis Christian	Inst. für Astrophysik Universität Kiel	Olshausenstr. 40 24098 Kiel	Germany
Thommes Eduard	MPI für Astronomie	Königstuhl 17 69117 Heidelberg	Germany
Tresse Laurence	DAEC Observatoire de Meudon	92195 Meudon Principal Cedex	France
Turner Michael S.	Department of Physics University of Chicago	Chicago IL 60637-1433	USA
White Simon	MPI für Astrophysik	Schwarzschild-Str. 1 85748 Garching	Germany
Williger Gerry M.	MPI für Astronomie	Königstuhl 17 69117 Heidelberg	Germany

Windhorst Rogier A.	Department of Physics Arizona State University	Box 871504 Tempe AZ 85287-1504	USA

The Hot Big Bang and Beyond

Michael S. Turner[1,2]

[1] Departments of Physics and of Astronomy & Astrophysics
Enrico Fermi Institute, The University of Chicago, Chicago, IL 60637-1433
[2] NASA/Fermilab Astrophysics Center
Fermi National Accelerator Laboratory, Batavia, IL 60510-0500

1 Overview

The hot big-bang cosmology provides a reliable account of the Universe from 10^{-2} sec after the bang until the present, as well as a robust framework for speculating back to times as early as 10^{-43} sec. Cosmology faces a number of important challenges; foremost among them are determining the quantity and composition of dark matter in the Universe and developing a detailed and coherent picture of how structure (galaxies, clusters of galaxies, superclusters, voids, great walls, and so on) developed. At present there is a working hypothesis—cold dark matter—which squarely addresses both issues. According to the cold dark matter theory, which is motivated by inflation, the Universe is flat, the density perturbations are almost scale invariant, and the bulk of the dark matter is in the form of slowly moving particles left over from the earliest moments (e.g., neutralinos or axions). If correct, cold dark matter would extend the big-bang model back to 10^{-32} sec and shed light on the unification of the forces. Many experiments and observations, from CBR anisotropy measurements to Hubble Space Telescope observations to experiments at Fermilab and CERN, are now putting the cold dark matter theory to the test. At present it appears that the theory is viable only if the Hubble constant is smaller than current measurements indicate (around $35\,\mathrm{km\,s^{-1}\,Mpc^{-1}}$), or if the theory is modified slightly, e.g., by the addition of a cosmological constant, a small admixture of hot dark matter (5 eV "worth of neutrinos"), more relativistic particles, or a tilted spectrum of density perturbations.

2 Successes

The success of the hot big-bang cosmology (or standard cosmology as it is known) is simple to describe: It provides a reliable and tested account of the Universe from a fraction of a second after the bang (temperatures of order a few MeV) until the present 15 Billion years later (temperature 2.726 K). When supplemented by the Standard Model of particle physics and various ideas about physics at higher energies (e.g., supersymmetry, grand unification, and superstrings) it provides a sound foundation for speculations about the Universe back to 10^{-43} sec after the bang (temperatures of 10^{19} GeV) and perhaps even earlier (for textbooks on modern cosmology see e.g., E.W. Kolb and M.S. Turner, *The Early Universe*, Addison-Wesley, Redwood City, CA, 1990, or P.J.E. Peebles, *Principles of Physical Cosmology*, Princeton University Press, Princeton, NJ, 1993).

The fundamental observational data that support the standard cosmology are: the universal expansion (Hubble flow of galaxies); the cosmic background radiation (CBR); and the abundance of the light elements D, ^{3}He, ^{4}He, and ^{7}Li. The Hubble law ($z \simeq v/c \simeq H_0 d$) has been tested to a redshift $z \sim 0.05$ (see e.g. Mould *et al.* 1991) and the highest redshift object is a QSO with $z = 4.90$. (One plus redshift is the size of the Universe today relative to its size at the time of emission, $1 + z = R_0/R_E$; R is the cosmic scale factor).

The surface of last scattering for the CBR is the Universe at an age of a few hundred thousand years ($T \sim 0.3$ eV and redshift $z \sim 1100$). COBE has determined its temperature to be 2.726 ± 0.005 K and constrains any deviations from a black-body spectrum to be less than 0.03% (Mather *et al.* 1994). The CBR temperature is very uniform: the difference between two points separated by angles from arcminutes to 90° is less than $300\mu K$, indicating that the Universe had a very smooth beginning.

According to the big-bang model the temperature of the CBR decreases as the Universe expands, and a recent measurement has confirmed this (Songaila *et al.* 1994). The relative populations of hyperfine states in neutral Carbon atoms seen in a gas cloud at redshift $z = 1.776$ indicated a thermodynamic temperature, 7.4 ± 0.8 K, which is consistent with the big-bang prediction for the CBR temperature at this earlier time $T(z) = (1 + z)2.726\,\mathrm{K} = 7.58\,\mathrm{K}$.

There is a dipole anisotropy in the CBR temperature of about 3 mK, due to our motion with respect to the cosmic rest frame (the "peculiar velocity" of the Local Group is $620\,\mathrm{km\,s^{-1}}$ toward the constellation Leo), and temperature differences on angular scales from 0.5° to 90° have been detected by about ten experiments at the level of about 30μK (see e.g., White *et al.* 1994).

The abundance of the light elements, which range from about 24% for ^{4}He to 10^{-5} for D and ^{3}He and 10^{-10} for ^{7}Li are consistent with the predictions of the hot big-bang model. The comparison between the predicted abundances and the light-element abundances measured today is not a simple matter; it is complicated by 15 Gyr of "chemical evolution" (astrophysical processes destroy D, produce ^{4}He, and destroy or produce ^{3}He and ^{7}Li). However, three decades of careful theoretical and observational work has put the comparison on a firm footing, and there is good agreement provided that the ratio of baryons to photons is between 2.5×10^{-10} and 6×10^{-10} (Copi *et al.* 1995); see Fig. 1. Since the synthesis of the light elements occurred when the Universe was of order seconds old and the temperature was of order MeV, big-bang nucleosynthesis is the earliest and perhaps most impressive test of the standard cosmology.

Finally, the standard cosmology provides a general framework for understanding how the very smooth early Universe evolved to the highly structured Universe today—galaxies, clusters of galaxies, superclusters, voids, great walls

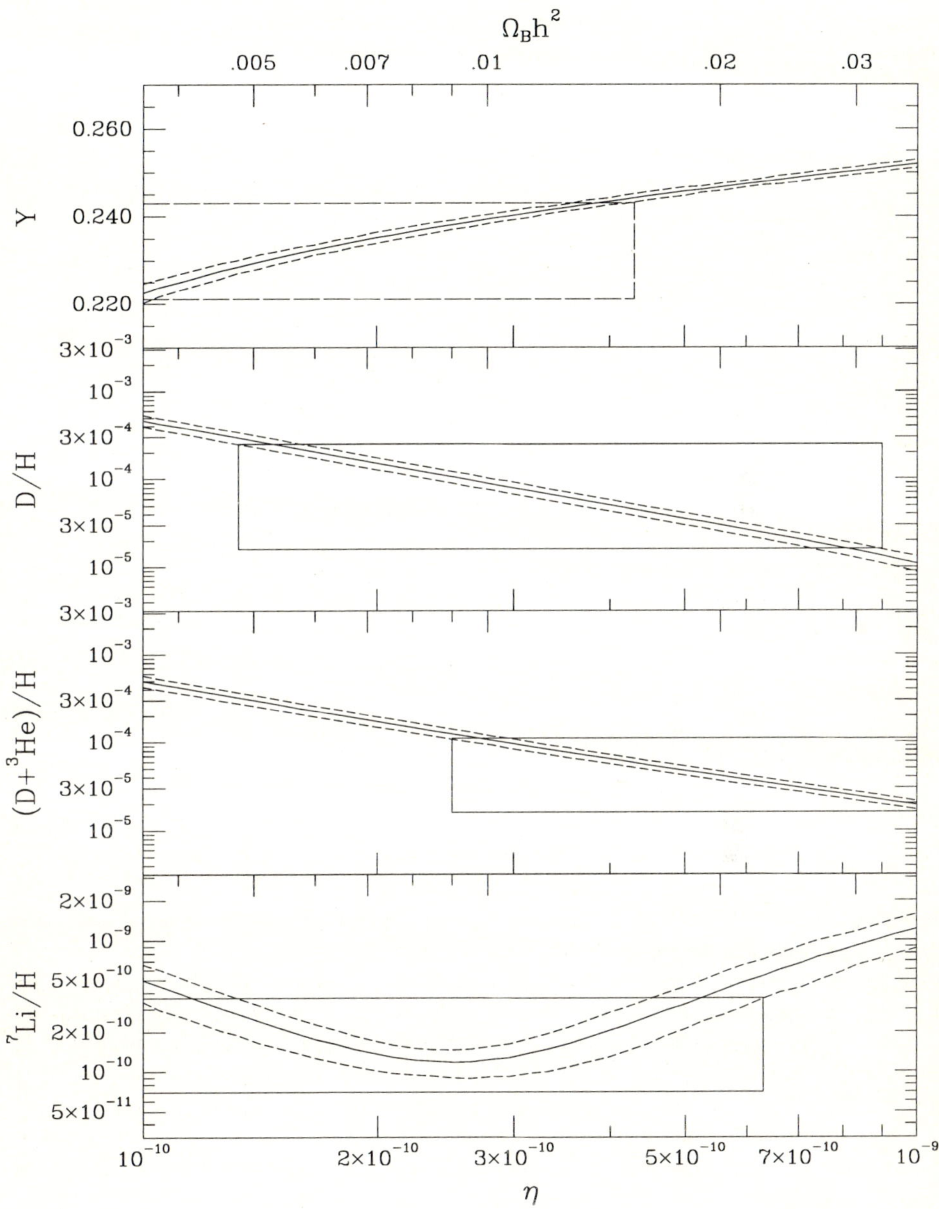

Fig. 1. The predicted light-element abundances as a function of the baryon-to-photon ratio η and $\Omega_B h^2$. The broken curves delineate the two-sigma theoretical uncertainties and the boxes delineate the acceptable range for η as allowed by measured abundances. The predictions for all four light elements are consistent with the observations for $\eta \simeq 2.5 \times 10^{-10} - 6 \times 10^{-10}$. (Figure courtesy of C. Copi.)

and so on. Small (primeval) variations in the matter density ($\delta\rho/\rho \sim 10^{-5}$) were amplified by gravity over the age of the Universe (the Jeans' instability in the expanding Universe) eventually resulting in the structure seen today (see e.g. Efstathiou 1992). The CBR temperature fluctuations detected on angular scales from 0.5° to 90° are strong evidence for the existence of these primeval density fluctuations; see Fig. 2.

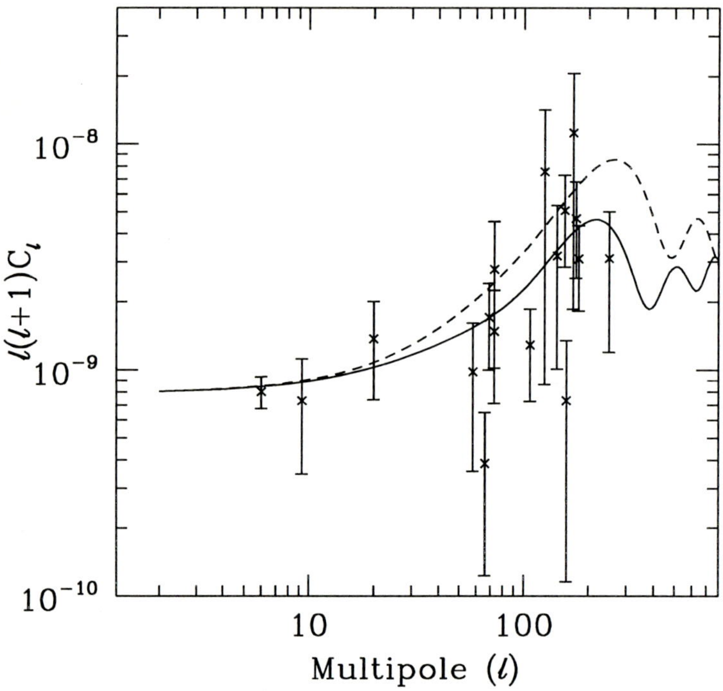

Fig. 2. Summary of current measurements of CBR anisotropy in terms of a spherical-harmonic decomposition, $C_l \equiv \langle |a_{lm}|^2 \rangle$. The rms temperature fluctuation measured between two points separated by an angle θ is roughly given by: $(\delta T/T)_\theta \simeq \sqrt{l(l+1)C_l}$ with $l \simeq 200°/\theta$. The curves are the cold dark matter predictions, normalized to the COBE detection, for Hubble constants of $50\,\mathrm{km\,s^{-1}}$ Mpc (solid) and $35\,\mathrm{km\,s^{-1}\,Mpc^{-1}}$ (broken). (Figure courtesy of M. White.)

According to the Standard Model of particle physics the fundamental particles are point-like quarks and leptons whose interactions are weak enough to treat perturbatively. The cosmological implications of this are profound: The Universe at temperatures greater than about 150 MeV (times earlier than 10^{-5} sec) consisted of a hot, dilute gas of quarks, leptons, and gauge bosons (photons, gluons, and at high enough temperatures W and Z bosons, the carriers of the electromagnetic, strong and weak forces). The Standard Model of particle physics, which has been tested up to energies of several hundred GeV, supplies the input microphysics needed for times as early as 10^{-11} sec. In addition, it provides a firm platform for speculations about the unification of forces and particles (e.g., supersymmetry and grand unification), and in turn, for extending cosmological speculations back to the Planck epoch. Earlier than the Planck time a quantum description of gravity is needed, and superstring theory is a good candidate for such.

While the hot big-bang cosmology and modern particle theory allow "sensible" — and very interesting — speculations about the early Universe, there is no evidence *yet* that any of these speculations is correct. However, contrast this with the situation before the early 1970s. The count of "elementary particles" (baryons and mesons) had exceeded 100 and was growing exponentially with mass; this, the strength of their interactions and their finite sizes precluded any sensible speculation about the Universe at times earlier than about 10^{-5} sec (see e.g. Weinberg 1972).

3 Challenges

Cosmology is not without its challenges. Because of its success, the hot big-bang model has allowed cosmologists to ask even deeper questions. They include: What is the quantity and composition of the ubiquitous dark matter? What is the nature and origin of the primeval density perturbations that seeded structure and precisely how did the structure arise? What is the origin of the cosmic asymmetry between matter and antimatter? Why is the observed portion of the Universe so smooth and flat? And does this mean that the entire Universe is the same? Are there observational consequences of the phase transitions that the Universe has undergone (transition from quarks to nucleons and related particles, electroweak symmetry breaking, and possibly others) during its earliest moments? Are there observable consequences of the quantum-gravity epoch? Why does the Universe have four dimensions? What caused the expansion in the first place?

The first two of these challenges, the nature of the dark matter and the details of structure formation, are in my opinion the most pressing—and could well be resolved soon. They offer an excellent opportunity for extending the big-bang cosmology back to much earlier times.

That is not to say that the other challenges are not important or do not have potential for advancing our understanding. In addition, there are "important practical problems;" for example, a precise determination of the three traditional parameters used to describe "our world model," the Hubble constant, the deceleration parameter, and the cosmological constant, or an explanation for the primeval magnetic fields required to seed the magnetic fields seen throughout the Universe today.

3.1 Discard the Big Bang?

There are a few who believe the big bang faces challenges of such enormity that they will led to its downfall (see e.g. Arp *et al.* 1990). One challenge involves the tension between the Hubble constant and stellar ages. If the Hubble constant is as large as some determinations indicate, say around $80\,\mathrm{km\,s^{-1}\,Mpc^{-1}}$ and the oldest stars are as old as some determinations indicate, say around 16 Gyr, then a real dilemma exists because *without recourse to a cosmological constant* the time back to the bang is less than 12 Gyr (Jacoby *et al.* 1992; Fukugita *et al.* 1993; Freedman *et al.* 1994). At present, the uncertainties in both the Hubble constant and stellar ages preclude drawing any firm conclusions.

Before the COBE discovery of CBR anisotropy in 1992 (Smoot *et al.* 1992), some had argued that the absence of anisotropy precluded inhomogeneity of a size large enough to seed the structure seen today. The big-bang has weathered that storm: Fluctuations in the CBR temperature have been detected and are now seen on scales from $0.5°$ to $90°$; see Fig. 2. In fact, careful calculations indicate that if anything the level of temperature fluctuations seen is slightly larger than expected (Ostriker 1993; White *et al.* 1994).

The only two competitors to the big bang are the quasi steady-state model (Hoyle *et al.* 1993, 1994). and the plasma-universe model. At the moment the problems that these models face seem far more daunting: thermalization of starlight to produce the 2.726 K black-body background with no spectral distortion (quasi steady-state) and the formulation of a model definite enough to be tested (plasma universe). Until these models (or another model) can account for the cosmological data that have been firmly established (expansion, CBR, light elements, and structure formation), the standard cosmology is without a serious competitor.

3.2 Dark Matters

An accurate inventory of matter in the Universe still eludes cosmologists. What we do know is: (i) luminous matter (i.e., matter closely associated with bright stars) contributes a fraction of the critical density that is about $0.003h^{-1}$ (Faber & Gallagher 1979); (ii) based upon big-bang nucleosynthesis baryons contributions a fraction of critical density between $0.009h^{-2}$ and $0.022h^{-2}$ (Copi *et al.* 1995), which for a generous range of the Hubble constant corresponds to between about 0.01 and 0.15 of the critical density; (iii) there are indications that the fraction of critical density contributed by all forms of matter is *at least* $0.1-0.3$ (Trimble 1987) —flat rotation curves of spiral galaxies, virial mass determinations of rich clusters—and perhaps around the critical density—the peculiar motions of galaxies, cluster-mass determinations based upon gravitational lensing and x-ray measurements (Trimble 1987; see e.g. Kaiser *et al.* 1991, Strauss *et al.* 1992, Dekel 1994). (Here the Hubble constant $H_0 = 100h\,\mathrm{km\,s^{-1}\,Mpc^{-1}}$ and the critical density $\rho_{\rm crit} = 3H_0^2/8\pi G = 1.88h^2 \times 10^{-29}\,\mathrm{g\,cm^{-3}} \simeq 1.05h^2 \times 10^4\,\mathrm{eV\,cm^{-3}}$.)

From this one concludes that: (i) most of the matter in the Universe is dark; (ii) most of the baryons are dark; (iii) the dark matter is not closely associated with bright stars, i.e., it is more diffusely distributed, e.g., in the extended halos of spiral galaxies; and (iv) if the total mass density is greater than about 20% of the critical density, then there must be another form of matter since baryons can at most account for 15% of the critical density (and only for a very low value of the Hubble constant: if $h \geq 0.6$, then $\Omega_B \lesssim 5\%$).

The case for $\Omega_0 \gtrsim 0.2$ and nonbaryonic dark matter receives additional support, albeit indirectly, from other lines of reasoning. First, it is difficult to reconcile all the data concerning the formation of structure in the Universe with a theory without nonbaryonic dark matter (the one model that *may* be able to do so is Peebles' primeval baryon isocurvature model or PBI; Peebles 1987a, b). Second, the most compelling and comprehensive theory of the early Universe, inflation (Guth 1981), predicts a flat Universe (total energy density equal to the critical density) and thus requires something other than baryons. Third, since the deviation of Ω from unity grows with time, if Ω_0 is not equal to unity, the epoch when Ω_0 just begins to deviate significantly from one is a special epoch and is today(!) (this is often called the Dicke-Peebles timing argument).

Last but not least, there are three compelling candidates for the nonbaryonic dark matter: an axion of mass between 10^{-6} eV and 10^{-4} eV; a neutralino of mass between 10 GeV and 1000 GeV; and a light neutrino species of mass between 10 eV and about 50 eV (Turner 1993). By compelling, I mean these particles arose out of efforts to unify the forces of Nature, and the fact that a particle was predicted whose relic mass density is close to critical is a bonus. This may be the "Great Hint"—or the "Grand Misdirection" if it proves to be wrong.

For the axion, the underlying particle physics is Peccei-Quinn symmetry which is the most attractive solution to the so-called strong-CP problem (the fact that Standard Model of particle physics predicts the electric dipole moment of the neutron to be almost ten orders of magnitude larger than the current upper limit). For the neutralino, it is supersymmetry, the symmetry that relates fermions and bosons and which helps to explain the large discrepancy between the weak scale (300 GeV) and the Planck scale and may hold the key to unifying gravity with the other forces. Unlike the axion or the neutralino, neutrinos are known to exist, come in three varieties, and have a relic abundance known to three significant figures (113 cm^{-3} per species). The only issue is their mass. Almost all attempts to unify the forces and particles of Nature lead to the prediction that neutrinos have mass, often in the "eV range" (meaning anywhere from 10^{-6} eV or smaller to keV).

The axion and neutralino are referred to as "cold dark matter" because they move very slowly (neutralinos because they are heavy and axions because they were produced coherently in the early Universe with very small momenta). Neutrinos on the other hand are referred to as "hot dark matter" because they move rapidly (due to their small mass). The distinction between the two is crucial for structure formation: at early times neutrinos can "run out" of overdense regions and into underdense regions, damping density perturbations on scales smaller than those corresponding to superclusters. This means that in the absence of additional seed perturbations that don't involve neutrinos (e.g., cosmic string) the sequence of structure formation in a hot dark matter universe proceeds from the "top down:" objects like superclusters form first and then fragment into smaller objects (galaxies and the like). Because there is now much evidence that "small objects" (galaxies, quasars, neutral hydrogen clouds, and clusters) were ubiquitous at redshifts from 1 to 4, and "large objects" are just forming today, hot dark matter is not viable.

To end on a sober note, at present the data can neither prove nor disprove: (i) $\Omega_0 = \Omega_B \simeq 0.15$; (ii) $\Omega_0 = 1$ with $\Omega_B \sim 0.05$ and $\Omega_{\rm CDM} \sim 0.95$. (In the first case the Hubble constant must be near its lower extreme since the nucleosynthesis constrains $\Omega_B = 0.009h^{-2} - 0.022h^{-2}$.) In any case, I will focus on the second, more radical, possibility.

3.3 Coherent Picture of Structure Formation

Because the energy densities of matter (baryons + CDM?) and radiation (photons, light neutrinos, and at early times all the other particles in the thermal plasma) evolve differently, R^{-3} for matter and R^{-4} for radiation, the energy density in radiation exceeded that in matter earlier at early times, $t \lesssim t_{\rm EQ} \sim 10^4\,$yr ($T \gtrsim T_{\rm EQ} \sim 5\,$eV and $R \lesssim R_{\rm EQ} \sim 3 \times 10^{-5} R_{\rm today}$). Moreover, matter density perturbations do not grow during the radiation-dominated era, and thus the formation of structure did not begin in earnest until the epoch of matter-radiation equality. After that, (linear) perturbations in the matter grow as the scale factor, for a total (linear) growth factor of around 30,000. This factor sets the characteristic amplitude of density perturbations, about $1/30,000 \sim few \times 10^{-5}$ (nonlinear structures have formed by the present) and thus the expected size of temperature fluctuations in the CBR (density perturbations lead to fluctuations in the CBR temperature of comparable size).

The detection of CBR anisotropy at the level of about 10^{-5} validates the gravitational instability picture of structure formation. This success should be viewed in the same way that the evidence for a large primeval mass fraction of ^{4}He validated the basic idea of primordial nucleosynthesis in the late 1960s. From this early success, big-bang nucleosynthesis developed into a coherent and detailed explanation for the abundances of D, ^{3}He, ^{4}He and ^{7}Li, and now provides the earliest test of the big bang, the most reliable determination of the baryon density, and an important probe of particle physics. It is not unreasonable to hope that a detailed and coherent picture of structure formation will develop, and that when it does, it will lead to similar advances in our understanding of the Universe and possibly even fundamental physics.

Two crucial elements underlay any detailed picture: specification of the quantity and composition of the dark matter and the nature of the density perturbations. With regard to the latter, what is wanted is a mathematical description of the spectrum of density perturbations. For example, the Fourier components δ_k of the density field and their statistical properties.

Presently there are three viable theories: cold dark matter models; topological-defect models (see e.g., Turok 1991); and the primeval baryon isocurvature model (PBI; Peebles 1987a, b). The effort being brought to bear—both experimental and theoretical—is great, and I am confident that *at least* two of these models, if not all three (!), will be falsified soon. It is my view that only cold dark matter will survive the next cut, but of course others may hold a different opinion.

Topological defect models, where the seeds are cosmic string, monopoles or textures produced in an early Universe phase transition and the dark matter is either neutrinos (cosmic string) or cold dark matter (textures), seem to predict CBR anisotropy on the degree scale that is significantly less than that measured. In addition, when normalized to the COBE measurements of anisotropy, they require a high level of "bias;" bias refers to the "mismatch" between the light and mass distributions, $b \simeq (\delta n_{\rm GAL}/n_{\rm GAL})/(\delta\rho/\rho)$, and b is generally believed to be of order $1-2$. Much of the difficulty in assessing the defect models is on the theoretical side; density perturbations are constantly being produced as the defect network evolves and thus cannot easily be described by Fourier components whose evolution is simple.

The basic philosophy behind the PBI model is to explain the formation of structure by using "what is here," rather then what early-Universe theorists (like myself) hope is here! The parameters for PBI are: $\Omega_0 = \Omega_B \sim 0.2$ and $H_0 \sim 70\,{\rm km\,s^{-1}\,Mpc^{-1}}$. An arbitrary power-law spectrum of fluctuations in the local baryon number (cut off at small scales to avoid difficulties with primordial nucleosynthesis) is postulated and its parameters (slope and normalization) are determined by the data (CBR fluctuations and large-scale structure). PBI has some serious problems: the baryon density violates the nucleosynthesis bound by a wide margin ($\Omega_B h^2 \sim 0.1 \gg 0.02$); it is difficult to make PBI consistent with the measurements of CBR anisotropy (Hu & Sugiyama 1995a, b). To wit, Peebles (1994) has considered variations on the basic theme (e.g., adding a cosmological constant, or even cold dark matter). At the very least PBI provides a useful model against which scenarios that postulate exotic dark matter can be compared; at its best, it may closely represent our Universe.

4 Inflation and Cold Dark Matter

Inflation represents a bold attempt to extend the standard big-bang cosmology to times as early as 10^{-32} sec and to resolve some of the most fundamental questions in cosmology. In particular, inflation addresses squarely both the dark matter and structure formation problems, as well as providing an explanation for the flatness and smoothness of the Universe. That is, it explains the apparent extreme specialness of the initial data required to produce a Universe qualitatively similar to ours. (The set of initial data that lead to a Universe qualitatively similar to ours is of measure zero; Collins & Hawking 1973.) If inflation is correct, it would represent a truly remarkable addition to the standard cosmology.

Two elements are essential to inflation: (1) accelerated ("superluminal") expansion and the concomitant tremendous growth of the scale factor; and (2) massive entropy production (Hu *et al.* 1994). Together, these two features allow a small, smooth subhorizon-sized patch of the early Universe to grow to a large enough size and contain enough heat (entropy in excess of 10^{88}) to easily encompass our present Hubble volume. Provided that the region was originally small compared to the curvature radius of the Universe it would appear flat then and today (just as any small portion of the surface of a sphere appears flat).

While there is presently no standard model of inflation—just as there is no standard model for physics at these energies (typically 10^{15} GeV or so)—viable models have much in common. They are based upon well posed, albeit highly speculative, microphysics involving the classical evolution of a scalar field. The superluminal expansion is driven by the potential energy ("vacuum energy") that arises when the scalar field is displaced from its potential-energy minimum, which results in nearly exponential expansion. Provided the potential is flat, during the time it takes for the field to roll to the minimum of its potential the Universe undergoes many e-foldings of expansion (more than around 60 or so are required to realize the beneficial features of inflation). As the scalar field nears the minimum, the vacuum energy has been converted to coherent oscillations of the scalar field, which correspond to nonrelativistic scalar-field particles. The eventual decay of these particles into lighter particles and their thermalization results in the "reheating" of the Universe and accounts for all the heat in the Universe today (the entropy production event).

Superluminal expansion and the tremendous growth of the scale factor (by a factor greater than that since the end of inflation) allow quantum fluctuations on very small scales ($\lesssim 10^{-23}$ cm) to be stretched to astrophysical scales ($\gtrsim 10^{25}$ cm). Quantum fluctuations in the scalar field responsible for inflation ultimately lead to an almost scale-invariant spectrum of density perturbations (Guth & Pi 1982; Hawking 1982; Starobinskii 1982; Bardeen *et al.* 1983), and quantum fluctuations in the metric itself lead to an almost scale-invariant spectrum of gravity-waves (Rubakov *et al.* 1982; Fabbri & Pollock 1983; Starobinskii 1983b; Abbott & Wise 1984). Scale invariance for density perturbations means scale-independent fluctuations in the gravitational potential (equivalently, density perturbations of different wavelength cross the horizon with the same amplitude); scale invariance for gravity waves means that gravity waves of all wavelengths cross the horizon with the same amplitude. Because of subsequent evolution, neither the scalar nor the tensor perturbations are scale invariant today.

Although there is no standard model of inflation, there are three predictions that almost all viable models make: (1) spatially flat Universe (Inflationary models have been constructed with $\Omega_0 < 1$; see e.g., Steinhardt 1990; Bucher *et al.* 1995); (2) nearly scale-invariant spectrum of gaussian density (scalar metric) perturbations (Guth & Pi 1982; Hawking 1982; Starobinskii 1982; Bardeen *et al.* 1983); (3) nearly scale-invariant spectrum of gravity waves (tensor metric perturbations: (Rubakov *et al.* 1982; Fabbri & Pollock 1983; Starobinskii 1983; Abbott & Wise 1984).

The prediction of a flat Universe means the total energy density (including matter, radiation, and the vacuum energy density associated with a cosmological constant) is equal to the critical density, that is $\Omega_0 = 1$. Coupled with our knowledge of the baryon density, this implies that the bulk of matter in the Universe (95% or so) must be nonbaryonic. The two simplest possibilities are hot dark matter and cold dark matter. Structure formation with hot dark matter has been studied, and, sadly, does not work; thus we are led to cold dark matter.

While inflation predicts that the density perturbations are nearly scale invariant and are described by gaussian statistics, the overall amplitude of the spectrum is model dependent. In that regard the COBE measurement of CBR anisotropy was crucial as it allows the spectrum to be normalized. The density perturbations and their statistical properties are described by the power spectrum, $P(k) \equiv \langle|\delta_k|^2\rangle$, which at early times is $P(k) = Ak^n$ ($n = 1$ corresponds to scale invariant). The *rms* density perturbation on a given scale is related to $P(k)$ by $(\delta\rho/\rho)^2 \sim k^3 P(k)/2\pi^2$.

When the Universe became matter dominated, the sizes of density perturbations were largest on small scales. For cold dark matter there is no damping of perturbations on small scales, and so small structures formed first ("bottom up"). Clumps of dark matter and baryons continuously merge to form larger objects. "Typical galaxies" are formed at redshifts $z \sim 1-2$; "rare objects" such as quasars and radio galaxies can form earlier from regions where the density perturbations have larger than average amplitude. Clusters form in the very recent past (redshifts less than order unity), and superclusters are just forming today. Voids naturally arise as regions of space are evacuated to form objects (see e.g. Blumenthal *et al.* 1984).

4.1 Metaphysical Implications

Inflation alleviates the "specialness" problem greatly, but does not eliminate all dependence upon the initial state (Starobinskii 1983b; Turner & Widrow 1986; Jensen & Stein-Schabes 1987). All open FRW models will inflate and become flat; however, many closed FRW models will recollapse before they can inflate. If one imagines the most general initial space-time as being comprised of negatively and positively curved FRW (or Bianchi) models that are stitched together, the failure of the positively curved regions to inflate is of little consequence: because of exponential expansion during inflation the negatively curved regions will occupy most of the space today. Nor does inflation solve the smoothness problem forever; it just postpones the problem into the exponentially distant future: We will be able to see outside our smooth inflationary patch and Ω will start to deviate significantly from unity at a time $t \sim t_0 \exp[3(N - N_{\rm min}]$, where N is the actual number of e-foldings of inflation and $N_{\rm min} \sim 60$ is the minimum required to solve the horizon/flatness problems.

Linde (1990) has emphasized that inflation has changed our view of the Universe in a very fundamental way (Linde 1990). While cosmologists have long used the Copernician principle to argue that the Universe must be smooth because of the smoothness of our Hubble volume, in the post-inflation view, our Hubble volume is smooth because it is a small part of a region that underwent inflation. On the largest scales the structure of the Universe is likely to be very rich: Different regions may have undergone different amounts of inflation, may have different laws of physics because they evolved into different vacuum states (of equivalent energy), and may even have different numbers of spatial dimensions. Since it is likely that most of the volume of the Universe is still undergoing inflation and that inflationary patches are being constantly produced (eternal inflation), the age of the Universe is a meaningless concept and our expansion age merely measures the time back to the end of our inflationary event!

4.2 Almost, But Is Something Missing?

The cold dark matter scenario is the most well motivated, most specific, and most successful scenario for structure formation yet proposed. It is so attractive that it has inspired a generation of observers to go out and prove it wrong! As I will describe, thus far it receives general support from a diversity of observations; however, there are indications that it does not have "all the truth" and requires some minor adjusting.

Broadly speaking, testing the cold dark matter scenario involves measuring the quantity, composition, and distribution of dark matter and determining the spectrum of density perturbations. I have already discussed the current state of our knowledge of dark matter. While a host of observations provide information about the primeval spectrum of density perturbations, measurements of the anisotropy of the CBR and mapping the distribution of matter today (as traced by bright galaxies) are perhaps most crucial. (For reference, perturbations on scales of about 1 Mpc correspond to galactic size perturbations, on 10 Mpc to cluster size perturbations, on 30 Mpc to the large voids, and 100 Mpc to the great walls. Fourier wavenumber is related to wavelength by $k = 2\pi/\lambda$)

CBR anisotropy probes the power spectrum on large scales. The CBR temperature difference measured on a given angular scale is related to the power spectrum on length scales $\lambda \sim (\theta/\text{deg})100h^{-1}\,\text{Mpc}$. Since the COBE detection, a host of ground-based and balloon-borne experiments have also detected CBR anisotropy, on scales from about $0.5°$ to $90°$, at the level of around 30μK, corresponding to $\delta T/T \sim 10^{-5}$. The measurements are consistent with the predictions of cold dark matter (see Fig. 2), though there are still large statistical uncertainties as well as concerns about contamination by foreground sources (see e.g., White *et al.* 1994). There is a great deal of experimental activity (more than ten groups), and measurements in the near future should improve the present situation significantly. The CBR contains important information on angular scales down to about 0.1 deg (anisotropy on smaller angular scales is washed out due to the finite thickness of the last scattering surface). A satellite-borne experiment with ten times greater resolution than COBE is being studied in both Europe and the US, and a variety of earth-based and balloon-based experiments should hopefully map CBR anisotropy on scales from about 0.1 deg to 90 deg in the next decade.

The COBE detection of CBR anisotropy not only provided the first evidence for the existence of primeval density perturbations, but also an unambiguous way to normalize the spectrum of density perturbations: Given the shape of the power spectrum (for cold dark matter, approximately scale invariant) the COBE measurement (on an angular scale of about $10°$ which corresponds to a length scale of about $10^3h^{-1}\,\text{Mpc}$) ties down the spectrum on all scales. This leads to definite predictions that can be tested by other CBR measurements and observations of large-scale structure.

The comparison of predictions for structure formation with present-day observations of the distribution of galaxies is very important, but fraught with difficulties. Theory most accurately predicts "where the mass is" (of course, only in a statistical sense) and the observations determine where the light is. Redshift surveys probe present-day inhomogeneity on scales from around one Mpc to a few hundred Mpc, scales where the Universe is nonlinear ($\delta n_{\rm GAL}/n_{\rm GAL} \gtrsim 1$ on scales $\lesssim 8h^{-1}$ Mpc) and where astrophysical processes undoubtedly play an important role (e.g., star formation determines where and when "mass lights up," the explosive release of energy in supernovae can move matter around and influence subsequent star formation, and so on). The distance to a galaxy is determined through Hubble's law ($d = H_0^{-1}z$) by measuring a redshift; peculiar velocities induced by the lumpy distribution of matter are significant and prevent a direct determination of the true distance. There are the intrinsic limitations of the surveys themselves: they are flux not volume limited (brighter objects are seen to greater distances and vice versa) and relatively small (e.g., the CfA slices of the Universe survey contains only about 10^4 galaxies and extends to a redshift of about $z \sim 0.03$). Last but not least are the numerical simulations which bridge theory and observation; they are limited in dynamical range (about a factor of 100 in length scale) and in microphysics (in the largest simulations only gravity, and in others only a gross approximation to the effects of hydrodynamics/thermodynamics).

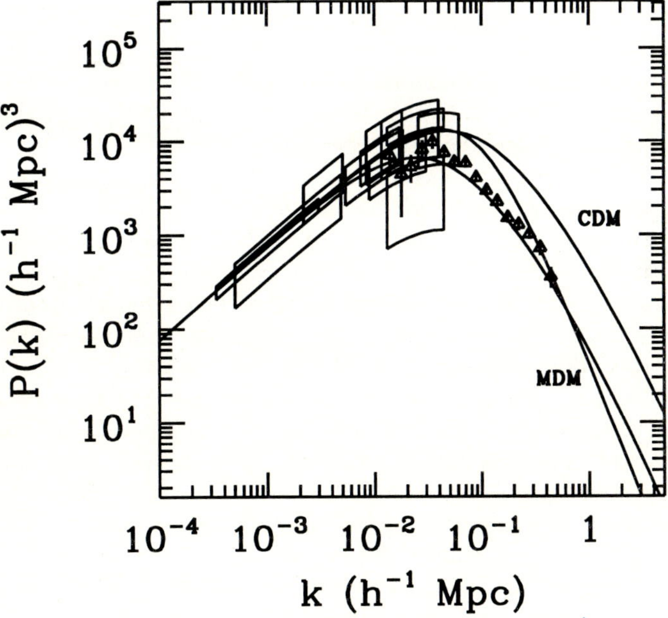

Fig. 3. Comparison of the cold dark matter perturbation spectrum with CBR anisotropy measurements (boxes) and the distribution of galaxies today (triangles). Wavenumber k is related to length scale, $k = 2\pi/\lambda$; error flags are not shown for the galaxy distribution. The curve labeled MDM is hot + cold dark matter ("5 eV" worth of neutrinos); the other two curves are cold dark matter models with Hubble constants of 50 km s^{-1} Mpc (labeled CDM) and 35 km s^{-1} Mpc. (Figure courtesy of M. White.)

This being said, redshift surveys do provide an important probe of the power spectrum on small scales ($\lambda \sim 1-300\,\mathrm{Mpc}$). Even with their limitations redshift surveys (as well as other data) indicate that while the simplest version of COBE-normalized cold dark matter is in broad agreement with the data, the shape of the power spectrum as well as its amplitude on small scales is not quite right (Ostriker 1993; Liddle & Lyth 1993). At least three possibilities come to mind: (i) the comparison of numerical simulations and the observations is still too primitive to draw firm conclusions; (ii) cold dark matter has much, but not all, of the "truth;" or (iii) cold dark matter has been falsified.

For three reasons I believe that it is worthwhile exploring the second possibility, namely that cold dark matter needs a little tinkering. First, cold dark matter is such an attractive theory and part of a bold attempt to extend greatly the standard cosmology. Second, many observations seem to point to the same problem (e.g., the abundance of x-ray clusters and the cluster-cluster correlation function). Third, there are other reasons to believe that the Universe is more complicated than the simplest model of cold dark matter.

4.3 The Cold Dark Matter Family of Models

Somewhat arbitrarily, standard cold dark matter has come to mean: precisely scale-invariant density perturbations; baryons + CDM only; and Hubble constant of $50\,\mathrm{km\,s^{-1}\,Mpc^{-1}}$ (to ensure a sufficiently aged Universe with a Hubble constant still within the range of observations). This is the vanilla or default model, which, when normalized to COBE has too much power on small scales and the wrong spectral shape on slightly larger scales.

The spectrum of density perturbations today depends not only upon the primeval spectrum (and the normalization on large scales provided by COBE), but also upon the energy content of the Universe. While the fluctuations in the gravitational potential were initially (approximately) scale invariant, the Universe evolved from an early radiation-dominated phase to a matter-dominated phase which imposes a characteristic scale on the spectrum of density perturbations seen today; that scale is determined by the energy content of the Universe, $k_{\mathrm{EQ}} \sim 10^{-1} h\,\mathrm{Mpc}^{-1}\,(\Omega_{\mathrm{matter}} h/\sqrt{g_*})$ (g_* counts the relativistic degrees of freedom, $\Omega_{\mathrm{matter}} = \Omega_B + \Omega_{\mathrm{CDM}}$). In addition, if some of the nonbaryonic dark matter is neutrinos, they reduce power on small scales somewhat through freestreaming (see Fig. 2). With this in mind, let me discuss the variants of cold dark matter that have been proposed to improve its agreement with observations.

1. **Low Hubble Constant + cold dark matter (LHC CDM)** (Bartlett *et al.* 1995). Remarkably, simply lowering the Hubble constant to around $30\,\mathrm{km\,s^{-1}\,Mpc^{-1}}$ solves all the problems of cold dark matter. Recall, the critical density $\rho_{\mathrm{crit}} \propto H_0^2$; lowering H_0 lowers the matter density and has precisely the desired effect. It has two other added benefits: the expansion age of the Universe is comfortably consistent with the ages of the oldest stars and the baryon fraction is raised to a value that is consistent with that measured in x-ray clusters. Needless to say, such a small value for the

Hubble constant flies in the face of current observations (Jacoby *et al.* 1992; Fukugita *et al.* 1993; Freedman *et al.* 1994); further, it illustrates that the problems of cold dark matter get even worse for the larger values of H_0 that are favored by recent observations.

2. **Hot + cold dark matter (νCDM)** (Shafi & Stecker 1984; Achilli *et al.* 1985; Ikeuchi *et al.* 1988; van Dalen & Schaefer 1992; Davis *et al.* 1992; Primack *et al.* 1995; Pogosyan & Starobinskii 1995). Adding a small amount of hot dark matter can suppress density perturbations on small scales; adding too much leads back to the longstanding problems of hot dark matter. Retaining enough power on very small scales to produce damped Lyman-α systems at high redshift limits Ω_ν to less than about 20%, corresponding to about "5 eV worth of neutrinos" (i.e., one species of mass 5 eV, or two species of mass 2.5 eV, and so on). This admixture of hot dark matter rejuvenates cold dark matter provided the Hubble constant is not too large, $H_0 \lesssim 55\,\mathrm{km\,s^{-1}\,Mpc^{-1}}$; in fact, a Hubble constant of closer to $45\,\mathrm{km\,s^{-1}\,Mpc^{-1}}$ is preferred.
3. **Cosmological constant + cold dark matter (ΛCDM)** (Turner *et al.* 1984); Turner 1991; Peebles 1984; Efstathiou *et al.* 1990; Kofman & Starobinskii 1985). (A cosmological constant corresponds to a uniform energy density, or vacuum energy.) Shifting 50% to 80% of the critical density to a cosmological constant lowers the matter density and has the same beneficial effect as a low Hubble constant. In fact, a Hubble constant as large as $80\,\mathrm{km\,s^{-1}\,Mpc^{-1}}$ can be accommodated. In addition, the cosmological constant allows the age problem to solved even if the Hubble constant is large, addresses the fact that few measurements of the mean mass density give a value as large as the critical density (most measurements of the mass density are insensitive to a uniform component), and allows the baryon fraction of matter to be larger, which alleviates the cluster baryon problem. Not everything is rosy; cosmologists have invoked a cosmological constant twice before to solve their problems (Einstein to obtain a static universe and Bondi, Gold, and Hoyle to solve the earlier age crisis when H_0 was thought to be $250\,\mathrm{km\,s^{-1}\,Mpc^{-1}}$). Further, particle physicists can still not explain why the energy of the vacuum is not at least 50 (if not 120) orders of magnitude larger than the present critical density, and expect that when the problem is solved the answer will be zero.
4. **Extra relativistic particles + cold dark matter (τCDM)** (Dodelson *et al.* 1994; Bond & Efstathiou 1991). Raising the level of radiation has the same beneficial effect as lowering the matter density. In the standard cosmology the radiation content consists of photons + three (undetected) cosmic seas of neutrinos (corresponding to $g_* \simeq 3.36$). While we have no direct determination of the radiation beyond that in the CBR, there are at least two problems: What are the additional relativistic particles? and Can additional radiation be added without upsetting the successful predictions of primordial nucleosynthesis which depend critically upon the energy density of relativistic particles? The simplest way around these problems is an unstable tau neutrino (mass anywhere between a few keV and a few MeV) whose decays

produce the radiation. This fix can tolerate a larger Hubble constant, though at the expense of more radiation.

5. **Tilted cold dark matter (TCDM)** (Cen *et al.* 1992; Davis *et al.* 1992; Lucchin *et al.* 1992; Salopek 1992; Liddle & D. Lyth 1992; Lidsey & Coles 1993; Souradeep & Sahni 1992). While the spectrum of density perturbations in most models of inflation is very nearly scale invariant, there are models where the deviations are significant ($n \approx 0.8$) which leads to smaller fluctuations on small scales. Further, if gravity waves account for a significant part of the CBR anisotropy, the level of density perturbations can be lowered even more. A combination of tilt and gravity waves can solve the problem of too much power on small scales, but seems to lead to too little power on intermediate and very small scales.

In evaluating these better fit models, one should keep the words of Francis Crick in mind (loosely paraphrased): A model that fits all the data at a given time is necessarily wrong, because at any given time not all the data are correct(!). ΛCDM provides an interesting/confusing example. When I discussed it in 1990, I called it the best-fit Universe, and quoting Crick, I said that ΛCDM was certain to fall by the wayside (Turner 1991). In 1995, it is still the best-fit model (Krauss & Turner 1995).

Let me end by defending the other point of view, namely, that to add something to cold dark matter is not unreasonable, or even as some have said, a last gasp effort to saving a dying theory. Standard cold dark matter was a starting point, similar to early calculations of big-bang nucleosynthesis. It was always appreciated that the inflationary spectrum of density perturbations was not exactly scale invariant (Steinhardt & Turner 1984) and that the Hubble constant was unlikely to be exactly $50\,\mathrm{km\,s^{-1}\,Mpc}$. As the quality and quantity of data improve, it is only sensible to refine the model, just as has been done with big-bang nucleosynthesis. Cold dark matter seems to embody much of the "truth." The modifications suggested all seem quite reasonable (as opposed to contrived). Neutrinos exist; they are expected to have mass; there is even some experimental data that indicates they do have mass. It is still within the realm of possibility that the Hubble constant is less than $50\,\mathrm{km\,s^{-1}\,Mpc^{-1}}$, and if it is as large as $70\,\mathrm{km\,s^{-1}\,Mpc^{-1}}$ to $80\,\mathrm{km\,s^{-1}\,Mpc^{-1}}$ a cosmological constant seems inescapable based upon the age problem alone. There is no data that can preclude more radiation than in the standard cosmology and deviations from scale invariance were always expected.

5 The Future

5.1 Testing and Discriminating

The stakes for cosmology are high: if correct, inflation/cold dark matter represents a major extension of the big bang and our understanding of the Universe, which can't help but shed light on the fundamental physics at energies of order $10^{15}\,\mathrm{GeV}$ or higher.

What are the crucial tests and when will they be carried out? Because of the many measurements/observations that can have significant impact, I believe the answer to when is sooner rather than later. The list of pivotal observations is long: CBR anisotropy, large redshift surveys (e.g., the Sloan Digital Sky Survey will have 10^6 redshifts), direct searches for nonbaryonic in our neighborhood (both for axions and neutralinos) and baryonic dark matter (microlensing), x-ray studies of galaxy clusters, the use of back-lit gas clouds (quasar absorption line systems) to study the Universe at high redshift, evolution (as revealed by deep images of the sky taken by the Hubble Space Telescope and the Keck 10 meter telescope), measurements of both H_0 and q_0, mapping of the peculiar velocity field at large redshifts through the Sunyaev-Zel'dovich effect, dynamical estimates of the mass density (using weak gravitational lensing, large-scale velocity fields, and so on), age determinations, gravitational lensing, searches for supersymmetric particles (at accelerators) and neutrino oscillations (at accelerators, solar-neutrino detectors, and other large underground detectors), searches for high-energy neutrinos from neutralino annihilations in the sun using large underground detectors, and on and on. Let me end by illustrating the interesting consequences of several possible measurements.

A definitive determination that H_0 is greater than $55\,\mathrm{km\,s^{-1}\,Mpc^{-1}}$ would falsify LHC CDM and νCDM. Likewise, if H_0 is shown to be $75\,\mathrm{km\,s^{-1}\,Mpc^{-1}}$ or larger a cosmological constant would be mandatory. A flat Universe with a cosmological constant has a very different deceleration parameter than one dominated by matter, $q_0 = -1.5\Omega_\Lambda + 0.5 \sim -(0.4 - 0.7)$ compared to $q_0 = 0.5$, and this could be settled by galaxy number counts or numbers of lensed quasars. The level of CBR anisotropy in τCDM and LHC CDM on the $0.5°$ scale is about 50% larger than the other models, which should be easily discernible. If neutrino-oscillation experiments were to provide evidence for a neutrino of mass 5 eV (or two of mass 2.5 eV) νCDM would seem almost inescapable.

Many more CBR measurements are in progress and there should many interesting results in the next few years. In the wake of the success of COBE there are proposals, both in the US and Europe, for a satellite-borne instrument to map the CBR sky with a factor of ten better resolution. A map of the CBR with $0.5° - 1°$ resolution could separate the gravity-wave contribution to CBR anisotropy and provide evidence for the third robust prediction of inflation, as well as determining other important parameters (Knox 1995), e.g., the scalar and tensor indices, Ω_Λ, and even Ω_0 (the position of the "Doppler" peak scales as $0.5°/\sqrt{\Omega_0}$) (Kamionkowski *et al.* 1994).

There are key measurements that can shed light on the quantity and composition of dark matter in the Universe. Steady progress is being made in the quest to measure Ω_0 and to detect particle dark matter. At the moment there are two particularly pressing issues involving measurements of the baryonic fraction of dark matter.

X-ray observations of rich clusters are able to determine the ratio of hot gas (baryons) to total cluster mass (baryons + CDM) (by a wide margin, most of the baryons "seen" in cluster are in the hot gas). To be sure there are assumptions and uncertainties; the data at the moment indicate that this ratio is $0.04h^{-3/2}$ –

$0.10h^{-3/2}$ (see e.g., Briel *et al.* 1992; White *et al.* 1993; White & Fabian 1995). If clusters provide a fair sample of the universal mix of matter, then this ratio should equal $\Omega_B/(\Omega_B + \Omega_{\rm CDM}) \simeq (0.009 - 0.022)h^{-2}/(\Omega_B + \Omega_{\rm CDM})$. Since clusters are large objects they should provide an approximately fair sample. Taking the numbers at face value, cold dark matter is consistent with the cluster gas fraction provided either: $\Omega_B + \Omega_{\rm CDM} = 1$ and $h \sim 0.3$ or $\Omega_B + \Omega_{\rm CDM} \sim 0.3$ and $h \sim 0.7$, favoring LHC CDM or ΛCDM. The cluster baryon fraction is an important test of and discriminator between cold dark matter models, and we haven't heard the final word yet.

If cold dark matter is correct, then a significant, if not dominant, fraction of the dark halo of our galaxy should be cold dark matter (the halos of spiral galaxies are not large enough to guarantee that they represent a fair sample). Direct searches for faint stars have failed to turn up enough to account for the halo (see e.g. Bahcall *et al.* 1994 and references therein). Over the past few years, microlensing has been used to search for dark stars (stars below the $0.08M_\odot$ limit for hydrogen burning). Five stars in the LMC have been observed to change brightness in a way consistent with their being microlensed by dark halo objects passing along the line of sight. While the statistics are small, and there are uncertainties concerning the size of dark halo, these results indicate that only a small fraction (5% to 30%) of the dark halo is in the form of dark stars (Gates *et al.* 1995, Alcock *et al.* 1995). So far, this one goes in the "plus column" for cold dark matter.

5.2 Reconstruction

If cold dark matter is shown to be correct, then a window to the very early Universe ($t \sim 10^{-34}$ sec) will have been opened. While it is certainly premature to jump to this conclusion, I would like to illustrate one example of what one could hope to learn. As mentioned earlier, the spectra and amplitudes of the the tensor and scalar metric perturbations predicted by inflation depend upon the underlying model, to be specific, the shape of the inflationary scalar-field potential. (Inflation involves the classical evolution of a scalar field ϕ rolling down its potential energy curve $V(\phi)$.) If one can measure the power-law index of the scalar spectrum and the amplitudes of the scalar and tensor spectra, one can recover the value of the potential and its first two derivatives around the point on the potential where inflation took place (Copeland *et al.* 1993a,b; Turner 1993a, 1993b; Knox & Turner 1994; Knox 1995). (Measuring the power-law index of the tensor perturbations in addition, allows an important consistency check of inflation.) Reconstruction of the inflationary scalar potential would shed light both on inflation as well as physics at energies of the order of 10^{14} GeV.

5.3 Concluding Remarks

We live in exciting times. We have a cosmological model that provides a reliable account of the Universe from 0.01 sec until the present. Together with the Standard Model of particle physics it provides a framework for both asking and addressing deeper questions about the Universe. Much progress has been made in the past fifteen years of intensive study of the earliest history of the Universe. With inflation and cold dark matter we may be on the verge of a very significant extension of the standard cosmology. Most importantly, the data needed to test the cold dark matter theory is coming in at a rapid rate. At the very least we should know soon whether we are on the right track or if it's back to the drawing board.

Acknowledgments. This work was supported in part by the DOE (at Chicago and Fermilab) and by the NASA through grant NAG 5-2788 (at Fermilab).

References

Abbott, L. & Wise, M. 1984, *Nucl. Phys. B* **244**, 541

Achilli, S., Occhionero, F., & Scaramella, R. 1985, *Astrophys. J.* **299**, 577

Alcock, C., *et al.* 1995, *Phys. Rev. Lett.* **74**, 2867

Arp, H., *et al.* 1990, *Nature* **346**, 807

Bahcall, J., *et al.* 1994, *Astrophys. J. Lett.* **435**, L51

Bardeen, J.M., Steinhardt, P.J., & Turner, M.S. 1983, *Phys. Rev. D* **28**, 697

Bartlett, J., *et al.* 1995, *Science* **267**, 980

Blumenthal, G.R. *et al.* 1984, *Nature* **311**, 517

Bond, J.R. & Efstathiou, G. 1991, *Phys. Lett. B* **265**, 245

Briel, U.G. *et al.* 1992, *Astron. Astrophys.* **259**, L31

Bucher, M., *et al.* 1995, *Phys. Rev. D* **52**, in press

Cen, R., Gnedin, N., Kofman, L., & Ostriker, J.P. 1992, *Astrophys. J. Lett.* **399**, L11

Collins, C.B. & Hawking, S.W. 1973, *Astrophys. J.* **180**, 317

Copeland, E.J., Kolb, E.W., Liddle, A.R., Lidsey, J.E. 1993a, *Phys. Rev. Lett.* **71**, 219

Copeland, E.J., Kolb, E.W., Liddle, A.R., Lidsey, J.E. 1993b, *Phys. Rev. D* **48**, 2529

Copi, C., Schramm, D.N. & Turner, M.S. 1995, *Science* **267**, 192.

Davis, R. *et al.* 1992, *Phys. Rev. Lett.* **69**, 1856

Davis, M., Summers, F., & Schlegel, D. 1992, *Nature* **359**, 393

Dekel, A. 1994, *Ann. Rev. Astron. Astrophys.* **32**, 371

Dodelson, S., Gyuk, G. & Turner, M.S. 1994, *Phys. Rev. Lett.* **72**, 3578

Efstathiou, B., *et al.* 1990, *Nature* **348**, 705

Efstathiou, G. 1992, in *The Physics of the Early Universe*, eds. J.A. Peacock, A.F. Heavens and A.T. Davies, Adam Higler, Bristol

Fabbri, R. & Pollock, M. 1983, *Phys. Lett. B* **125**, 445

Faber, S. & Gallagher, J. 1979 *Ann. Rev. Astron. Astrophys.* **17**, 135

Freedman, W., *et al.* 1994, *Nature* **371**, 757

Fukugita, M., Hogan, C.J., & Peebles, P.J.E. 1993, *Nature* **366**, 309

Gates. E., Gyuk, G., & Turner, M.S. 1995, *Phys. Rev. Lett.* **74**, 3724

Guth, A. 1981, *Phys. Rev. D* **23**, 347

Guth, A.H. & Pi, S.-Y. 1982, *Phys. Rev. Lett.* **49**, 1110

Hawking, S.W. 1982, *Phys. Lett. B* **115**, 295

Hoyle, F., Burbidge, G. & Narlikar, J.V. 1993, *Astrophys. J.* **410**, 437

Hoyle, F., Burbidge, G. & Narlikar, J.V. 1994a, *Mon. Not. R. Astron. Soc.* **267**, 1007

Hoyle, F., Burbidge, G. & Narlikar, J.V. 1994b, *Astron. Astrophys.* **289**, 729

Hu, Y., Turner, M.S., & Weinberg, E.J. 1994, *Phys. Rev. D* **49**, 3830

Hu, W. & Sugiyama, N. 1995a, *Astrophys. J.* **444**, 489
Hu, W. & Sugiyama, N. 1995b, *Phys. Rev. D* **51**, 2599
Ikeuchi, S., Norman, C., & Zahn, Y. 1988, *Astrophys. J.* **324**, 33
Jacoby, G.H., *et al.* 1992, *Proc. Astron. Soc. Pacific* **104**, 599
Jensen, L. & Stein-Schabes, J. 1987, *Phys. Rev. D* **35**, 1146
Kaiser, N., *et al.* 1991, *Mon. Not. R. Astr. S.* **252**, 1
Kamionkowski, M., *et al.* 1994 *Astrophys. J. Lett.* **426**, L57
Kofman, L. & Starobinskii, A.A. 1985, *Sov. Astron. Lett.* **11**, 271
Knox, L. & Turner, M.S. 1994, *Phys. Rev. Lett.* **73**, 3347
Knox, L. 1995, *Phys. Rev. D* **52**, in press
Krauss, L. & Turner, M.S. 1995, *Gen. Rel. Grav.*, in press
Liddle, A.D. & Lyth, D. 1993, *Phys. Repts.* **231**, 1
Liddle, A.D. & Lyth, D. 1992, *Phys. Lett. B* **291**, 391
Lidsey, J.E. & Coles, P. 1993, *Mon. Not. R. Astron. Soc.*
Linde, A.D. 1990, *Inflation and Quantum Cosmology,* Academic Press, San Diego, CA
Lucchin, F., Mattarese, S., & Mollerach, S. 1992 *Astrophys. J. Lett.* **401**, L49
Mather, J., *et al.* 1994. *Astrophys. J.* **420**, 439
Mould, J., *et al.* 1991, *Astrophys. J.* **383**, 467
Ostriker, J.P. 1993, *Ann. Rev. Astron. Astrophys.* **31**, 689
Peebles, P.J.E. 1984, *Astrophys. J.* **284**, 439
Peebles, P.J.E. 1987a, *Nature* **327**, 210
Peeples, P.J.E. 1987b, *Astrophys. J. Lett.* **315**, L73
Peebles, P.J.E. 1994, *Astrophys. J. Lett.* **432**, L1
Pogosyan, D. & Starobinskii, A. 1995, *Astrophys. J.*, in press
Primack, J., *et al.* 1995, *Phys. Rev. Lett.* **74**, 2160
Rubakov, V.A., Sazhin, M., & Veryaskin, A. 1982, *Phys. Lett. B* **115**, 189
Salopek, D. 1992, *Phys. Rev. Lett.* **69**, 3602
Shafi, Q. & Stecker, F. 1984, *Phys. Rev. Lett.* **53**, 1292
Smoot, G., *et al.* 1992, *Astrophys. J. Lett.* **396**, L1
Songaila, A., *et al.* 1994, *Nature* **371**, 43
Souradeep. T. & Sahni, V. 1992, *Mod. Phys. Lett. A* **7**, 3541
Starobinskii, A.A. 1982, *Phys. Lett. B* **117**, 175
Starobinskii, A.A. 1983a, *JETP Lett.* **37**, 66
Starobinskii, A.A. 1983b, *Sov. Astron. Lett.* **9**, 302
Steinhardt, P.J. & Turner, M.S. 1984, *Phys. Rev. D* **29**, 2162
Steinhardt, P.J. 1990, *Nature* **345**, 47
Strauss, M.A., *et al.* 1992, *Astrophys. J.* **397**, 395
Trimble, V. 1987, *Ann. Rev. Astron. Astrophys.* **25**, 425
Turner, M.S., Steigman, G., & Krauss, L. 1984, *Phys. Rev. Lett.* **52**, 2090
Turner, M.S. & Widrow, L.M. 1986, *Phys. Rev. Lett.* **57**, 2237
Turner, M.S. 1991, *Physica Scripta* **T36**, 167
Turner, M.S. 1993 *Proc. Natl. Acad. Sci. (USA)* **90**, 4822
Turner, M.S. 1993a, *Phys. Rev. D* **48**, 3502
Turner, M.S. 1993b, *Phys. Rev. D* **48**, 5539
Turok, N. 1991, *Physica Scripta* **T36**, 135
van Dalen, A. & Schaefer, R.K. 1992, *Astrophys. J.* **398**, 33
Weinberg, S. 1972, *Gravitation and Cosmology,* J. Wiley, New York
White, S.D.M. *et al.* 1993 *Nature* **366**, 429
White, M., Scott, D., & Silk, J. 1994, *Ann. Rev. Astron. Astrophys.* **32**, 319
White, D.A. & Fabian, A.C. 1995, *Mon. Not. R. Astron. Soc.* **273**, 72

Cosmological Structure Formation in Hot and Cold Dark Matter Scenarios

Gerhard Börner, H.J. Mo, Y.P. Jing

Max–Planck–Institut für Astrophysik,
Karl–Schwarzschild–Str. 1, 85740 Garching, Germany

1 Introduction

The luminous matter in the universe, as it has been observed up to now, shows an extremely inhomogeneous distribution: from the sizes and shapes of individual galaxies to the clustering hierarchy of galaxies into groups, clusters and superclusters the astronomers have discovered a wealth of interesting structures. There is no evidence yet of a largest scale in the architecture of the galaxy distribution. Even the statistical description of these observations is not yet completed. While on scales just a few times the size of clusters of galaxies ($\sim 10h^{-1}$ Mpc, where h parametrizes the value of the Hubble constant) the clustering is well measured by a two–point correlation function following a power law $\propto r^{-1.8}$ (e.g. Peebles 1980; Mo & Börner 1990; Jing *et al.* 1991), there seems to be evidence for large characteristic scales of $60h^{-1}$ Mpc and $130h^{-1}$ Mpc, probably corresponding to the average separation of superclusters (Mo *et al.* 1991). Recently completed large surveys lend support to this result and also give hints of an intricate topology of the large–scale structure with galaxies located predominantly in sheets and filaments surrounding large empty regions of almost spherical shape. It is the aim of most modern theories to explain these patterns as a result of the growth by gravitational interaction of initially small perturbations in a smooth background density.

Clearly there must be some scale, below which local physics — radiation hydrodynamics — is the dominant effect in shaping the objects. But it is not clear what this scale is. Is it at the size of individual galaxies, or at the size of clusters? As long as cosmological effects determine the objects initial conditions and boundary conditions are decisive. But once local physics takes over, there is the chance that some universal properties appear which are independent of the arbitrariness of initial and boundary conditions.

Two different general schemes have been proposed. In some theories, the formation of structures started on small scales and proceeded step by step to the larger structures. Galaxies cluster hierarchically to build groups; groups form clusters, and perhaps clusters merge to form superclusters. This "hierarchical clustering model" (Peebles 1980) has recently found a successful representation in models making use of "Cold Dark Matter" (CDM; see section 3). In other models, the largest structures formed first via gravitational instability, followed by fragmentation processes to shape the smaller structures. It can be shown that the largest inhomogeneities collapsed anisotropically forming thin, dense sheets of matter. This aspect has given to this scenario its popular name, the "pancake picture" (Zel'dovich 1970; Doroshkevich *et al.* 1980; Buchert 1989). The pancake scenario has its natural representation in "Hot Dark Matter" models (HDM; see section 3). In the following we want to present an overview of these different dark matter models, and to discuss observational tests which may help to discriminate between the different scenarios.

The assumption of a smooth background of matter has its justification in the successes of the standard model of cosmology, although such an assumption seems not very natural in view of the inhomogeneous and well-structured distribution of the luminous matter. The most solid and convincing evidence to most (with a few tenacious exceptions) for a homogeneous and isotropic cosmic background comes from the measurements of a radiation field in the microwave range with a Planckian spectrum. The satellite COBE (COsmic Background Explorer) launched by NASA on Nov. 18, 1989 has measured this radiation field which was discovered by Penzias and Wilson in 1964. One instrument aboard the satellite has measured the spectrum with remarkable precision. It turned out to be of almost perfect Planck-shape with a mean temperature of $T_\gamma = 2.726 \pm 0.01$K. The universe seems to be the ultimate black-body! (Mather *et al.* 1994) This smooth radiation field strongly supports the idea of a hot big bang. It gets more and more difficult to find other explanations, besides the simple hypothesis that the microwave background is the remnant of a homogeneous and structureless primeval explosion. The error limits on the black-body shape of the spectrum constrain various processes in the early universe quite tightly.

The radiation field is also very isotropic. Two differential radiometers (DMR A and B) scan the sky for temperature differences. The first anisotropy discovered was the dipole pattern of the Doppler shift due to the peculiar motion of the sun, or equivalently the Local Group of galaxies with respect to the rest system of the background radiation. The amplitude of (3.3 ± 0.3)mk in the direction $(\alpha, \delta) = (11.2^h \pm 0.2^h,\ -7^\circ \pm 2^\circ)$ agrees reasonable well with previous measurements (Wright 1991). After subtraction of the dipole fluctuations appear at a level of $\frac{\Delta T}{T} \approx 10^{-5}$ (Fig. 1).

The analysis of the first year of data provided evidence for the rms sky variation of $\Delta T \simeq (30 \pm 5)\mu$K ($\frac{\Delta T}{T} = 1.1 \times 10^{-5}$). The rms-quadrupole amplitude was $(13 \pm 4)\mu$K ($\frac{\Delta T}{T} \simeq 6 \times 10^{-5}$). The data are smoothed out with a Gaussian window function of 7° FWHM which together with the $\approx 7^\circ$ FWHM antenna beam results in a $\sim 10^\circ$ Gaussian smoothing on the sky. The power spectrum of $C(\theta) = \langle \Delta T(\underline{n}) \Delta T(\underline{n}') \rangle$ is derived by an expansion in terms of Legendre polynomials

$$C(\theta) = \frac{1}{4\pi} \Sigma\, \Delta T_\ell^2\, P_\ell(\cos\theta)) \ .$$

In Figure 2 the power spectrum is displayed in terms of the 53 GHz plus 90 GHz data. Structure on all scales can be seen, from the beam size (7°) to the

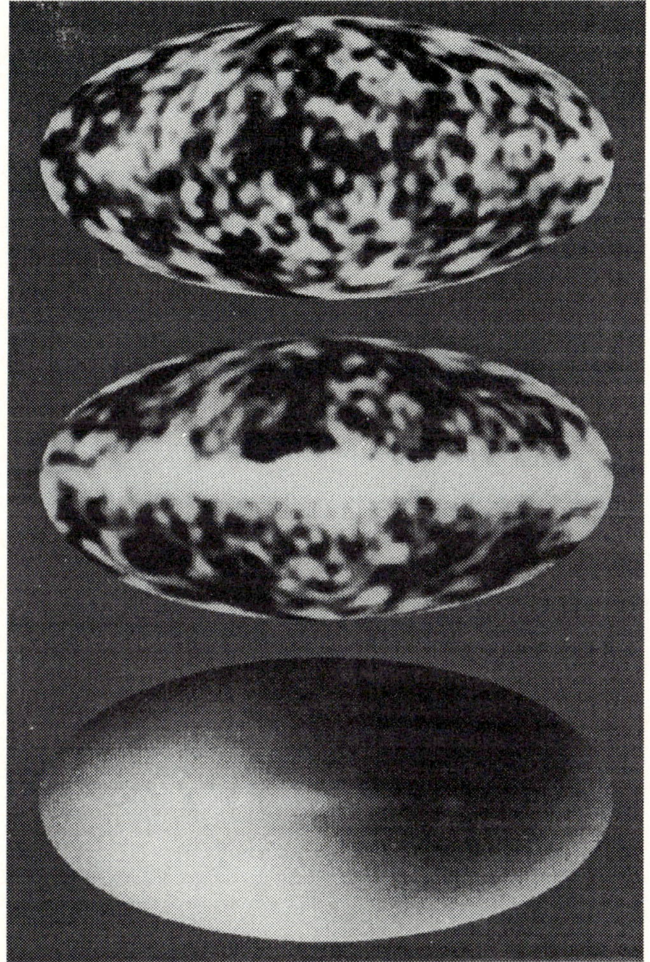

Fig. 1. Sky maps of the COBE experiments. The fluctuations seen on scales of $\sim 10^\circ$ correspond to temperature variations $\frac{\Delta T}{T} \simeq 10^{-5}$ (by courtesy of the COBE team).

quadrupole (90°). The spectrum is consistent with a flat Harrison–Zel'dovich spectrum ($C(\theta) \sim \theta^{1-n}$ with $n = 1$; see section 2).

The implications of this result for the problem of galaxy formation are obvious: The gravitational instability picture with dark matter components has now firm support. The initial conditions can be fixed.

It seems that other causes for these measurements such as systematic instrument errors and Galactic emission can be excluded with some confidence. First positive results from ground–based observations have also been reported.

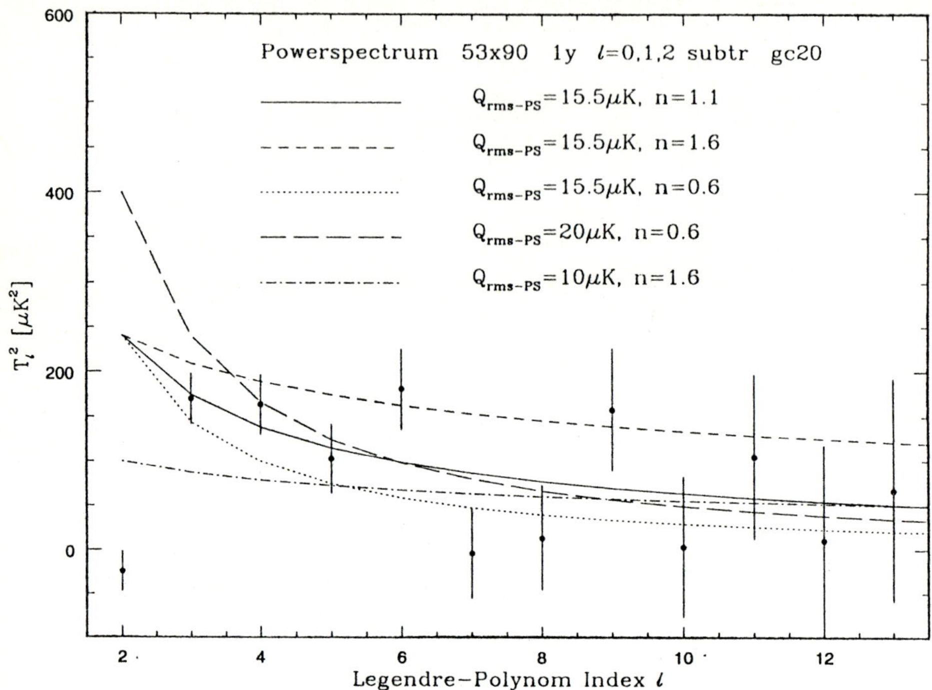

Fig. 2. Power spectrum of the CMB fluctuations in terms of an expansion in Legendre polynomials. The index ℓ corresponds to a scale $L = 2H_0^{-1}/\ell$ or an angular scale $\frac{180}{\ell}$ degrees.

2 Linear Perturbation Theory

2.1 The Smooth Background

The homogeneous Hubble expansion and the isotropic Cosmic Microwave Background (CMB) find an adequate representation in the Friedmann–Lemaître models of cosmology which can be described by the metric

$$ds^2 = +dt^2 - R^2(t)\left\{dr^2 + f^2(r)(d\theta^2 + \sin^2\theta d\varrho^2)\right\} \tag{1}$$

where

$$f^2(r) = \begin{cases} \sin^2 r & \text{for K} = +1 \\ r^2 & \text{for K} = 0 \\ \sin^2 hr & \text{for K} = -1 \ . \end{cases}$$

The Gaussian curvature of the 3–spaces at constant time is $\frac{K}{R^2}$, and here K has been normalized to ± 1 or 0. Einstein's field equations for this highly symmetric

metric can be obtained for an ideal fluid for which $(r, \theta, \varrho = \text{const.})$ are the flow lines. In these "comoving coordinates" one finds

$$\left(\tfrac{1}{R}\tfrac{dR}{dt}\right)^2 = \tfrac{8\pi}{3} G\bar{\varrho}_b(t) - \tfrac{K}{R^2} + \tfrac{1}{3}\Lambda \ ; \quad (2a)$$

$$\tfrac{d}{dt}\left(\bar{\varrho}R^3\right) + p\,\tfrac{d}{dt}\left(R^3\right) = 0 \ ; \quad (2b)$$

$\bar{\varrho}_b(t)$ and $p(t)$ are the homogeneous density and pressure of the fluid.

An equation of state has to be specified to close the equation s(2). For collisionless dark matter with negligible velocity dispersion and for cold baryonic matter, we may set $p = 0$. Then from (2b)

$$\bar{\varrho}\,R^3 = \text{const.}$$

Using the redshift z instead of R,

$$1 + z \equiv R_0/R \ , \quad (3)$$

we may write

$$\bar{\varrho} = \bar{\varrho}_0(1+z)^3 \ , \quad (4)$$

for pressureless matter.

For radiation, or for relativistic particles $p = 1/3\bar{\varrho}$ (here units are used in which the velocity of light $c = 1$), and then

$$\bar{\varrho} = \bar{\varrho}_0(1+z)^4 \ . \quad (5)$$

The ratio of pressureless matter $\bar{\varrho}_m$ and radiation $\bar{\varrho}_\gamma$ is

$$\frac{\bar{\varrho}_m}{\bar{\varrho}_\gamma} = \frac{\bar{\varrho}_{m_0}}{\bar{\varrho}_{\gamma_0}}\,\frac{1}{(1+z)} \ .$$

At present the matter density $\bar{\varrho}_{m_0}$ dominates. But at early times the density of radiation and relativistic matter increases faster by a factor $(1+z)$. There is a time, when $\bar{\varrho}_r = \bar{\varrho}_m$ namely

$$(1 + z_{eq}) = \frac{\bar{\varrho}_{m_0}}{\bar{\varrho}_{\gamma_0}} \ . \quad (6)$$

The value of z_{eq} deepens on the radiation and matter density at the present epoch, i.e. also on the amount of dark matter.

The universe is surely dominated by collisionless, nonrelativistic matter after the time of recombination, when hydrogen atoms have formed. This happens around $1 + z \simeq 1500$, about 3×10^5 years after the big bang.

G is the gravitational constant, and Λ the cosmological constant which has an interesting history of being set to zero, and being reintroduced alternatingly. For easy reference later on we want to introduce here a set of special variables, as follows: $H(t) \equiv \frac{1}{R(t)}\frac{dR}{dt}$, the Hubble parameter; at the present epoch $t = t_0$

it becomes $H(t_0) \equiv H_0$, the Hubble constant. $\lambda(t) \equiv \frac{\Lambda}{3H(t)^2}$; $\Omega(t) \equiv \frac{8\pi G\bar{\varrho}(t)}{3H^2}$; the density parameter. Rewrite equ.(2a) to read

$$\Omega + \lambda - 1 = \frac{K}{R^2 H^2} \ ; \tag{7}$$

at the present epoch we have

$$K = R_0^2 H_0^2 (\Omega_0 + \lambda_0 - 1) \ . \tag{8}$$

Since we have normalized K to ± 1, the value of R_0 is fixed.

Various astronomical observations limit the values of the cosmic parameters. For the purpose of this review we need not discuss in detail the observationally favourable values for Ω_0, λ_0, and H_0, as well as t_0, the age of the universe. From the range of values discussed (cf. Börner 1993), we can set limits

$$\begin{gathered} 0.005 \leq \Omega_0 \leq 2 \ ; \\ \lambda_0 \leq 0.7 \ ; \\ 12 \cdot 10^9 \text{years} \leq t_0 \leq 20 \cdot 10^9 \text{years} \ ; \\ 50 \leq H_0 \leq 100 \text{km s}^{-1} \text{Mpc}^{-1} \ . \end{gathered}$$

We write $H_0 = h \cdot 100 \left[\text{km s}^{-1}\text{Mpc}^{-1}\right]$, and to account for the uncertainty in H_0 we let h vary between 1/2 and 1. Then we restrict our consideration to either spatially open ($\Omega_0 < 1$, $\lambda_0 = 0$) or flat ($\Omega_0 \leq 1$, $\lambda_0 = 1 - \Omega_0$) models. The simplest model, the Einstein–deSitter model, with $K = 0$, $\lambda_0 = 0$, $\Omega_0 = 1$ is often used for comparison, although it may already be at variance with the astronomical data. In this model $R(t) \propto t^{2/3}$, and $t_0 H_0 = 2/3$.

2.2 Newtonian Perturbation Theory

For a Newtonian ideal fluid approximation the continuity equation, Euler's equation and the Poisson equation for the gravitational field are analysed for a smooth background on which small perturbations are superimposed (cf. Peebles 1980; Weinberg 1971; Börner 1993). The density contrast δ and the peculiar velocity $\underline{v}$ are introduced as the perturbation variables

$$\begin{aligned} \delta &\equiv \frac{\varrho(\underline{x}, t)}{\bar{\varrho}(t)} - 1 \\ \underline{v} &\equiv R(t)\dot{\underline{x}} \ . \end{aligned} \tag{9}$$

The evolution equation for δ is then

$$\ddot{\delta} + 2\frac{\dot{R}}{R}\,\dot{\delta} = \frac{\nabla^2 p}{\varrho R^2} + \frac{1}{R^2}\underline{\nabla} \cdot [(1+\delta)\underline{\nabla}\phi] + \frac{1}{R^2}\left[(1+\delta)v^i v^j\right]_{,ij} \tag{10}$$

Here ϕ is the gravitational potential

$$\nabla^2 \phi = 4\pi G \bar{\varrho} R^2 \delta \ .$$

These equations can in general only be solved by a numerical approach. As a first step it is, however, quite useful to consider the evolution of these quantities in the linear regime $|\delta| \ll 1$. If the pressure p is neglected — which may be a good approximation for large-scale structures — the linear equation for δ is

$$\ddot{\delta} + 2\frac{\dot{a}}{a}\dot{\delta} - 4\pi G\bar{\varrho}\delta = 0 \ . \tag{11}$$

This second-order differential equation has two independent solutions which are conventionally denoted by $D_1(t)$ and $D_2(t)$. $D_1(t)$ is a growing mode (δ increases with time), $D_2(t)$ a decaying one.

In Fig. 3 $D_1(t)$ is plotted for various sets of cosmological parameters. For $\Omega_0 = 1$ one simply has $D_1(t) \propto R(t)$. We can see from the curves that for $\Omega_0 < 1$ the linear growth is strongly reduced, as $R_0/R = 1 + z$ approaches 1.

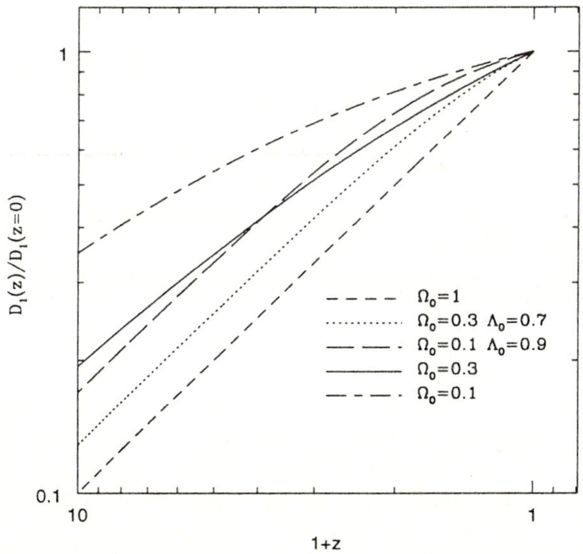

Fig. 3. The linear growth factor $D_1(z)$ for different cosmological models as a function of z. Various combinations of Ω_0, λ_0 are displayed. All models have been normalised to the same value at the present epoch $z = 0$.

Even for the strongest growth of $\delta(t)$, there is the following problem: Perturbations in the baryonic matter–density can only grow from the time of recombination t_R. The linear growth for $\Omega_0 = 1$ is thus $\frac{D_1(t_0)}{D_1(t_R)} = (1 + z_R) \simeq 1500$.

At recombination $\delta(t_R)$ is limited by the anisotropies of the CMB. If we take the COBE result as typical $\frac{\Delta T}{T} = 10^{-5}$, then for adiabatic perturbations $\delta(t_R) = 3|\frac{\Delta T}{T}| = 3\times10^{-5}$, and $\delta(t_0) = (1+z_R)\,\delta(t_R) = 1500\cdot3\cdot10^{-5} = 4.5\times10^{-2}$. The amplitude of the density perturbations stays small, does not grow to the nonlinear regime $\delta \simeq 1$. This means that the universe stays almost homogeneous, and that the structures we observe could not have formed.

One way out of this dilemma is to postulate dark, nonbaryonic matter as an additional component of cosmic matter. Perturbations of nonbaryonic dark matter can grow from z_{eq} which often is larger than z_R, and they couple to photons only through the gravitational potential wells they form, the "Sachs–Wolfe" effect (Sachs & Wolfe 1967). Thus the CMB anisotropy limits can be met, and also structures may form in the dark matter component. This approach will be described in the following.

2.3 Statistical Description of Density Perturbations

For a statistical description of the cosmic density the Fourier components of δ are widely used. The basic assumption is that the distribution of mass is periodic, when sufficiently large cubic volumes V_u, fixed in comoving coordinates, are considered. The power spectrum of density fluctuations is defined as

$$P(\underline{k}) \equiv \langle |\delta_k|^2 \rangle \ , \tag{12}$$

where δ_k is the Fourier component with wave number k of the density contrast:

$$\delta_k \equiv 1/v_u \int \delta(\underline{x}) \exp(i\underline{k} \cdot \underline{x})\, dx \tag{13}$$

Here we wrote $\delta(\underline{x})$, since we consider the complete density contrast $\delta(\underline{x}, t)$ as consisting of 2 factors: $\delta(\underline{x}, t) = D_1(t)\delta(\underline{x})$.

The brackets $\langle \ \rangle$ denote an ensemble average. The background is homogeneous and isotropic, and therefore the ensemble average is replaced by an average over a fair sample, or the spatial average within it.

Another statistical measure is the two–point correlation function $\xi(\underline{x}_1, \underline{x}_2)$. It is for point–like objects defined by the joint probability to find an object in a comoving volume δV_1 centered at $\underline{x}_1$ and another one in δV_2 at $\underline{x}_2$:

$$\delta P_{12} = \bar{n}^2 R^6 (1 + \xi(\underline{x}_1, \underline{x}_2)\, dV_1\, dV_2 \ ,$$

where $\bar{n}$ is the mean number density of objects. In terms of a density field $\varrho(\underline{x})$ ξ can be written as

$$\xi(\underline{x}_1, \underline{x}_2) = \frac{< \varrho(\underline{x}_1)\, \varrho(\underline{x}_2)}{\bar{\varrho}^2} - 1 \ . \tag{14}$$

Because of rotational and translational invariance $\xi(\underline{x}_1, \underline{x}_2) \equiv \xi(|\underline{x}_1 - \underline{x}_2|)$ $\xi(x)$ and $P(k)$ are a Fourier transform pair

$$P(k) = \frac{4\pi}{V_u} \int \xi(x) \frac{\sin kx}{kx}\, x^2\, dx \ ;$$
$$\xi(x) = \frac{V_u}{\pi^2} \int P(k) \frac{\sin kx}{kx}\, k^2\, dk \ . \tag{15}$$

$P(k)$ and $\xi(x)$ give equivalent information. For the practical data analysis, however, $\xi(x)$ is more convenient since it can more easily be adapted to observational requirements, such as the shape of the survey region or selection effects. This is best achieved by using the estimator

$$\xi(x) = \frac{DD(x)}{RR(x)} \frac{n_B}{n_D} - 1 \ , \tag{16}$$

where n_D is the mean number density of data points, n_R the mean density of points in a random distribution in the same volume.

$P(k)$, as well as $\xi(x)$, are defined through the amplitudes $|\delta_k|$ only. But $\delta_k \equiv |\delta_k| \exp(i\,\phi_k)$ is a complex field, and it might be interesting to recover information on the phase ϕ_k too. This question has not been dealt with in great detail so far.

To set up acceptable initial conditions for galaxy formation the statistical distribution of the density fluctuations must be prescribed. Usually a random Gaussian field is assumed which leads to a probability function for the Fourier components, such as

$$\Pi(|\delta_k|\,, \phi_k)\, d\,|\delta_k|\ d\phi_k = \frac{2\,|\delta_k|}{P(k)} \exp\left(-\frac{|\delta_k|^2}{P(k)}\right) d\,|\delta_k|\ \frac{d\Phi_k}{2\pi} \ . \tag{17}$$

The distribution function is completely determined by $P(k)$ as expected. The ansatz for $\delta(\underline{x})$ as a Gaussian random field becomes invalid, when the fluctuations become nonlinear. $\delta(x) \geq -1$ by definition, but for a random Gaussian field one must formally have $-\infty < \delta < +\infty$. In the linear regime $|\delta| \ll 1$ there is no problem, but for nonlinear fluctuations the distribution functions will loose their Gaussian nature.

2.4 The Primeval Fluctuation Spectrum

The power spectrum $P(k)$ must be derived from physical processes in the early universe. Up to now only the model of the inflationary universe (Linde 1990) has produced at least an idea, how quantum fluctuations just after the big bang can lead to density perturbations later on. These schemes can certainly not be viewed as an exact derivation, but they make it plausible that a specific type of power spectrum arises: $P(k) \propto k$. The amplitude of the perturbations is fixed by the gradient of the effective potential energy $\frac{dV}{d\varphi}$ of the Higgs–field φ which drives inflation (cf. Linde 1990). The potential $V(\varphi)$ is adapted such that the CMB anisotropy limits can be met. It seems that the CMB measurements favour a power spectrum $\propto k$. This has become therefore the favourite input for models of structure formation.

It has already been pointed out earlier (Harrison 1970) that for a power law ansatz

$$P_i(k) = A_i k^n \ , \tag{18}$$

("i" for initial) the index n is constrained, because for $n \gg 1$ there would be an overproduction of small–mass objects, like black holes, while for $n \ll 1$ the large scales would have too high amplituded which could destroy the global isotropy and homogeneity of the universe. Harrison concluded that $n = 1$ was a preferred value for the index.

The value $n = 1$ has the consequence that in the matter dominated epoch fluctuations of different wavelengths $\frac{2\pi}{k}$ have the same amplitude at the time when their scale $k^{-1}R(t)$ is equal to the horizon scale $\lambda_H = H^{-1}(t)$.

The amplitude A_i can only be predicted, when the physical mechanisms and the cosmic epoch of the generation of primordial density fluctuations are known. Since we have only vague glimpses of the generating processes the general attitude is to view the initial density fluctuations as free parameters which can be chosen subject only to some weak constraints from observations. For simplicity one takes

$$P_i(k) = A_k k^n \ , \tag{19}$$

with $n = 1$ as a preferred choice. This ansatz is not compulsory, however, and we should keep in mind that in principle physical processes determine the initial power spectrum. We choose it freely at present only because of our lack of knowledge.

3 Matter Content of the Universe and Time Evolution of the Power Spectrum

On each scale the primordial fluctuations larger than the horizon size just follow the evolution equations of a FL–model. The spectrum remains unchanged. But once the horizon scale encompasses the fluctuation, the spectrum is changed, because the density perturbations of non–relativistic matter ϱ_m do not grow, as long as the energy density in relativistic particles ϱ_r dominates. From the time t_{eq}, when $\varrho_m = \varrho_r$ the fluctuations grow. This defines a characteristic scale, namely the size of the horizon λ_{eq} at cosmic time t_{eq}. The changes of the primeval spectrum depend on the components of the cosmic matter. We shall consider possible matter models in section (3.1).

3.1 Cosmic Matter Components

Dark matter in the universe can be baryonic or nonbaryonic. Evidence from nucleosynthesis limits the value of the baryonic mean density

$$\Omega_B h^2 \leq 0.015 \ . \tag{20}$$

The analysis of the dynamics of galaxies and clusters, and the large–scale velocity fields (cf. Börner 1993; Peebles 1993) points to a much larger value of Ω, perhpas even $\Omega \simeq 1$. Thus, in addition to the constraints on the density fluctuations there is a hint from astronomical observations that nonbaryonic dark matter exists.

There are many theoretical candidates, but so far no positive detection of a massive dark matter particle. Dark matter particles are in thermal equilibrium at early stages, but eventually they decouple from the rest of the universe. A useful terminology seems to classify them according to their random velocity at the time of their decoupling (Bond *et al.* 1983). "Hot dark matter (HDM)" are light particles ($m \leq 100$eV) which are relativistic when they decouple. The relativistic free streaming motion erases all fluctuations on scales smaller than the horizon scale at the epoch when these particles have cooled down to non–relativistic velocities. A typical example would be a hypothetical massive neutrino with a mass $m_\nu \simeq 10$eV. Neutrinos of this kind have a number–density comparable to the density of photons.

"Cold dark matter" (CDM) particles decouple very early. They are either very heavy particles (such as photinos, gravitinos or neutralinos) which are non–relativistic very early, or they appear with essentially zero random velocity (a Bose condensate of axions, quark lumps). No CDM particle is know experimentally, while for HDM we know the neutrinos, we just do not know whether they have a mass.

a) Hot dark matter

Massive neutrinos ($m_\nu \leq 100$eV) decouple when the temperature drops below 1 MeV. For each ν–species we have today

$$n_{\nu 0} = \frac{3}{4}\frac{4}{11}\, n_{\gamma_0} = 109\, \vartheta^3 \text{cm}^{-3} \left(\vartheta \equiv \frac{T_{\gamma_0}}{2.7K}\right) .$$

The contribution to the cosmological density is — if we consider one species with mass m_ν

$$\varrho_{\nu_0} = m_\nu\, n_{\nu_0} = 3270 \text{eV cm}^{-3} \left(\frac{m_\nu}{30\text{eV}}\right) \vartheta^3 .$$

Since the critical density is $\varrho_c = 10500 h^2 \text{eV cm}^{-3}$ we write

$$\Omega_\nu \equiv \varrho_\nu / \varrho_c = 0.31 h^{-2} \vartheta^3 \left(\frac{m_\nu}{30\text{eV}}\right) . \tag{21}$$

Thus massive neutrinos can give $\Omega_\nu = 1$. They can be the dominating cosmic matter component. The neutrinos are relativistic until the temperature has dropped below m_ν (i.e. $kT \leq m_\nu c^2$). At that time the horizon size is

$$\lambda_H = d_\nu = ct(T = m_\nu) \simeq (G\varrho)^{-1/2} \simeq M_{p\ell} m_\nu^{-2} , \tag{22}$$

where we have set $(G)^{-1/2} = M_{p\ell} \simeq 10^{19}$GeV ($\hbar = c = 1$), and $\varrho = am^4$ at $T = m_\nu$.

All fluctuations on a scale less than $d_\nu \sim M_{pe} m_\nu^{-2}$ are damped by the "free–streaming" of the relativistic collisionless neutrinos. The free–streaming scale is (at present)

$$d_\nu = 13(\Omega_\nu h^2)^{-1} \vartheta^{-2} \text{Mpc} . \tag{23}$$

This scale is comoving with the expanding universe, i.e.

$$d_\nu \propto (1+z)^{-1} .$$

b) Cold dark matter

For cold dark matter the free–streaming is unimportant, because the scale is smaller than any scale of astronomical interest. CDM fluctuations can only grow, when the energy density of the nonrelativistic CDM component ("X–particles") becomes larger than the relativistic energy density. With N_ν types of neutrinos we have

$$\varrho_{\rm rel} = \varrho_\gamma \left(1 + N_\nu 7/8 \left(\frac{T_\nu}{T_\gamma}\right)^4\right) \simeq \varrho_\gamma (1 + 0.227 N_\nu) \ . \tag{24}$$

Then $\varrho_x = \varrho_{\rm rel}$ at t_{eq}, and

$$1 + z_{eq} = 4.2 \times 10^4 \ \Omega \times h^2 \ (1 + 0.227 N_\nu)^{-1} \ . \tag{25}$$

The horizon size $\lambda_{eq} \simeq \int_0^{t_{eq}} \frac{dt}{R(t)} \propto \Omega^{-1} h^{-2}$ determines a typical scale.

c) Baryons
Fluctuations in the baryonic component can only grow on scales larger than the Jeans length $\lambda_J = \frac{2\pi}{k_J}$, with $k_J = \left(\frac{4\pi G \bar{\varrho}}{c_s^2}\right)^{1/2}$. λ_J is very large–comparable to d_ν of the neutrinos before recombination, and after recombination it is rather small $\lambda_J^3 \simeq (10^5 M_\odot (\Omega_B h^2)^{-1/2} / \bar{\varrho})$; $\lambda_J \simeq 5 {\rm kpc} (\Omega_B h^2)^{-1/2}$.

After recombination the baryons just follow the gravitational potential created by the DM components.

3.2 Changes of the Spectrum

The inital power spectrum $P_i(k)$ will be changed by the time evolution of the DM components around t_{eq} (or $T = m_\nu$). A precise calculation must solve the linearized Boltzmann equations for photons, baryons and DM distribution functions (cf. Bardeen *et al.* 1986). The dark matter particles are taken as collisionless, noninteracting except by gravity. Then the properties of each component can be described separately. The power spectrum

$$P(k,t) = c(t) P_i(k) T^2_{\rm linear}(k) \tag{26}$$

for the linear evolution can be described by a linear transfer function.

For HDM

$$T_{\rm lin.} = \begin{cases} 1 & \text{for } k < \frac{2\pi}{d_\nu} \\ e^{-d_\nu k} & \text{for } k > \frac{2\pi}{d_\nu} \end{cases} ; \tag{27}$$

for CDM

$$T_{\rm lin.} = \begin{cases} 1 & \text{for } k < \frac{2\pi}{\lambda_{eq}} \simeq 0.64 (\Omega h^2) {\rm Mpc}^{-1} \\ k^{-2} & \text{for } k > \frac{2\pi}{\lambda_{eq}} \end{cases} . \tag{28}$$

The clustering behaviour in these two models is quite different. A fluctuation on the physical scale r has an rms–mass fluctuation amplitude of

$$\langle \left(\frac{\delta M}{M}\right)^2 \rangle \propto r^{-(3+n)} \ , \quad \text{if } P(k) \propto k^n \ .$$

Starting from $P_i(k) \propto k$, we found that in the CDM–model $P(k) \propto k^{-3}$ for $k > k_{eq}$. Actually this is a gradual transition from $P_i \propto k$. The amplitude

$\langle (\frac{\delta M}{M})^2 \rangle$ decreases monotonically with the physical scale r. Thus small scale structures will grow fastest, collapse first, and then cluster to form successively larger structures. This is the pattern of hierarchical clustering.

In the HDM–model, the exponential cut–off $e^{-d_\nu k}$ dominates completely, and for any $P_i \propto k^n$ the first structures to form have scales proportional to d_ν.

Smaller structures must then originate via fragmentation processes of these large initial perturbation which encompasses typically masses of $\sim 10^{16} M_\odot$.

For baryons the typical horizon size is ct_R, the horizon at recombination $(k_R = 0.05(\Omega\ ^2)\mathrm{Mpc}^{-1})$ h

$$T_{\mathrm{lin}} = \begin{cases} 1 & \text{for } k < k_R \\ k^2 & \text{for } k_R < k < k_s \\ e^{-(k/k_s)^2} & \text{for } k > k_s \end{cases} \quad . \tag{29}$$

k_s is a damping scale which is produced by photon diffusion at epochs before recombination, when baryonic perturbation oscillate like acoustic waves. The mass–scale corresponding to this "Silk damping" (Silk and Efstathiou 1983) is

$$M_s = 3 \times 10^{12} (\Omega h^2)^{-5/4} M_\odot \quad .$$

In Figure 4 the transfer function is plotted for various matter models. We can already see that on large scales the models are very similar. This is an effect of the normalisation procedure.

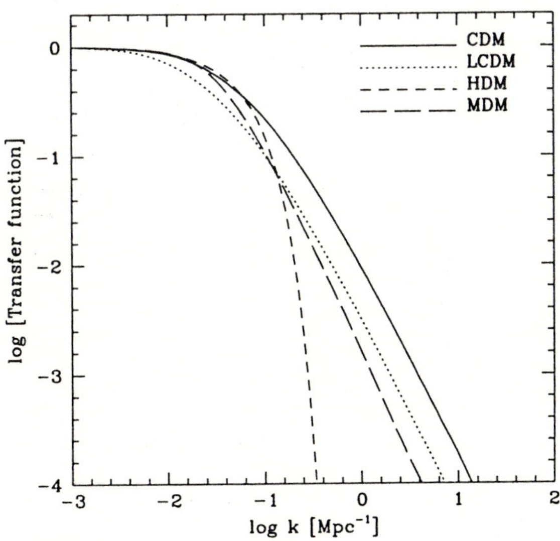

Fig. 4. The logarithm of the transfer function is plotted against wave number k. The scale λ_{eq} corresponding to the equality of matter and radiation energy densities is indicated.

3.3 Normalisation

The amplitude $C(t)A$ must be fixed ($P_i(k) = Ak$), and this is usually done by a comparison with astronomical data. Under the assumption that the linear theory is valid until the present epoch, we write $C(t) = \left[\frac{D_1(t)}{D_1(t_o)}\right]^2$, where $D_1(t)$ is the linear growth rate. To determine A, one way is to use the COBE results which apply to very large scales, such that even at present $T_{\rm lin} = 1$.

For $\Omega + \lambda = 1$ models

$$A_{\rm COBE} = \left(\frac{96\pi^2}{5}\right) \Omega_0^{-1.54} H_0^{-4} \left(\frac{Q_{\rm rms}}{T_{\gamma_0}}\right)^2 , \tag{30}$$

where $Q_{\rm rms}$ is the COBE rms–quadrupole fluctuation. Then at large scales, where $T_{\rm lin} = 1$, all the models have the same $P(k)$.

Another popular way is to use the amplitude of the mass fluctuation in randomly placed spheres of radius S.

$$\langle \left(\frac{\delta M}{M}\right)^2 (S) \rangle = \int_0^\infty W^2(kR)\, P(k)\, \frac{k^2 dk}{2\pi^2} . \tag{31}$$

W is a filter function. For a so–called top–hat window

$$W_{TH}(x) = \frac{3}{x^3} (\sin x - \cos x) . \tag{32}$$

The corresponding mass fluctuations for galaxies can be derived from the two–point correlation function ξ_{gg}. At a scale of $8h^{-1}$Mpc one finds that the quantity

$$\sigma_{8,g} \equiv \left(\frac{3}{4\pi S^3}\right) \left(\int_V dV_1 dV_2 \xi_{gy}(r_{12})\right)^{1/2} \simeq 0.96 \tag{33}$$

(Peebles 1980).

It has turned out that good fits to the obervations can only be achieved in DM models of galaxy formation, if δ_8 from the models is set equal to $b^{-1}\delta_{8,g}$, where b is called a biasing parameter (e.g. Davis *et al.* 1985).

In a CDM model with $P_i(k) \propto k$, $\Omega = 1$ a value of $b \simeq 3$ is required to achieve a reasonable fit to ξ_{gg}. The physical basis for the bias is the idea that the galaxies do not trace the DM distribution, but rather that peaks in the DM density field correspond to galaxies. The value of b is introduced ad hoc, however, and the uncomfortable feeling remains that here is another quantity which is treated as a free parameter, but which in fact is determined by the physics of galaxy formation.

3.4 A Few Remarks on the Status Quo

Linear perturbation theory is simple and reliable once the conditions of a model have been set down, and as long as the density contrast is smaller than unity. The evolution of structures in the universe can only be described in detail, if non-linear gravitational clustering is understood. Before we look into that problem it may be worthwhile to sit back for a moment, and consider what we have achieved in our model building. Several quantitites are needed for a model of structure formation: The amplitude and spectrum of the initial density perturbations, the type and fraction of dark matter, and the bias factor have to be specified to define a model, whose linear evolution can be computed unambiguously. In addition, of course, the parameters of the cosmological model $(H_0, \Omega_0, \lambda_0, T_{\gamma_0})$ are needed. Usually some limits are imposed on these quantities from observations. But there is a wide range of possible models. It is quite a task to explore the parameter space that seems available, and to narrow down the possibilities by a careful consideration of the observational consequences. It is important to realize that the freedom in model building is not intrinsic, but is due to our lack of knowledge of the basic physical processes, as well as to the large uncertainties in the measurements of the cosmic parameters. In fact, looking at the situation from first principles, we find that the cosmic parameters, as well as the type and fraction of dark matter can be determined from observations. The initial density perturbations as well as the bias factor should be derived from physical mechanisms operative in the very early universe, and in galaxy formation resp. Thus there is really just one correct model of structure formation, but since we do not have the necessary information, we can only try out one model after another, and see how it fares with respect to the observations.

4 Some Remarks on Nonlinear Gravitational Clustering

The evolution of a system of N self–gravitating particles is best studied by numerical simulations. But the various N–body computer codes are all limited in speed and resolution. To gain some understanding of the physical processes and to study the dependence on cosmological parameters the numerical methods may be supported by analytic considerations.

The Zel'dovich pancake model (cf. section 4.1) and the Press–Schechter method to study the size distribution of the condensed masses (cf. section 4.3) will be briefly discussed in the following.

4.1 The Zel'dovich Pancake Model

In a neutrino dominated universe structures of mass $M_{p\ell}^2 m_\nu^{-2} \sim 10^{15} M_\odot$ are the first to start growing, the first to reach the non–linear regime, the first to collapse. Subsequent fragmentation of such supercluster structures should then be the cause of galaxy formation. Although the fragmentation process is not well understood, it seems that the simple adiabatic collapse cannot produce a sufficient number of condensed objects on smaller scales (cf. Fig. 4). Thus without additional ingredients the "pancake" model cannot serve as a model for galaxy formation, but it gives an excellent account of the large-scale structure observed (Geller and Huchra 1989) — the arrangement of galaxies in sheets and filaments surrounding large empty regions.

It was shown some time ago (Zel'dovich 1970) that in general this collapse will proceed asymmetrically. The theory was originally proposed for baryonic adiabatic fluctuations with a large scale below which all substructure was suppressed. The theory therefore applies nicely to neutrino structures. The main point is that particle trajectories are considered of the following form: If $\underline{q}$ is the Lagrange coordinate of a particle its position $\underline{r}$ in Eulerian coordinates is expressed as

$$\underline{r} = R(t)\left(\underline{q} - \frac{b(t)}{R(t)}\nabla\psi(\underline{q})\right) . \tag{34}$$

$\psi(\underline{q})$ is a function of the comoving coordinates $\underline{q}$ describing the effect of an inhomogeneity on the motion.

The density evolution is derived from the continuity equation

$$dm = \bar{\varrho}\, d^3q\, R^3 = \varrho(\underline{q},t)\, dxdydz$$

as

$$\varrho(\underline{q},t) = \bar{\varrho}(t)\left\{\det\left[\delta_{ij} - \frac{b}{R}\frac{\partial^2\psi}{\partial_{q_i}\partial_{q_j}}\right]\right\}^{-1} . \tag{35}$$

If the eigenvalues of the matrix are $\alpha > \beta > \gamma$, then

$$\varrho = \bar{\varrho}(1-\alpha b/R)^{-1}(1-\beta b/R)^{-1}(1-\gamma b/R)^{-1} . \tag{36}$$

In the linear limit ($ab/R \ll 1$) it reduces to

$$\delta \propto (\alpha+\beta+\gamma)b/R \ ;$$

$b(t)$ can be determined (for $K=0$, $\Omega=1$: $b/R \propto t^{2/3}$ i.e. $b \propto R^2$).

b/R grows as a power of the time t; $\alpha b/R$ will approach one, and the density becomes infinite. The collapse of an inhomogeneity thus occurs first in one particular direction, leading to transient sheet-like or "pancake" structures. The validity of these extrapolations into the non-linear regime must be justified. In one dimension the form (35) is an exact solution of the Euler-Poisson equations. Despite interesting attempts to justify (35) in general as an approximation (Buchert 1989) the best argument in favour of this scheme is the excellent agreement with numerical N-body simulations (Buchert, Melott, Weiß 1994). In Fig. 5 a high resolution Zel'dovich approximation is shown (Buchert & Bartelmann 1991). The HDM free-streaming length d_ν was implemented as a cut-off scale in a flat initial spectrum. Several million particles have been used to simulate the density field of a section of the universe. A cellular pattern of the large-scale structure clearly emerges, where pancakes as filamentary or sheet-like objects surround large areas of very low density. The qualitative similarity to the CfA results is quite remarkable.

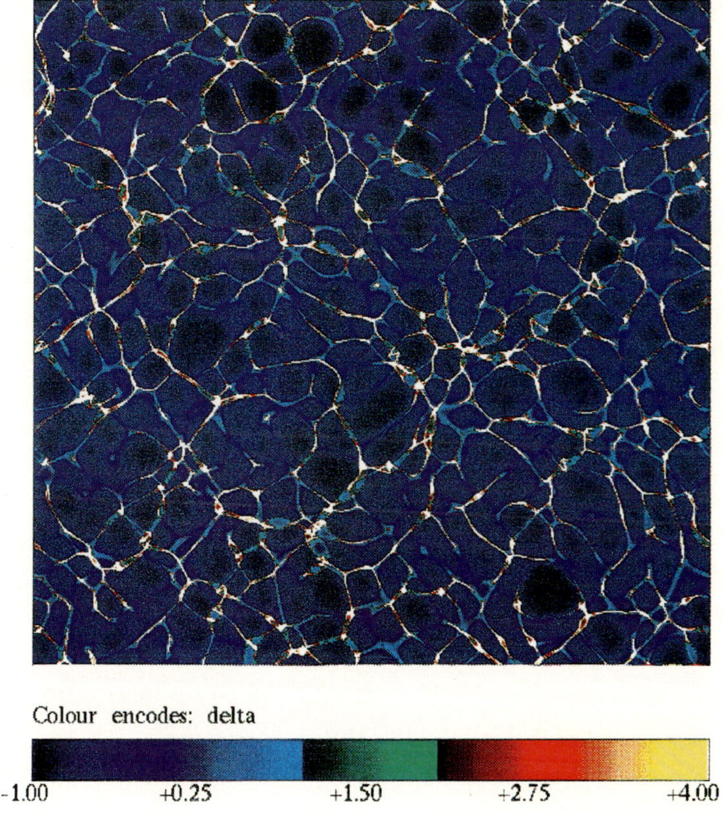

Fig. 5. A numerical representation of a slightly improved Zel'dovich approximation (Buchert & Bartelmann 1991). Several million particles were used in this representation. The density field is shown in a 2D–slice with 512×512 pixels. Dark regions have low density, light regions high density. The initial $P(k)$ was flat with a cut–off length λ_c. The side–length of the square is $12\lambda_c$.

4.2 Spherical Inhomogeneities

It is quite instructive to consider the evolution of a spherical inhomogeneity in a FL universe filled with dust, i.e. zero–pressure matter. This problem can be solved analytically (Novikov & Zel'dovich 1983). The spherical, overdense region reaches its maximum expansion at an epoch t_m, it separates from the background expansion after that, and recollapses. The kinetic energy is zero at the moment of turn–around, and therefore the total energy is equal to the potential energy at turn–around (per unit mass)

$$E_{\text{tot}} = -\frac{1}{2} \frac{GM}{r_m} \quad ;$$

after the collapse, we assume that a virialised state is reached with radius $r_{\rm vir}$. The total energy is conserved, and the virial theorem tells us that the new potential energy is equal to $2E_{\rm tot}$. Therefore

$$r_{\rm vir} = r_m/2 \ , \quad \text{and} \quad \varrho_{\rm vir} = 8\varrho_m \ .$$

At the time t_m of maximum expansion we use the equations of a $k = +1$ FL–model to find

$$r_m = \frac{2t_m}{\pi} \ ; \quad \frac{2GM}{r_m} = 1 \ .$$

In a background $K = 0$ model $\bar\varrho(t) = (6\pi G t^2)^{-1}$, and thus $\varrho_m = \frac{3\pi}{32 G t_m^2}$. The density contrast at the time of maximum expansion can be computed most simply in a flat background model (Einstein–deSitter, $K = 0$)

$$\frac{\varrho_m}{\bar\varrho} = \frac{9}{16}\pi^2 \simeq 5.6 \ , \quad \text{i.e. } \delta_m = 4.6 \ . \tag{37}$$

This value is independent of initial conditions, and thus it gives a very useful criterion for the turn–around and collapse. If one extrapolates linear theory the density contrast at turn–around is $\delta_m \sim \frac{3}{20}(6\pi)^{2/3} \approx 1.06$. If we assume that the collapse proceeds to a virialized state the sphere shrinks to a radius $r_{\rm vir} = \frac{1}{2} r_m$ at turn–around at a time $t_c = 2t_m$. Since $\bar\varrho(t_m) = 4\bar\varrho(t_c)$ the density of the virialized clumps is

$$\bar\varrho_{\rm vir} = 8\varrho_m = 18\pi^2 \bar\varrho(t_c) = 177.7\bar\varrho(t_c) \ , \tag{38}$$

about 180 times above the background density. This is again a result which characterizes collapsed virialized clumps of matter independent of initial conditions. Again an extrapolation of linear theory gives for this epoch

$$\delta_{\rm lin,vir} \approx 1.69 \equiv \delta_c \ . \tag{39}$$

Whether this value is a reasonable estimate is unclear. Obviously linear theory can be valid only for $\delta \ll 1$.

4.3 The Press–Schechter Mass Function

An important application of the spherical collapse calculation is an estimate of the number density of objects of mass M which have turned around and collapsed at redshift z. The simplifying assumption that spherical random Gaussian density fluctuations collapse and form bound objects when their density contrast smoothed over the mass scale M and computed in linear theory reaches δ_c (equation (39)), leads to the differential mass function per unit comoving volume $n(M,z)dM$ at redshift z (Press & Schechter 1974; PS)

$$\begin{aligned} M = & -\left(\tfrac{2}{\pi}\right)^{1/2} \tfrac{\varrho_0}{M} \tfrac{\delta_c}{\Delta^2(M,z)} \tfrac{d\Delta(M,z)}{dM} \\ & \times \exp\left\{-\tfrac{\delta_c^2}{2\Delta^2(M,z)}\right\} dM \end{aligned} \tag{40}$$

$n(M,z)$ is the comoving number density at redshift z of objects with mass between M and $M+dM \cdot \varrho_0$ is the present mass density, and $\Delta(M,z)$ is the density fluctuation computed from linear theory

$$\Delta^2(M,z) = \frac{D^2(z)}{2\pi^2} \int_0^\infty P(k,t_0)\, W^2(kr_0)\, k^2 dk \qquad (41)$$

$D(z)$ is the linear growth factor $D(z) = \frac{D_1(R(z))}{D_1(R_0)}$. $W(kr_0)$ is the top–hat window function, with $\frac{4\pi}{3}\, \varrho_0\, r_0^3 \equiv M$.

This expression has been originally derived in the Einstein–deSitter universe, but it is also valid in $\Omega_0 < 1$, and $\lambda_0 \neq 0$ models (Lilje 1992). It is not clear, how serious one should take this Press–Schechter formalism, because its physical foundation is a bit shaky. Studies with N–body simulations (Lacey & Cole 1994) show that the PS estimate is quite a good approximation provided δ_c is close to the critical value 1.69. It seems that this model can be used quite successfully to build an evolution model of galaxy formation, where smaller objects aggregate to form successively larger structures as time progresses.

The formula (40) for $n(M,z)$ can be used to estimate the mass density of objects $\Omega(M,z)dM$:

$$\Omega(M,z)dM = 3.60h^4 \frac{M}{10^{12}M_\odot}\, n(M,z)\, dM\; (\mathrm{Mpc})^3 \;. \qquad (42)$$

The mass function $n(M,z)$ can also be written in terms of the circular velocity of bound objects

$$V_c = \sqrt{\frac{GM}{S}} \;.$$

If we take S to be the virial radius $r_{\mathrm{vir}} = 1/2 r_m(M,z)$ of dissipationless spherical collapse, it may be related to the comoving radius $r_0 = r_0(M)$.

With the definition $\delta_{\max}(z) \equiv \frac{r_0(M)}{r_m(M,z)}$, we find

$$V_c = (GM/1/2r_m)^{1/2} = H_0 \Omega_0^{1/2} \delta_{\max}^{1/6} r_0(M) \qquad (43)$$

and finally

$$n(V_c,z)dV_c = \frac{3}{(2\pi)^{3/2}}\, \delta_c H_0^2 \Omega_0 \delta_{\max}^{1/3} \Delta^{-2}(M,z)\, \frac{d\Delta}{dr_0} \times \exp\left\{-\frac{\delta_c^2}{2\Delta^2}\right\} dV_c \;. \qquad (44)$$

For arbitrary Ω_0 and λ_0, the collapse factor $\delta_{\max}(z)$ has to be found numerically on the basis of the spherical nonlinear model (see e.g. Suto 1993).

4.4 Numerical Simulations

The effects of nonlinear gravitational clustering in generic situations can only be studied using numerical codes which integrate the equations of motion for N particles in an expanding universe. The ideal situation would be to include the radiation hydrodynamics of baryonic matter on galactic scales in addition to the purely gravitational interaction of the dark matter. This is not possible, however, since observations show structures from 100 kpc to 100 Mpc suggesting a dynamical range of at least 10^3. Therefore to resolve the structures on all scales about 10^9 particles should be simulated just to cover the purely gravitational interaction adequately. In addition probably at least 10^2 particles per galaxy are needed to give a rough idea of the hydrodynamic effects. Altogether the straightforward model would need more than 10^{11} particles, whereas sophisticated codes on available computers describe about 10^6 particles.

It is clear that we are a long way from the ideal situation. There must be approximations to the physics in each type of numerical simulations.

As a consequence several open questions, such as the biasing mechanism, cannot yet be addressed by theoretical models. If purely dark matter simulations are considered, it is quite tricky to identify individual galaxies, for instance. (cf. Davis *et al.* 1985; Suginohara *et al.* 1991; Suto 1993)

Some way to find galaxies as peaks in the density field has to be implemented. Then the simulated section of the universe can be compared to the observed universe, and quantities such as the galaxy and cluster 2–point correlation function, the velocity field, the Hubble sequence, and the rotation curve of galaxies can be used to test the models. As a result a preferred set of cosmological parameters (Ω_0, λ_0, H_0; power spectrum index n; DM composition) might be found.

We will briefly comment on this in section 5; but especially the many excellent reviews on this subject should be consulted (see e.g. Suto 1993; White *et al.* 1993; Peebles 1993; Efstathiou *et al.* 1992). Let us just mention a model that consists of a mixture of CDM and HDM (Jing *et al.* 1994). The idea of such a MDM ("mixed dark matter") model is to combine the small–scale perturbations of a CDM component with the large–scale power of an HDM component. In Fig. 6 some linearly evolved power spectra $P(k)$ are shown.

Fig. 7 gives a visual impression of the galaxy distribution in an MDM model. The galaxies trace the structures in the mass distribution, and the galaxy distribution is clumpier (the biasing factor is $b \sim 1.5$). The morphology of the large–scale structures resembles that revealed by the observations (e.g. Geller & Huchra 1989).

5 Observational Tests

The voids and sheets of the large–scale distribution of the galaxy which give such spectacular visual impressions are difficult to describe with quantitative statistical methods. Statistical measures of the topology have not been applied to such an extent that useful tests of models can be done. There are several

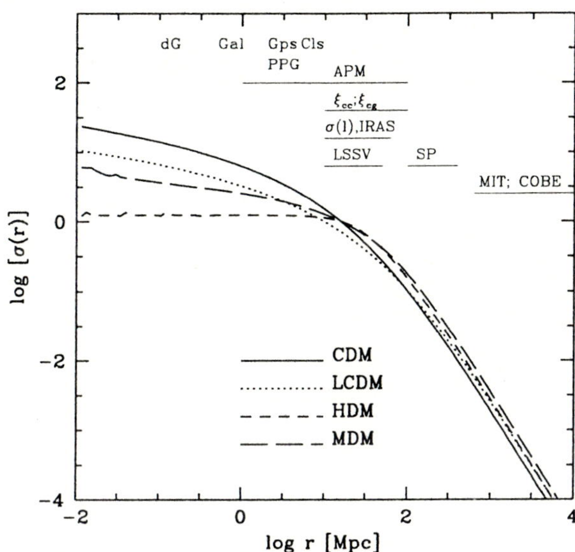

Fig. 6. The power spectrum $P(k)$ for CDM, MDM, HDM models is plotted.

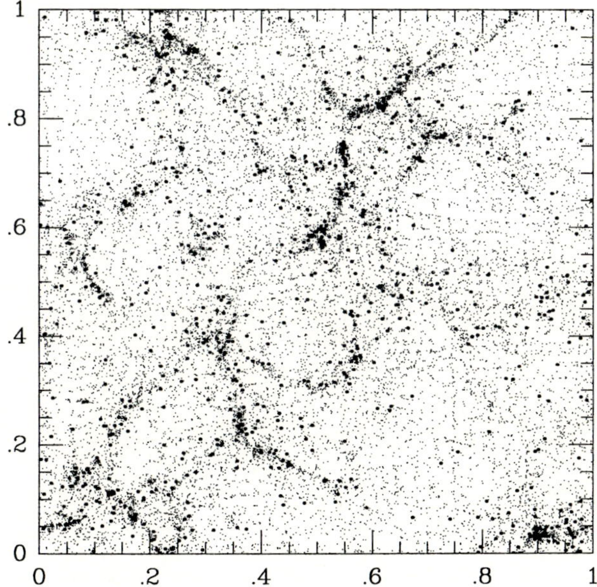

Fig. 7. The galaxy distribution extracted from one realization of a MDM model (Jing *et al.* 1994). A rectangular slice in real space with dimension $60 \times 60 \times 18h^{-3}\,\mathrm{Mpc}^3$ is shown. 847 'galaxies' are plotted with thick points; 21490 mass particles (30% of the total number) in the slice are plotted with thin points. $b = 1.5$ from a correlation analysis.

other quantitative tests, and we want to describe some of them briefly in the following.

5.1 2–Point Correlation Functions

The 2–point correlation function of galaxies is the most widely used tool. An approximate power–law has been derived for the galaxy–galaxy correlation (cf. Peebles 1980)

$$\xi(r) = \left(\frac{r_0}{r}\right)^{\gamma} \tag{45}$$

for $0.1h^{-1}$ Mpc $< r < 10h^{-1}$ Mpc; $\gamma = 1.77 \pm 0.04$; $r_0 = (4.7 \pm 0.3)h^{-1}$ Mpc.

All models reproduce this power law, because it is actually used as a condition to identify galaxies in the models. Galaxies are density peaks which have a 2–point correlation function

$$\xi_{gg}(r) = b^2 \xi_m(r) \tag{46}$$

with a bias parameter b. b depends on the model, and the galaxy catalogue used. Typically values of b in the range 1.5 to 3 are introduced. The 2–point correlation function of galaxy clusters is the same power law with $r_0 = 20h^{-1}$Mpc. The correlation length for clusters in standard CDM–models ($\Omega = 1$, $n = 1$) is too small, but MDM models and modifications of CDM ($\Omega_0 = 0.2$, $\lambda_0 = 0.8, h = 1 : \Lambda$CDM; etc.) can reproduce this correlation for the rich clusters in the simulations.

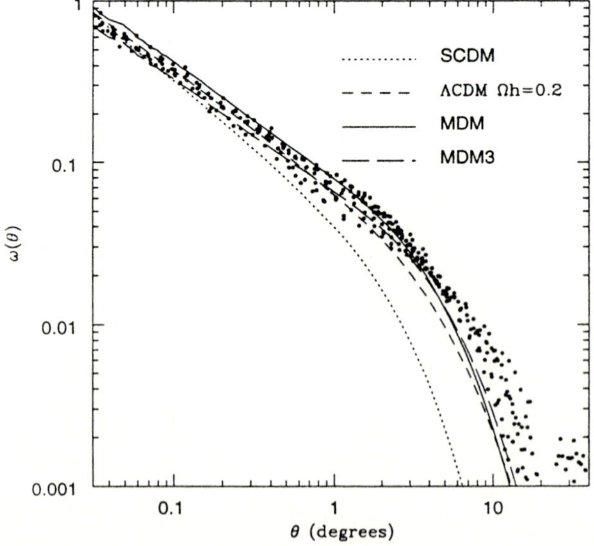

Fig. 8. The angular two–point correlation function ($w(\theta)$) of various models are compared to the data of the APM survey (Maddox *et al.* 1990).

The most stringent constraint on the models comes from the angular–correlation function $w(\theta)$ of galaxies of the APM survey (Maddox *et al.* 1990). In Fig. 8 the observations have been plotted together with various model fits. It seems that the standard CDM–model (Davis *et al.* 1985) does not produce big enough correlations on angular scales above 1°, while ΛCDM or an MDM model with a 30% admixture of hot DM can reproduce the observations quite well (Jing *et al.* 1994). It has been this result which created doubts about the standard CDM model which was quite successful in explaining various small–scale features of the galaxy distribution. Since then numerous modifications of the standard CDM model have been explored. We should note that only one specific set of parameters has been eliminated ($\Omega = 1$, $n = 1$), and e.g. changes of the inital power spectrum $P_i(k) \propto k$ have not been seriously explored so far. There is, of course, the suspicion that the CMB anisotropies will narrow down the possible range of values to $n \sim 1$.

The cross–correlation of IRAS galaxies with Abell clusters and radio galaxies has revealed that on large scales again the standard CDM–model does not have a sufficiently large correlation function (Mo *et al.* 1993). In Fig. 9 the observed average cross–correlations are shown divided by the correlation function predicted by the standard CDM model. Above $r \sim 10h^{-1}$Mpc all observations — despite some internal variations — consistently increase above the standard CDM model.

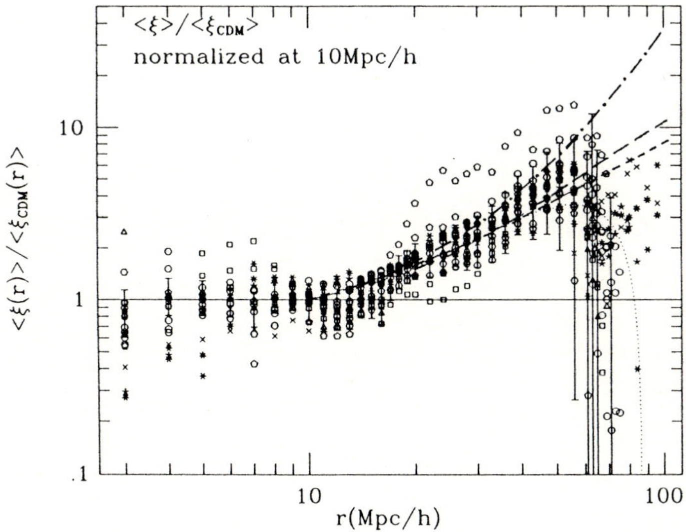

Fig. 9. The observed average cross–correlation functions of galaxies from the QDOT sample (IRAS) with Abell clusters and radio galaxies are shown divided by the correlation function predicted by the standard CDM model. For details see (Mo, Peacock & Xia 1993). All the observations agree quite well and show an increase over the SCDM correlations at separations larger than $10h^{-1}$Mpc.

5.2 Velocity Fields

Measurements of the peculiar velocity of galaxies are an important tool to analyse the mass distribution — luminous and dark matter together. The peculiar motion depends only on the gradient of the gravitational potential, i.e. on the total mass distribution. It is independent of the biasing.

a) Bulk flow

Measurements of a bulk motion for a sample of 200 elliptical galaxies (Dressler *et al.* 1987) have motivated several groups to investigate the redshift vs. distance relation. The peculiar velocities

$$v_p = cz - H_0 d$$

can only be determined, if reliable distance indicators can be found. The true uncertainties in the use of semi–empirical relations such as the infrared Tully–Fisher relation for spirals (luminosity vs. line width) or the $D_n - \sigma$ (diameter vs. velocity dispersion) relation are difficult to assess.

A comparision of the mass density field derived from a reconstruction of the bulk velocity field with galaxy counts from the IRAS survey gave

$$\Omega_0^{0.6}/b = 1.0 \pm 0.3 \ ; \tag{47}$$

(see Dekel 1994), consistant both with high and low Ω_0, and therefore not very exciting. But the method has great potential, since it can lead to a measurement of the bias factor.

b) Pairwise velocity dispersion of galaxies

The pairwise galaxy velocity dispersion (Davis & Peebles 1983) has been widely used to put constraints on numerical simulations. The method for estimating this quantity from redshift–catalogues is a slightly involved statistical analysis which starts out from defining a two–point correlation function $\xi(r_p, \pi_e)$ as a function of two variables, the separations parallel and perpendicular to the line of sight of two galaxies

$$\pi_e \equiv v_1 - v_2 \ ; \quad r_p = [(v_1 + v_2)/H_0] \ \tan(\theta_{12}/2) \tag{48}$$

$\xi(r_p, \pi_e)$ is then defined by the probability of finding a galaxy in the volume δV (located at a distance r) at separation (r_p, π_ℓ) from another galaxy

$$\delta P = n\,[1 + \xi(r_p, \pi_\ell)]\,\delta V \ \phi(r) \ , \tag{49}$$

where n is the mean number density of galaxies, and $\phi(r)$ is the selection function which gives the relative probability that a galaxy at distance r is included in the sample. $\xi(r_p, \pi_\ell)$ can be estimated from redshift catalogues. On the other hand it is the convolution of the real space correlation function $\xi_r(r)$ with the distribution function $f(v)$ of the pairwise peculiar velocity.

Now

$$1 + \xi(r_p, \pi_\ell) = H_0 \int_0^\infty dy \left\{ 1 + \xi_r \left[(r_p^2 + y^2)^{1/2} \right] \right\} f(v) \ , \tag{50}$$

and

$$V = \pi_\ell - H_0 y \left\{ 1 - \left(1 + \frac{r_p^2 + y^2}{r_0^2} \right)^{-1} \right\} \ ,$$
$$\xi_r(r) = (r_0/r)^\gamma \ .$$

For $f(v)$ a simple model is assumed

$$f(v) \equiv C \exp \left[-2^{1/2} |V| / \sigma_g(r_p) \right] \ . \tag{51}$$

For a given r_p, we derive the pairwise velocity dispersion $\sigma_g(r_p)$ by fitting the model (C, σ_g) to the data $(\xi(r_p, \pi_\ell)^{\cdot})$.

For $r_p = 1$Mpc one finds for various models the following values:

model	*bias*	δ_p(1Mpc)[km s^{-1}]
CDM	$b = 1$	1000
CDM	$b = 2$	500
ΛCDM	$b = 1$	450
MDM	$b = 1.5$	500

Previous analysis of the observations (Davis and Peebles 1983) gave a value

$\sigma_p = 340 \pm 40$km s^{-1}which was almost impossible to meet by the numerical models. But a recent analysis (Mo *et al.* 1993) showed that the value of Davis and Peebles is more like a lower limit (see Fig. 10). From Fig. 10 we conclude that a better estimate is

$$\sigma_p = 450 \pm 50 km\ s^{-1} \ , \tag{52}$$

and thus the discrepancy between models and observations is removed.

5.3 Counts of High Redshift Objects

The Press–Schechter formula $n(M, z)$ can be used to estimate the comoving number density of objects in different redshift ranges. The application to clusters (White *et al.* 1993; Carlberg *et al.* 1994) is very promising, but does not givevery strong constraints yet. The rms–mass normalisation turns out to be between

$$0.6 \le \sigma_g \le 0.9 \ .$$

The PS–scheme applied to quasars (Haehnelt 1993) again does not produce any strong restrictions.

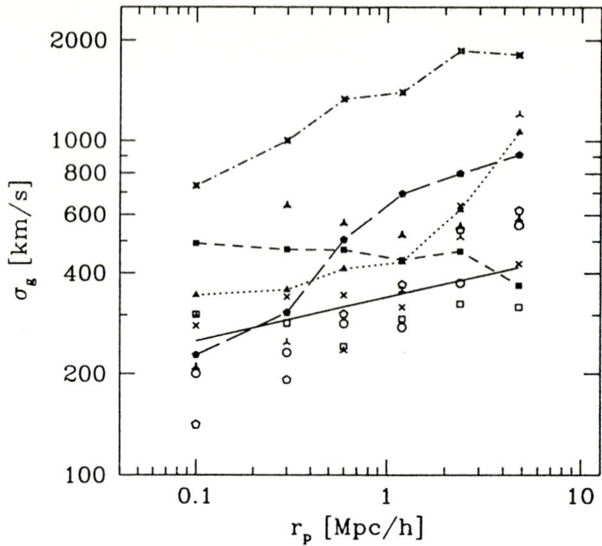

Fig. 10. The pairwise velocity dispersion σ_g as a function of the projected separations r_p for different galaxy samples (for details see Mo *et al.* 1993). The solid line is a fit to the result of Davis & Peebles (1983): $\sigma_g = 340 r_p^{0.13}$ (km s^{-1}, r_p in h^{-1}Mpc). This result seems to be a lower limit to the data.

The most promising bounds come from the consideration of damped Lyman–alpha systems (Wolfe 1993) which seem to be the best probe right now to study galaxy formation and evolution at high redshifts. DL systems are detected as absorption line systems along the line of sight to a high redshift quasar. They have high H I column densities of $10^{19-22}\mathrm{cm}^{-2}$. The column density distribution $f(N_{HI}, z)dN_{HI}$ at a certain redshift can be used to estimate the mass density of neutral hydrogen in DL systems

$$\Omega_D = H_0 \frac{\mu m_H}{\varrho_c} \int N_{HI} f(N_{HI}, z) dN_{HI} \tag{53}$$

where ϱ_c is the critical density, $\mu = 1.33$. The data points for the DL–system in Fig. 11 are determined in this manner (Lanzetta *et al.* 1993).

These data are compared to model values derived from an estimate of $n(M, z)$ from the PS–scheme (eq. 40 and following). The curves in Figure 11 (Mo and Miralda–Escudé 1994) are for different circular velocities of halos, and for each model the fraction of baryons in dark halos is taken equal to the baryonic fraction in each model. The plot shows the density of baryons in virialized halos as a function of redshift. These should be upper limits on the observed H I mass in DL–systems. Thus when the observational points come close to or even are above the theoretical curves the models do not provide a good fit. We see from Fig. 11 that all the models have some problems to account for the high–redshift data,

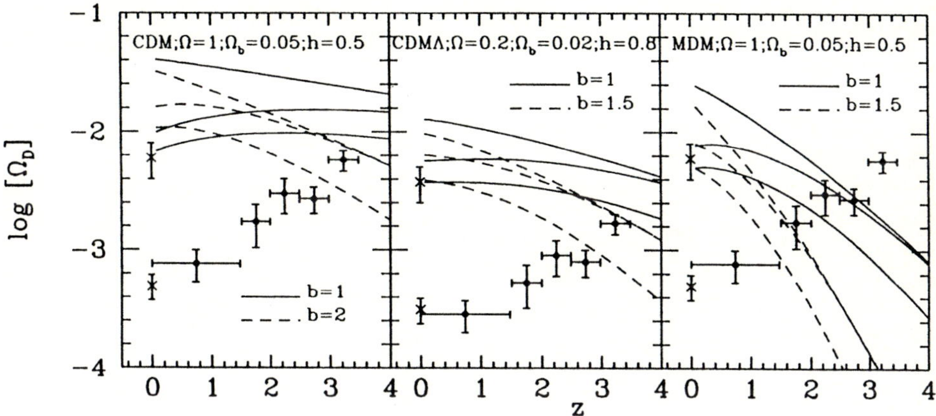

Fig. 11. Ω_D is the density parameter of baryons (scaled to $z = 0$, in units of the present critical density) in virialised halos. Ω_D is shown as a function of redshift, as it is predicted by CDM, COMΛ, and MDM models. For each model results for two bias parameters are shown, and for $b = 1$, resp. $b = 1.5$, three different intervals of halo circular velocities: from top to bottom $v_c = 50 - 2000$, $50 - 250$, $100 - 250$km s^{-1}. The data points for DL systems (solid circles) are adapted from Lanzetta *et al.* (1993). The upper cross–bar shows Ω in stars at the present epoch, the lower cross–bar shows the mass inferred for H I in present–epoch spiral galaxies.

but the MDM model with $\Omega_\nu = 0.3$ seems definitely to be ruled out. In this model there are obviously not enough collapsed halos of DM at high z.

Considering that the neutral hydrogen may form disk–like structures (Mo and Miralda–Escudé 1994) there are indications that for simple models the discrepancy gets larger.

5.4 Clusters of Galaxies

The evolution of galaxy clusters also influenced by the cosmological model. At P^3M simulation of clusters in MDM, SCDM, and ΛCDM (Jing & Fang 1994) can illustrate this.

In Fig. 12 the temperature function $n(T)$ is plotted versus T. Here $n(T)$ is the mean number of clusters expected per unit volume and per unit gas temperature (in keV^{-1}). In DM simulations the temperature is just not equal to the virial temperature, i.e. the velocity disperson σ_v in the cluster core,

$$T \propto \sigma_v^2 \ . \tag{54}$$

The MDM model (CHDM in the figure) shows a rapid evolution with redshift, whereas the ΛCDM model (LCDM) has almost no evolution since $z = 0.5$. The

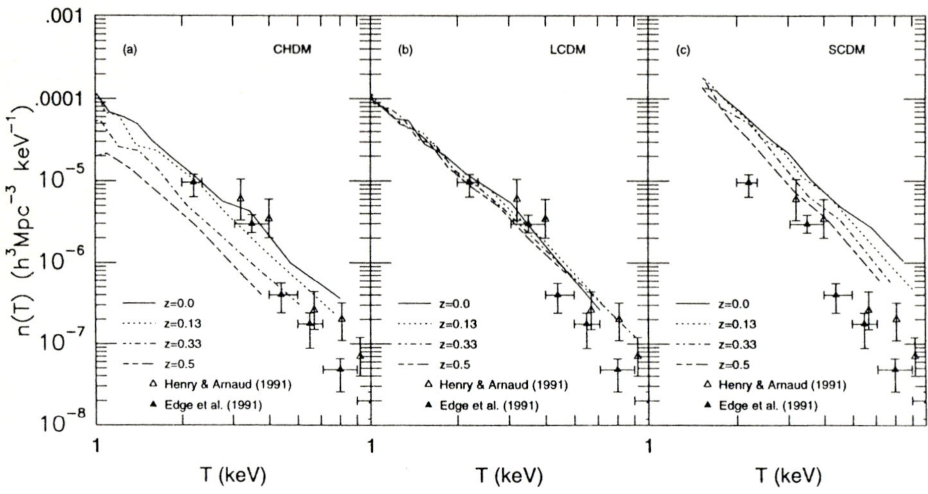

Fig. 12. Temperature functions $n(T)$ of galaxy clusters at three different epochs from $z = 0.5$ to $z = 0$ are shown for 3 different cosmological models.

ΛCDM model ($h = 0.75$, $\Omega_0 = 0.3$, $\lambda_0 = 0.7$) seems to give the best fit to the data.

Another interesting test could stem from the comparison of high-resolution simulations with the data with respect to cluster density profiles and morphologies. Although there may be useful cosmological tests, our results (Jing *et al.* 1995) show that only very low density models ($\Omega_0 = 0.1$) seem to run into a conflict with observations, because they do not show much substructure, and have very steep density profiles (see Fig. 13).

References

Bardeen, J., Bond, J.R., Kaiser, N., Szalay, A.S. 1986, *Astrophys. J.* **304**, 15

Bond, J.R., Szalay, A.S. 1983, *Astrophys. J.* **274**, 443

Börner, G. 1993, *The Early Universe,* (Springer–Heidelberg, 3rd ed.)

Buchert, T. 1989, *Astron. Astrophys.* **223**, 9

Buchert, T., Bartelmann, M. 1991, *Astron. Astrophys.* **251**, 389

Buchert, T, Melott, A.L., Weiß A.G. 1994, *Astron. Astrophys.* **289**, 349

Carlberg, R.G., *et al.* 1994, *J. R. Astron. Soc. Can.* **88**, 39

Davis, M., Peebles, P.J.E. 1983, *Astrophys. J.* **267**, 465

Davis, M., Efstathiou, G., Frenk, C.S., White S.D.M. 1985, *Astrophys. J.* **292**, 371

Dekel, A. 1994, *Ann. Rev. Astron. Astrophys.* **32**, 371

Doroshkevich, A.G., Khlopov, M., Yu., Sunyaev, R.A., Szalay, A.S., Zel'dovich, Ya.B. 1980, *Ann. N.Y. Acad. Sci.* **375**, 32

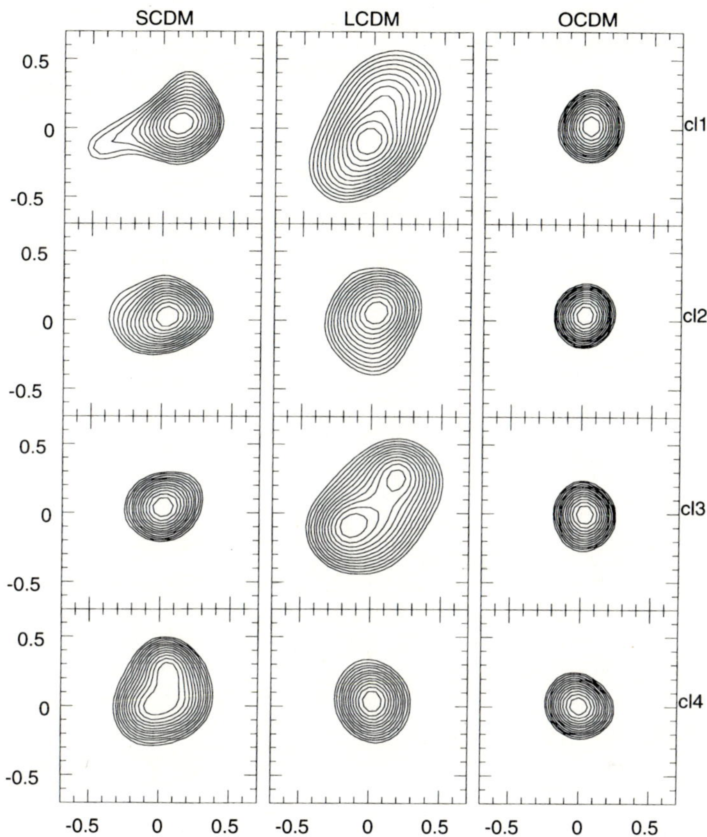

Fig. 13. The projected density–square contours for twelve clusters at $z = 0$ are shown. Contour level i is chosen to be $10^{-0.1i}$ times the maximum value. Scale units are in h^{-1}Mpc.
OP01: $\Omega = 0.1, \lambda_0 = 0$, FL03: $\Omega_0 = 0.3, \lambda_0 = 0.7$, SCDM: $\Omega_0 = 1.0, \lambda_0 = 0$

Dressler, A., *et al.* 1987, *Astrophys. J. Lett.* **313**, L37
Edge, A.C., *et al.* 1990, *Mon. Not. R. Astr. Soc.* **245**, 559
Efstathiou, G., Bond, J.R., White, S.D.M. 1992, *Mon. Not. R. Astr. Soc.* **258**, 1p
Efstathiou, G., Silk, J. 1983, *Fund. Cosmic Phys.* **9**, 1
Geller, M.Y., Huchra, J. 1989, *Science* **246**, 897
Haehnelt, M.G. 1993, *Mon. Not. R. Astr. Soc.* **265**, 727
Harrison, E.R. 1970, *Phys. Rev. D* **1**, 2726
Henry, J.P., Arnaud, K.A. 1991, *Astrophys. J.* **372**, 410
Jing, Y.P., Mo, H.J., Börner, G. 1991, *Astron. Astrophys.* **252**, 449
Jing, Y.P., Mo, H.J., Börner, G., Fang, L.Z. 1994, *Astron. Astrophys.* **284**, 703
Jing, Y.P., Fang, L.Z. 1994, *Astrophys. J.* **432**, 438
Jing, Y.P., Mo, H.J., Börner, G., Fang, L.Z. 1995, *Mon. Not. R. Astr. Soc.*, in press

Lacey, C., Cole, S. 1994, *Mon. Not. R. Astr. Soc.* **271**, 676

Lanzetta, K.M., Turnshek, D.A., Wolfe, A.M. 1993, *Astrophys. J. Suppl.* **84**, 1

Lilje, P.B. 1992, *Astrophys. J.* **386**, 33

Linde, A.D., 1990, *Perticle Physics and Inflationary Cosmology*, (Harvard Acad. Publ.)

Maddox, S.J., Sutherland, W.J., Efstathiou, G., Loveday, J. 1990, *Mon. Not. R. Astr. Soc.* **243**, 692

Mather, J.C., *et al.* 1994, *Astrophys. J.* **420**, 439

Mo, H.J., Börner, G. 1990, *Astron. Astrophys.* **224**, 1

Mo, H.J., Deng, Z.G., Xia, X.Y., Schiller, P., Börner, G. 1992, *Astron. Astrophys.* **257**, 1

Mo, H.J., Jing, Y.P., Börner, G. 1993, *Mon. Not. R. Astr. Soc.* **264**, 825

Mo, H.J., Miralda–Escudé, J. 1994, *Astrophys. J. Lett.* **430**, L25

Mo, H.J., Peacock, J.A., Xia, X.Y. 1993, *Mon. Not. R. Astr. Soc.* **260**, 121

Peebles, P.J.E. 1980, *The Large–Scale Structure of the Universe*, (Princeton Univ. Press)

Peebles, P.J.E. 1993, *Physical Cosmology*, (Princeton Univ. Press)

Press, W.H., Schechter, P. 1974, *Astrophys. J.* **187**, 425

Sachs, R.K., Wolfe, A.M. 1967, Astrophys. J. **147**, 73

Suginohara, T., Suto, Y., Bouchet, F.R., Hernquist, L. 1991, *Astrophys. J. Suppl.* **75**, 631

Suto, Y. 1993, *Prog. Theor. Phys.* **90**, 1173

Weinberg, S., 1972, *Gravitation and Cosmology*, (Wiley, New York)

White, S.D.M., Efstathiou, G., Frenk, C.S. 1993, *Mon. Not. R. Astr. Soc.* **262**, 1023

Wolfe, A.M. 1993, *Ann. N.Y. Acad. Sci.* **688**, 281

Wright, E.L., *et al.* 1992, *Astrophys. J. Lett.* **396**, L13

Zel'dovich, Ya.B. 1970, *Astron. Astrophys.* **5**, 84

Zel'dovich, Ya.B., Novikov, I.D. 1983, *The Structure and Evolution of the Universe*, (Univ. Chicago Press)

Observations and Cosmological Models

George F. R. Ellis

Applied Mathematics Department, University of Cape Town,
Rondebosch 7700, South Africa.

1 Introduction

My topic is the nature of different approaches to distinguishing between cosmological models on the basis of observations. The thesis[1] is that one should recognize two major possibilities: the *Direct Approach* and the *Inverse Approach*, giving alternative ways of undertaking cosmological modelling. A third approach, the *Fitting Approach*, to some extent lies between these two. Each has advantages and disadvantages, hence the suggestion is one should try to follow all three as far as possible: for each has something to give us in terms of understanding cosmological models, and what we can learn from them. Considering this range of possibilities helps us determine the different roles played by various types of observations, and so identify which can play the most crucial role (see Stoeger 1987).

2 The Direct Approach

Here one starts by assuming the nature of the geometry of the universe: one chooses a particular geometry for the cosmological model, most often assuming it belongs to the Friedmann-Lemaître-Robertson-Walker (FLRW) family (see e.g. Weinberg 1972, Ellis 1987). One must also specify the assumed matter content of the universe. A two stage process then occurs: firstly, one attempts to determine the major parameters describing this model, and secondly to describe the statistics of fluctuations about it. Although conceptually independent, these turn out to be in fact closely linked.

Thus in the standard case, in terms of the FLRW scale parameter $a(t)$, the parameters are the Hubble constant $H_0 = (\dot{a}/a)(t_0)$, deceleration parameter $q_0 = -(\ddot{a}/a)(t_0)H_0^{-2}$, spatial curvature K_0 and density parameter Ω_0, related to each other by the Friedmann and Raychaudhuri equations. Given the specified matter content (baryons, photons, scalar fields, cosmological constant, etc.) one can solve the dynamical equations for the model and then the observational

[1] cf. the discussion by Matravers *et al.* (1995)

equations, leading to predictions of the number count-redshift and magnitude-redshift relations and their close variants (angular diameter - redshift, radio source counts, and so on). These can then be compared with observations, as can the predicted light element abundances and the full spectrum of the cosmic background radiation (CBR), including in particular the black-body microwave background radiation (CMBR) that is interpreted as relic radiation from the hot big bang. Making detailed predictions depends on understanding well matter-radiation interactions, the mass to light ratio, luminosity functions, and source evolution properties. Further, physical models of the growth of structure, based on assumptions about the hot and cold dark matter present, give statistical predictions for the structures that should be present as well as the background radiation anisotropies. Comparing this array of predictions with observations one can hope to select a particular FLRW model as the best one to choose, determining the famous two numbers of classical cosmology (Sandage, 1968, 1988) as well as the nature of the matter content (in particular, the dark matter). The basic picture is well-supported by observations (Peebles *et al.* 1991). Nevertheless, a series of problems arise.

2.1 Problems Within the Family

These are all well known and are familiar to our participants: the problem of evolution effects, and our inability to find a reliable standard candle; the problem of selection effects, and our inability to directly observe dark matter; the problem of unknown physics at very early stages in the history of the universe, because the processes at work then are beyond any presently possible or even conceivable laboratory tests. In view of the vast amount of recent discussion of these topics, I will not consider them further except to make two remarks.

Firstly, one thing which is clear is that selection effects cannot be fully described in terms of any single parameter (Ellis, Perry & Sievers 1984); consequently for example the idea of a 'magnitude limited sample' should be treated with considerable caution (what about the faint surface-brightness galaxies that may be out there?, c.f. Disney 1976).

Secondly, although there are now a few analyses of large-scale structure suggesting a high density parameter, nevertheless the weight of evidence favours a value of Ω_0 of about 0.2 to 0.3 rather than the critical density 1 preferred by many theorists (Coles & Ellis 1994). If we could really tie down the Hubble constant this would be very helpful in constraining Ω_0; currently a tendency to favour higher values of H_0 pushes the balance towards lower values of Ω_0.

2.2 Problems of Description

The problem here is what description to use for space-time. There are two aspects. Firstly, these relate to local inhomogeneities.

Lumpiness: Here the issue is, what is the relation of the completely smooth FLRW universe models and the 'lumpy' reality of the universe? The clusters, voids, walls, streaming motions, great attractors out there do not look much like a FLRW model. How can we relate them? Specifically,

* At what scales is the universe well-described by a FLRW geometry?

* What description should we use to relate smaller scale structure to the large-scale smooth geometry?

Secondly, one can be more radical and consider large-scale variations from the initially chosen model.

Globally different geometries: Here the issue is, there are other models that give similar observations to the FLRW universes, but differ radically in their geometry (perhaps at early or late times, or in distant places): they cannot be considered just as perturbed FLRW models, even if they are quite similar for substantial periods of time (and give very similar observations). Specifically, one can consider here at least

1: *Almost Steady-State models* (Hoyle & Narlikar 1966), which hanker back to the idea of an unchanging universe in the large,

2: *Globally Inhomogeneous models*, such as on one hand the spherically symmetric models that have been proposed from time to time (Omer 1949, Ellis, Maartens & Nel 1978), and on the other *hierarchical* or *fractal* models (de Vaucoleurs 1970, Coleman *et al.* 1988), and now arising in new forms in some of the inflationary universe scenarios (Linde 1990, 1994);

3: *Spatially homogeneous* (Bianchi & Kantowski Sachs) universe models (Ellis & MacCallum 1969), allowing shear and vorticity, and hence the possibility of quite different singularity structures at early times and anisotropic expansion at late times, but which can have lengthy FLRW-like periods (Ellis & Wainwright 1995).

4: *Small Universe models*, with the same local structure as the standard FLRW models, but quite different global topology: they close up spatially with a sufficiently small scale that we could have seen right round the universe one or more times (Ellis & Schreiber 1986).

The problem highlighted by these examples is that despite its uniqueness, the FLRW family of models is not the only possibility for describing the observed universe. Thus we need to face,

* What is the range of models that might describe well the geometry of the observed universe?

We can if we wish test only our favourite models, altering this only where absolutely necessary (we are forced to do so by insistent data); or we can face this problem of alternative geometries at the outset, trying to determine the full phase space of possibilities for observational cosmology, by determining the full range of models that could be compatible with our observations. One should emphasize here that the Ehlers-Geren-Sachs (EGS) theorem (Ehlers Geren & Sachs 1968) and its generalisation to more realistic (almost-FLRW) geometries (Stoeger, Maartens & Ellis 1995) give good grounds for believing the universe is almost-FLRW within the observable domain (that is, within a spatial distance of a Hubble radius and back to the time of decoupling). The idea is that because we measure the CBR to be nearly isotropic, and we believe we are not privileged observers, then all fundamental observers reasonably near to us should also measure almost isotropic radiation; it then follows that the universe is almost FLRW in that region. However (Ellis 1975, 1980),

(a) Despite its plausibility, this argument rests on an *a priori* Copernican principle (the assumption we are not in a privileged position), which might or might not be true; this assumption should be tested in whatever ways are possible;

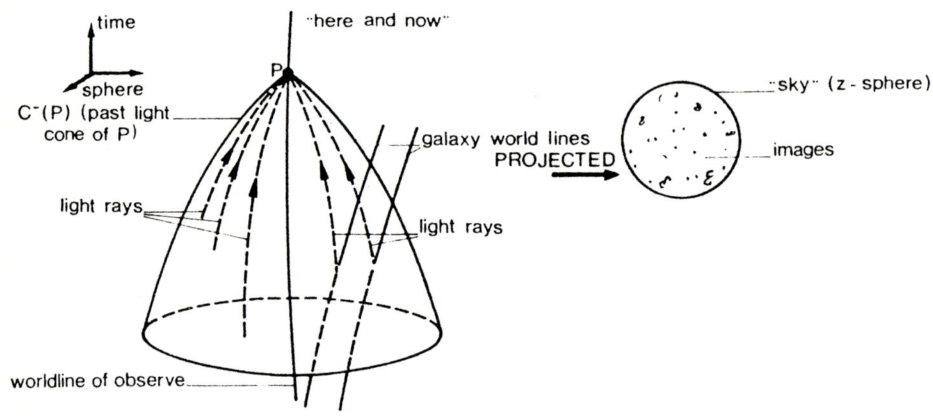

Fig. 1. Our past light cone $C^-(P)$ — the null-cone of our space-time point P ("here and now") — on which the light rays corresponding to all our actual and potential astronomical observations travel. From our perspective, furthermore, they are all projected onto a single 2-sphere, which is identifiable with the plane of the sky.

(b) Its plausible domain of applicability is within our past light cone, in the region between the present ('here and now') and the time of decoupling of the background radiation (about a redshift of 1200); it does not apply to earlier or later times (the universe can be very like a FLRW geometry in this domain, but be very anisotropic at earlier or later times);

(c) This argument cannot tell us *observationally* about the situation outside our light cone (that is, more than a Hubble distance away), despite various claims in the literature suggesting this is possible (these rely on analyticity assumptions which may or may not be true). Indeed in many model universes, e.g. 'chaotic inflationary' models, the universe is quite unlike a FLRW model on very large scales (that is, many times a Hubble size).

In summary, the usual 'Direct Approach' is theory driven, and its main aim is explanatory power. Models adopted *a priori* are used to predict observational results, and then a few remaining parameters adjusted to give a best fit to the observations. One gets no idea of how unique the model attained is, unless the approach explicitly also considers a whole range of other models such as those listed above. The EGS result gives plausible reason for assuming an almost-FLRW geometry within our past light cone for a limited period in the past and future, but is not by itself conclusive because of its *a priori* Copernican assumption, and also despite recent development to give quantitative results (Maartens, Ellis & Stoeger 1995) it has not yet been refined to the stage where it can help us determine on what scales the almost-FLRW result is applicable. Given these caveats we attain remarkable predictive power from this approach, in a synthesis that explains what we see on the largest scales in a basically satisfactory way, but still without resolutions of the values of some of the most basic parameters.

3 The Inverse Approach

The idea here is to start with the observations, and see what is out there, without making previous assumptions about the geometry of space-time or distribution of matter (but one does have to assume a knowledge of the nature of matter - the kinds of sources of radiation we are observing). The questions asked are,

* what is the real distribution of matter, and its motion?
* what is the real space-time geometry?

Now this has been the standard approach from the observational side from the beginning - it is observationally driven, rather than theory-driven. Recent examples are the Durham-AAT pencil beam surveys and the Harvard-Smithsonian 'slice of the sky' surveys. They have the obvious advantage that you look to see what is actually there, and so can find things your models have not predicted (voids and great walls, for example).

Now cosmologically speaking, those surveys are still relatively small-scale; they have not been taken as determining the nature of the universe model itself. The possibility of doing that was initially broached by McCrea in the late 1930's, and then developed systematically in the classic paper by Kristian and Sachs (1966). However even this was in a certain sense local (it developed the observable relations in power series). The extension of this approach to the fully relativistic regime - and to far times and places - has been carried out in a series of papers by Ellis, Stoeger, Maartens, and Nel (see e.g. Ellis *et al.* 1985), defining the theoretical nature of the problem: how to construct an accurate model of space-time geometry directly from astronomically feasible observations. The controversial aspect is in relating this theory to the complex reality of observations, and in particular facing up to dealing with the problem of source evolution. One viewpoint is that the approach is hopelessly naive because it does not give a definitive solution to that problem. Furthermore, it loses out heavily in comparison to the direct approach, in terms of its lack of explanatory power. However I suggest that despite these aspects, this approach is useful to us.

What is made clear by the approach is the full data needed to determine space-time geometry, and the regions of space time whose geometry we can determine from astronomical observations (Ellis 1975, 1980). In particular it makes clear how predictions to the future depend on unobservable data - unless we live in a small universe (Ellis 1984).

3.1 Null Data

The data we can actually observe at any distance is seen down our past null-cone (looking away to larger distances also corresponds to seeing further back in the past). What we need to determine, at each point down the past light cone that is accessible to observation, are,

(1) The cosmological redshift z_C, corresponding to measuring radial velocities. The prime problem here is splitting the observable redshift z into a local component z_D, presumed Doppler in nature, and its cosmological part z_C. This involves determining which galaxies observed in a region of sky are in the same cluster and which are not, enabling one then to determine the average redshift of the cluster and the deviations from this average. The whole argument then (Arp *et al.* 1990) depends on whether or not one has correctly identified the cluster members.

(2) Transverse velocity components u^θ, u^ϕ. These are determined by measuring galactic proper motions. Now these are almost impossible to measure, but the point is that they are essential data for determining the past and future positions of the galaxies we observe. One can assume these velocities are zero (because we have not measured non-zero values yet), or determine their values from the radial velocities by assuming the vorticity is zero (as in the POTENT analyses); such assumptions may or may not be correct. Whether we can measure their values or not,they are essential cosmological data. The fact they are so small that they are difficult to determine is already important evidence.

(3) Number counts of different objects and the associated energy densities. Here is where we encounter the problems of dark matter, the mass to light ratio, and selection effects. Through the selection effects, luminosity and Size evolution have an important effect on these counts. The need is to measure this density locally out at large distances; apart from source counts, which only gives partial information, dynamical methods (galaxy rotation curves, virial theorem for clusters) and gravitational lensing offer ways of doing so.

(4) Area/Angular diameter distances, or the closely related luminosity distances corresponds to the determinant of the metric of our past light cone. Here we run into the problem of a lack of standard candles (and of standard size objects), compounded by the effects of luminosity and size evolution. In principle a sensitive test of the space-time geometry is the distance of minimum apparent angular size for fixed-scale objects, in each direction; but even this is difficult to determine (Ellis & Tivon 1985).

(5) Distortion of images corresponds to the anisotropic part of the metric of our past light cone, caused by tidal gravitational forces (and, temporarily, by gravitational waves). An extreme form of such distortion is the formation of multiple images of a single object (the phenomenon of gravitational lensing, see Schneider, Ehlers & Falco 1992). Multiple images could also be caused by the

universe being spatially closed on such a scale that we could see round it several times (a 'small universe')

(6) The background radiation spectrum and anisotropy (at all angular scales and wavelengths) give us integrated information concerning the matter emitting the radiation and the intervening space-time geometry through which it has passed. The problem is to disentangle these two effects; a current view - which is disputed - is that a variant of the standard Sachs-Wolfe formula (Sachs & Wolfe 1967), modulated by pressure effects (see Stoeger, Ellis, & Xu 1994) represents the conditions at the surface of decoupling with all the other effects (intervening gravitational waves, vorticity perturbations, and effects of local structure formation traversed by the light on its way to us) implicitly contained in that formula. This analysis applies to all null radiation: electromagnetic radiation at all wavelengths, massless neutrinos, gravitons; the principle difference in the latter case being that the corresponding surface of decoupling was much earlier (allowing us to see back much further through such radiation, if ever we could measure it with sufficient accuracy).

3.2 Geological and Physical Data

All this is supplemented by data that does not pertain to the past light cone; rather it reflects conditions near our past world line, a long time ago (Ellis 1971):

(7) Element and particle abundances reflect baryosynthesis and nucleosynthesis near our world line a long time ago;

(8) Local structures - stars, galaxies, clusters, and so on - grew locally from initial perturbations in our vicinity with seeds in the every distant past. We can analyse them to determine the corresponding range of possible initial conditions, and in particular can try to determine age limits from them that we can then compare with the probable dynamical history of the universe near our world line in the past.

Other data with cosmological significance are the set of issues related to

(9) Boundary conditions for local physics: e.g. the dark night sky, the local arrow of time, the local compass of inertia (Ellis & Sciama 1972). However these are of a controversial nature, and it is not clear we can reliably extract cosmological information from them - although it is there (Penrose 1989).

3.3 Problems

There is in principle a problem with the shear volume of data needed to produce a detailed model, which has great descriptive power but little explanatory power; however modern computer technology makes this aspect technically feasible. The more fundamental problems with this approach are, firstly, the sparseness of real data compared to the ideal data needed (we in principle need data at each point on our past light cone, but can only observe it at isolated spots where some source happens to be situated; however we can handle this reasonably satisfactorily with well established interpolation procedures; and secondly, the particular problem associated with the need in many types of analysis to understand the evolutionary history of observed sources, which are poorly based in this kind of model (precisely because one is as far as possible avoiding introduction of a background model as a basis for such evolutionary analysis).

The basic response is that the same problems arise in the usual approach, but are hidden from view; the analyses one is led to by this observationally based approach bring out into the open issues that are there in all cases, but are covered over when one assumes a FLRW geometry from the start. Also the clear delineation of what the needed data are help to consider approaches that try to avoid such evolutionary problems, and hence are less model dependent than usual. Nevertheless, one would like to somehow relate this view more closely to the usual predictive models.

4 The Fitting Approach

Here we adopt an intermediate road: we use a basic background model (usually a FLRW model) but now try to fit this model to the actual space-time geometry in a detailed way (rather than just in a statistical way, which is what is usually done). Thus we have a basic explanatory model, but try to relate it to the detailed reality of the universe (at appropriate averaging scales) rather than just its typical properties.

This leads to new issues of some interest (Ellis & Stoeger 1987), analogous to problems in geodesy where one obtains successive approximations to the shape of the earth:

* How do we obtain a best-fit of an idealised universe model to the lumpy reality out there?

* How do we then characterise, the deviation of reality at each space-time point from the idealised model, and place a measure on the overall difference between them?

While there is not yet a fully developed relativistic theory of how this should be done, this is in fact being undertaken in a number of recent analyses, specifically the studies of streaming motions and the Great Attractor by the Seven Samurai, the IRAS analyses by Rowan Robinson *et al.* and the POTENT analyses of Dekel *et al.* , each leading to detailed maps of matter density and motion (albeit based on simplifying assumptions designed to get around the lack of true information on transverse motions). Because the fit depends on the parameters of the background model, these studies give estimates of both the FLRW parameters and the local deviations from a FLRW geometry simultaneously. What is needed here is,

* a well-defined 'optimal fit' procedure, leading to

* a 'goodness of fit' assessment, and then application of

* 'acceptance of fit' criteria (is the universe in fact well described by the proposed model, or not?)

The point is that without a clear procedure of this kind for defining the relation of the idealized model to the real universe, we do not have a proper basis for accepting or rejecting the model: for we have no criterion of acceptability that we can check. The fundamental further point here is that the whole analysis is completely dependent on the implicit averaging scale used in the process; so that scale should be made quite explicit throughout. In particular the conclusion ('the universe is indeed well -described by a FLRW model in the observable region since decoupling') should state not only the region where this is claimed to be true (as above) and the derived parameter values for the background model ('the Hubble constant is 80 $\mathrm{km\,sec^{-1}\,Mpc^{-1}}$ and the density parameter Ω is 0.24') but also the scales on which this is supposed to be true ('These results hold for averaging scales from 500 Mpc to 1500 Mpc') and the general accuracy of the statement ('On these scales, the deviation from a FLRW model, in terms of the fractional density variation magnitude, is at most 0.4').

One cannot make such detailed statements without carrying out a fitting procedure, which in turn depends on determining the real space-time geometry. However the advantage to a purely observational approach is that one is in effect using a successive approximation procedure, starting with the idealised smoothed-out background model; so one can use that model to help with age and evolution estimates, which one cannot do in a strictly observational approach. Thus this is a sensible and useful blend of the other two approaches, enabling one to make much more precise and meaningful statements than is otherwise the case. As has been commented it is in fact already in quite widespread use, but usually not articulated in quite these terms, and also usually this is envisaged as happening on rather a local scale; whereas eventually we should use these methods over the entire visible region of the universe. Then it can usefully lead to an interaction of complementary approaches, where the model–based explanatory approach and the observationally–based descriptive approach each support and modify the other, leading to a better overall understanding of the complexity of the true universe yet with its simplicity of underlying structure.

This kind of approach is by now quite widely used in terms of investigating local velocity fields and their relation to density inhomogeneities. The need is to extend its theoretical base to the fully relativistic regime, and to make more clear in these analyses the choice of best-fit criterion used in fitting the background FLRW model to the observations. We also have not yet seen attempts to define the overall deviation from a FLRW model in quantitative terms, or a good determination of the scale where we start to attain an accurate fit by the FLRW model (the scale at which the universe in fact attains spatial homogeneity).

5 What We Can Learn from This?

Apart from motivating the concepts of well-defined detailed fitting as discussed in the last section, one can use this discussion to discern some important features and opportunities in terms of cosmological observations and modelling.

5.1 Observational Data

We need to develop the theory in terms of real observational data, not theoretically produced quantities (such as the 'apparent magnitude' m) which are in fact unobservable, being deduced by manipulations of the actual data on the basis of a whole set of hidden assumptions (Ellis & Perry 1979). Further, we need as far as possible to choose data which probes conditions down the light cone (i.e. at large redshift), as well as close by (Ellis 1984). Together with the analysis above of the data needed for the pure observational approach, this gives some suggestions as to fruitful ways to pursue observations.

Direct observations at a distance: We need to as far as possible find data that depends only on intrinsic properties of the objects observed (determined by local physics and their characteristic nature) and so is independent of attempts to trace their evolutionary history . The recent extension of Cepheid observations to more distant galaxies is obviously a great improvement in this kind of analysis, but will still not take us very far. The great improvement in distance measurements in recent decades using the Tully-Fisher relation is another example of what one has in mind here, but one would still like to base the analysis either on more clearly physically based relations, or on direct measurements essentially independent of the source properties.

There are two particularly promising avenues here.The first is use of supernovae as standard candles; great strides in this regard have been made recently, and clearly a long-term programme of systematic observations of supernovae in distant galaxies is well worth while. The second is the possibility of direct parallax measurements being possible through interferometric techniques, bearing in mind that it is now technologically possible - should it be helpful - to construct interferometers based on satellites up to something like 1 A.U. apart. Considerable thought needs to be put into determining the optimal wavelength and size for new generation interferometers capable of measuring H_0 directly, and conceivably even q_0, taking into account the problems of distortion by lensing and scattering by any intergalactic medium. Insofar as measurement of such trigonometric parallaxes (Rosquist 1988) is possible, it has the inestimable advantage of giving results independent of the nature of the source, provided it is sufficiently compact. Problems arise here not only in terms of accuracy of position measurements and timekeeping in the satellite array used, but also in finding distant sources that are compact enough; and here again it could well be that supernovae are the best answer. If sources are time-varying on a short enough time scale, that time variation (which has to come from a compact region) can be used to determine such wave front distances without actually having to construct interference patterns.

Transverse velocities: It has been emphasized above that transverse velocities, observed as apparent motions, are a key element of cosmological data. In principle they are directly observable, but one runs here into major technical problems, particularly in terms of defining a stabilised non-rotating reference frame over a period of many decades or even centuries, independent of observations of distant galaxies. Two new possibilities are interesting here. The first (G Smoot, private communication) is that the Sunyaev-Zel'dovich effect in principle provides a way of measuring the motion of distant galaxies relative to the Cosmic Background Radiation, through polarisation of the scattered radiation due to such motion; not a very sensitive test, but important because it is independent of any direct measurement of apparent proper motions.

The second is that if one were to set up a satellite-based set of interferometers as mentioned in the previous section, this in principle gives a way of directly determining transverse motion of the source through measuring the spatial curvature of the wave-front (at least 5 receivers are required for this purpose; I. Kovner, unpublished). Hence long-term plans for astronomy in the future could include interferometers to undertake such measurements and give much improved limits on transverse motions.

Direct limits on null shear: Great strides have been made in recent years in detecting the shear induced in null rays by inhomogeneities, both in terms of direct measurements of image distortions (Tyson *et al.* 1984, Blandford *et al.* 1991) and through gravitational lensing observations (Schneider Ehlers and Falco 1992). What is needed here is to systematise this to an all-sky observational programme as implied in Kristian and Sachs (1966), but trying as far as possible to detect the contributions to shear from objects at different redshifts.

Local density measurements: This is part of the larger issue of trying to eventually deconvolve from observations the local density down our past light cone. Apart from lensing observations, the main tool here is dynamical observations (virial theorem, large scale flows) and again much progress has been made here in the local region through the POTENT, IRAS, and allied analyses. Again the issue is to systematise this to an all-sky programme that will eventually probe to higher and higher redshift systematically.

5.2 Consistency Tests

Putting together the data available through these various avenues (lensing, velocities, and matter densities) one can examine such questions as,

* Is there vorticity?
* Are there non-zero decaying modes?

Current wisdom has it that these are both zero, or at least completely negligible. However it would be most extraordinary if they were *exactly* zero (as they are in exact FLRW universes); and if they are non-zero, their magnitude is of considerable interest. If either were to presently exceed rather low values, their magnitude at decoupling would be rather large, and that would imply substantial deviation from a FLRW geometry then (which might well in turn cause measurable anisotropies in the background radiation temperature). Thus one can test the almost-FLRW picture by searching for non-zero vorticity or decaying modes (Haines and Ellis, unpublished), checking the validity of the standard picture.

This raises the importance of more general consistency tests: one can probe 'obvious' known 'facts' about the universe to see if they are in fact true; and propose a series of probes of the standard FLRW model, where we in principle know the answers (if the standard model is true). Thus we can probe its veracity, basically because in that model there are many correlations of conditions in different regions, because of the assumed spatial homogeneity. The point is that whenever we get zero (to within the accuracy of the measurements) we confirm the standard model at that scale and so help put it on firmer foundations; whenever we obtain non-zero values, these are deviation measurements from the standard model that can be used in the fitting process discussed in the last section, helping locate the precise differences between reality and the idealised standard models we propose.

Ages: One of the oldest concerns in cosmology, is the determination of ages of local objects and relation of these ages to the age of the universe (determined by the Hubble constant and Ω_0 in the FLRW case; there is more latitude in more generic geometries). Currently the tightness of the fit here tends to push us towards low Ω_0 (Coles & Ellis 1994), although one can always try to attain consistency by a non-zero cosmological constant Λ. In that case, a whole series of other tests must be used, related to number counts and gravitational lensing statistics in particular, to try to restrict the allowed range for Λ.

CMBR temperature and spectrum: One of the most powerful tools now available is testing not only the CMBR anisotropy but also its local temperature down the light cone, and its spectrum. As to the first, the point is that one can try to measure the background radiation temperature through measuring the excited states of molecules, as in the first detection of the 3K radiation by CN excitations, at points down the past null cone (Songaila *et al.* 1994)]. If the redshift of the molecules is z, the excitation temperature predicted by the standard model is $3(1+z)K$; thus we can check the thermal history of this radiation, in a spatially homogeneous universe model, through such measurements. As regards the spectrum, this is again a very sensitive measurement and if we can detect the spectrum of the background radiation down the light cone, at a redshift z, we can test conditions there. Here is where the Sunyaev-Zel'dovich effects comes in, for the scattering in hot intergalactic gas at a redshift z mixes the temperature of the incoming radiation there; so spectral distortions will result if the radiation is not closely isotropic. This is what enables us to turn the almost-EGS argument (see above) from one based purely on *a priori* and unverified supposition of a Copernican principle, to one that is to some degree tested and confirmed.

Number count anisotropies: A powerful check on our current view of the universe comes from the fact that we interpret the dipole CMBR anisotropy as due to our motion relative to the cosmological rest frame defined by that radiation. In the case of a FLRW universe model, the implication is that we are moving relative to the rest frame of that universe; consequently we must inevitably measure number count anisotropies also, at the 2% level (Ellis & Baldwin 1984). This is on the threshold of detectability; it is an important test to carry out, particularly because it is independent of selection effects and detection limits, as long as these are the same in the forward and backward directions (which they should be if the FLRW assumption is accurately true).

Local element abundances: A further test we can apply is checking local light element abundances down the light cone, at high redshift. This helps us to check conditions far out from our past world line, at very early times; thus testing the homogeneity of the universe then (the light element production would be spatially inhomogeneous in an inhomogeneous universe) and so providing powerful checks on the standard model. The feasibility of such observations has recently been demonstrated (Jacobsen *et al.* 1994). This can be regarded as part of a larger programme: namely to test homogeneity by testing for uniform thermal histories of matter in distant regions (Bonnor and Ellis 1986), and - more fundamentally - to try to check if the laws of physics (and the constants of nature) are the same out there? This can be done to a certain extent (e.g. testing if the fine structure constant in distant regions is the same as nearby).

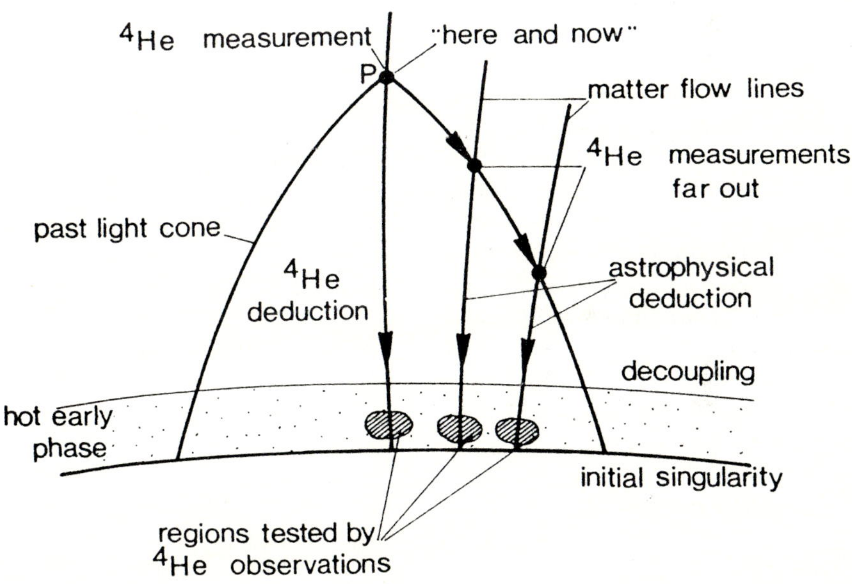

Fig. 2. The importance of determining element abundances (for example, that of He^4) on $C^-(P)$ far out from our world line. By astrophysical deduction these measurements can, in principle, constrain conditions in widely separated regions at very early times.

6 Discussion and Further Issues

This paper makes the case that complementary approaches to cosmology are possible and indeed should be undertaken. As well as the standard approach, based on a FLRW model, we should use a strictly observational approach, as far as that is possible, in an independent attack on understanding what is out there; and then should see how the results of the two approaches best fit together, by trying the 'fitting' programme discussed above. Exploring this range of options helps us see what is the best way to approach observations in as unbiased a way as possible - on the one hand applying detailed modelling of cosmological conditions from the outset; on the other, delaying doing so for as long as possible. Thus we recommend both the careful analysis of real information in as model-free a way as possible, as well as use of detailed models and simulations.

6.1 Effect of Lumpy Universe

It must be emphasized that the above only begins to touch on one important and as yet poorly understood area of cosmology; namely the effect of true inhomogeneity at small scales on both observations and dynamics at larger scales. These effects are due to the difference between Ricci and Weyl focussing effects, the former being directly caused by matter whereas the latter is mediated by distortions and lensing effects. The difference may be substantial in its effects both on dynamics (Futamase 1989, Zotov and Stoeger 1992) and on observations (Dyer and Oates 1968). Some useful approximations are available, for example the Dyer-Roeder area distance formula used in lensing analyses (Schneider *et al.* 1992); the nature and magnitude of the effects are not easy to determine when the perturbations are non-linear - as they certainly are, on small scales.

This has been a topic of discussion for decades, but is often ignored in observational analyses, and almost always in studying the dynamics of cosmology. This may be a substantial error (Ellis 1984); it may for example be one of the reasons different estimates of Ω appear to vary substantially with the length scale of averaging associated with the method of observation. There is no space to pursue these issues further here, other than to point out this is an area that needs attention.

References

Arp, H.C., Burbidge, G., Hoyle, F. , Narlikar, J.F., & Wickramasinghe, N.C. 1990, *Nature* **346**, 425

Blandford, R.D., Saust, A.B., Brainerd, T.G., & Villunsen, J.V. 1991, *Mon. Not. R. Ast. Soc* **251**, 600

Bonnor, W.B., & Ellis, G.F.R. 1986, *Mon. Not. R. Ast. Soc* **218**, 605

Coleman, P.H., Pietroneiro, L. & Sanders, R.H. 1988, *Astron. Astrophys.* **200**, L32

Coles, P. & Ellis, G.F.R. 1994, **Nature 370**, 609.

de Vaucoleurs, G. 1970, *Science* **157**, 1203

Disney, M. 1976, *Nature* **263**, 573

Dyer, C. & Oates, L. 1968, *Astrophys. J.* **326**, 50

Ehlers, J., Geren, P. & Sachs, R.K. 1968, *J. Math. Phys.* **9** 1344

Ellis, G.F.R. 1971, in *General Relativity and Cosmology*, Ed. R.K. Sachs (Academic Press), 104

Ellis, G.F.R. 1975, *Qu. Journ. Roy. Ast. Soc.* **16**, 245
Ellis, G.F.R. 1980, *Ann. NY. Acad. Sci.* **336**, 130
Ellis, G.F.R. 1984, in *General Relativity and Gravitation*, ed. B. Bertotti *et al.* (Reidel), 215
Ellis, G.F.R. 1987, in *Vth Brazilian School on Cosmology and Gravitation.* ed. M. Novello (World Scientific), 83
Ellis, G.F.R. & Baldwin, J. 1984, *Mon. Not. R. Ast. Soc.* **206**, 377
Ellis, G.F.R. & MacCallum, M.A.H. 1969, *Comm. Math. Phys.* **12**, 108
Ellis, G.F.R., Maartens, R. & Nel, S.D. 1978, *Mon. Not. R. Ast. Soc.* **184**, 439
Ellis, G.F.R., & Perry, J.J. 1979, *Mon. Not. R. Ast. Soc.* **187**, 357
Ellis, G.F.R., Perry, J.J. & Sievers, A. 1984, *Astron. J.* **89**, 1124
Ellis, G.F.R. & Schreiber, G. 1986, *Phys. Lett.* **A115** , 97
Ellis, G.F.R. & Sciama, D.W. 1972, in *General Relativity*, ed. L. O'Raifeartaigh (Oxford University Press), 35
Ellis, G.F.R. & Stoeger, W.R. 1987, *Class. Qu. Grav.* **4** 1679
Ellis, G.F.R., Nel, S.D., Stoeger, W., Maartens, R., & Whitman, A.P. 1985, *Phys. Reports* **124**, 315
Ellis, G.F.R. & Tivon, G. 1985, *Observatory* **105**, 189
Ellis, G.F.R., & Wainwright, J. (Ed) 1995, *The dynamical systems approach to cosmology.* (Cambridge University Press, in preparation)
Futamase, T. 1989, *Mon. Not. R. Ast. Soc.* **237**, 187
Hoyle, F. & Narlikar, J.V. 1966, *Proc. Roy. Soc.* **A290**, 162
Jacobsen, P., Boksenberg, A., Deharveng, J.M., Greenfield, P., Jedrzewski, R. & Paresce. F. 1994, *Nature* **370**, 35
Kristian, J. & Sachs, R.K. 1966, *Astrophys .J.* **143**, 379
Linde, A. 1990, *Particle Physics and Inflationary Cosmology*, (Harwood Academic Publishers)
Linde, A. 1994, *Scientific American*, November, p.32
Maartens, R., Ellis, G.F.R. & Stoeger, W.J. 1995, *Phys. Rev. .D* (in press)
Matravers, D.R., Stoeger, W.R. & Ellis, G.F.R. 1995, *Qu. Jour. Roy. Ast. Soc.* **36**, 47
Omer, G.C. 1949, *Astrophys. J.* **109**, 164
Peebles, P.J.E., Schramm, D.N., Turner, E.L. & Kron, R.G. 1991, *Nature* **352**, 769
Penrose, R. 1989, *The Emperor's New Mind* (Oxford University Press)
Rosquist, K. 1988, *Astrophys .J.* **331** 648
Sachs, R.K. & Wolfe, A.M. 1967, *Astrophys. J.* **147**, 73
Sandage, A. 1968, *Observatory* **88**, 91
Sandage, A. 1988, *Ann. Rev. Astron. Astrophys.* **26**, 561
Schneider, P., Ehlers, J. & Falco, E.E. 1992, *Gravitational Lenses*, (Springer, Berlin)
Songaila, A., Cowie, L.L., Vogt, S., Keane, W., Wolfe, A.M., Hu, E.M., Oren, A.L., Tytler, D.R. & Lanzetta, K.M. 1994, *Nature* **371**, 43
Stoeger, W. (Ed) 1987, *Theory and Observational Limits in Cosmology.* (Specola Vaticana - Vatican Observatory)
Stoeger, W., Ellis, G.F.R. & Xu, C. 1994, *Phys. Rev. D* **49**, 1845
Stoeger, W., Maartens, R. & Ellis, G.F.R. 1995, *Astrophys. J.* **442**, 1
Tyson, J.A., Valdes, F., Jarvis J. & Mills, A. 1984, *Astrophys. J. Lett.* **281**, L59
Weinberg, S. 1972, *Gravitation and Cosmology* (Wiley).
Zotov, N.V. & Stoeger, W.R. 1992, *Class. Qu. Grav.* **9**, 1023

Fluctuation Spectra and High-Redshift Objects

John Peacock and Stephen McNally

Royal Observatory, Blackford Hill, Edinburgh EH9 3HJ, UK.

1 Introduction

Observations of large-scale galaxy clustering have produced something of a crisis for theories of cosmological structure formation. The paradigm for the past decade has been a model where scale-invariant adiabatic primordial fluctuations cause clustering to grow in dark matter which is collisionless, and which has negligible thermal velocities; in many ways, observations have supported the basic elements of this picture. The clustering power spectrum is smooth and featureless, with no sign of the oscillatory features that would be expected if normal baryonic material was dynamically dominant (e.g. Peacock & Dodds 1994; hereafter PD). The detection of microwave-background anisotropies on a variety of angular scales favours a fluctuation spectrum which is indeed close to adiabatic and scale-invariant for wavelengths above about 100 Mpc (e.g. White, Scott & Silk 1994).

Despite these encouraging features, there has emerged a consensus that there is a problem with the shape of the fluctuation spectrum. The density of the universe should be written on the sky in the form of a break in the spectrum at around the comoving horizon scale at matter-radiation equality

$$r_{\rm H} = 16.0\,[\Omega h^2]^{-1}\ \mathrm{Mpc} \tag{1}$$

(as usual, $h \equiv H_0/100\ \mathrm{km\,s^{-1}\,Mpc^{-1}}$). This number, and all others in the paper, assume $T = 2.726$ K for the cosmic microwave background (CMB) temperature (Mather *et al.* 1994). Since observed wavenumbers come in units of $h\ \mathrm{Mpc}^{-1}$, the combination Ωh is measurable, and in practice is estimated by fitting a model of a scale-invariant spectrum modified by the cold dark matter (CDM) transfer function. According to PD, an approximate 95 per cent confidence range for the apparent value of the density is

$$0.22 < \Omega h|_{\rm apparent} < 0.29 \tag{2}$$

(allowing for effects of nonlinearity and redshift-space distortions). The problem is that this number appears to be inconsistent with current estimates of h and

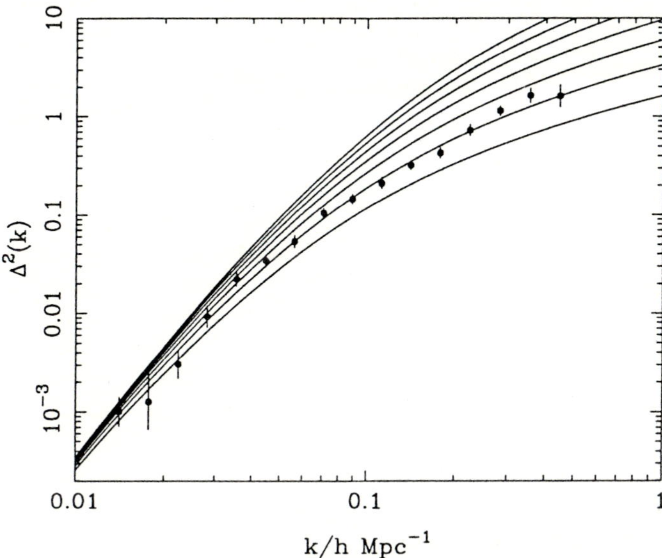

Fig. 1. The averaged linear power-spectrum data of PD, compared to various CDM models. These assume scale-invariant initial conditions, with the same large-wavelength normalization. Different values of the fitting parameter $\Omega h = 0.5, 0.45, \ldots 0.25, 0.2$ are shown; power is an increasing function of Ωh, so that the 'standard' $\Omega h = 0.5$ model has the highest power, whereas $\Omega h = 0.2$ has the lowest. $\Omega = 1$ and IRAS bias of unity are assumed; the normalization is to COBE with $Q_{\rm rms} = 19\ \mu$K.

an $\Omega = 1$ Einstein-de Sitter universe, and a variety of possible solutions have been suggested.

The least contrived escape routes seem to be those involving modifications either of the dark matter or of the relativistic content, and it on is this latter possibility that we wish to concentrate. It turns out that the characteristic prediction of such a model is of an enhanced amplitude of fluctuations on small scales, and we use this feature to place limits on the decaying particle involved in the model.

2 CDM with a decaying particle

Structure formation scenarios incorporating decaying particles have a long history (e.g. Davis *et al.* 1981; Bardeen, Bond & Efstathiou 1987). However, this model was given a strong boost by experimental work suggesting the possibility of neutrino eigenstates with $m \simeq 17$ keV (e.g. Simpson 1985). Such a neutrino cannot be stable, otherwise it would close the universe many times over. An acceptable present density can be achieved if the massive neutrino decays to products which are relativistic today – either to other massless neutrinos or possibly to some exotic new species. Decays to photons are not allowed for two reasons. First, the COBE results on the lack of CMB spectral distortions severely limit the allowed energy injection prior to recombination (Mather *et al.* 1994). Second, the relativistic density in photons is observed, and a total relativistic density is conventionally obtained by multiplying by a factor 1.68 to allow for three species of massless neutrinos. The possibility we wish to consider here, however, is that the true relativistic density is higher. If the decays producing this enhanced background occurred at redshifts greater than conventional matter-radiation equality ($z \simeq 24000\,\Omega h^2$), the onset of matter domination would be delayed, and we would have an explanation of the large-scale structure problem. The analysis presented in this paper follows on from work by Bond & Efstathiou (1991); hereafter BE. They calculated density fluctuations in $\Omega = 1$ models dominated by CDM, in which 17 keV neutrinos having lifetimes between 1 and 10^4 years decayed to relativistic products. BE derived power spectra which differ from standard CDM in two key ways (note that hereafter we use units where m is mass in keV and τ is lifetime in years):

(1) The decay increases the density of relativistic degrees of freedom ('radiation' for short) and so delays the onset of matter-radiation equality. The length scale associated with the Hubble radius at this epoch is correspondingly modified. This modification is parameterized by θ (Bardeen *et al.* 1987) which is the ratio of energy density in relativistic species with a decaying particle to that without. Equation (1) now reads

$$r_{\rm H} = 16.0\,[\Omega h^2]^{-1}\,\theta^{1/2}\ {\rm Mpc} \tag{3}$$

and as a result the apparent value of Ωh (the effective shape parameter for CDM transfer functions) is dependent on the mass and lifetime of the decaying particle.

(2) If the decay time exceeds $\sim 10m^{-2}$ years the universe can pass through *two* periods of matter domination, the first occuring when the density of undecayed particles exceeds that of relativistic species.

The horizon scale at the the time at which the non-relativistic density of the massive neutrinos first becomes dominant is given by equation (3). Now, a massive neutrino has $\Omega h^2 = m/0.095$ in our units (e.g. Kolb & Turner 1990, but scaling to the COBE $T = 2.726$ K); $\theta_{\rm eq1} = 1.45/1.68$; and $\theta_{\rm eq2}$ is boosted by the extra radiation:

$$r_{\rm H}^{\rm eq1} = 1.41\,m^{-1}\ {\rm Mpc} \tag{4}$$

$$r_{\rm H}^{\rm eq2} = 14.9\,h^{-2}\left[1 + x(m^2\tau)^{2/3}\right]^{1/2}\ {\rm Mpc} \tag{5}$$

with x a dimensionless constant. By a numerical solution of the full equations describing the problem, BE obtained $x \simeq 0.15$.

The effect of this is to yield a power spectrum with two characteristic break wavenumbers, of order the reciprocal of the appropriate $r_{\rm H}$. The result can be modelled as the sum of two CDM spectra with differing power-law break lengths and different amplitudes, i.e.

$$\Delta^2(k) = \Delta^2_{\rm LSS}(k) + \alpha^2\Delta^2_{\rm LSS}(k/\beta) \tag{6}$$

Throughout, we shall express power spectra in dimensionless form:

$$\Delta^2(k) \equiv d\sigma^2/d\ln k = k^3 P(k)/2\pi^2. \tag{7}$$

This expression says that the small-scale power spectrum looks like the CDM model which fits large-scale structure, but with a 'bump' superimposed which is a copy of the large-scale spectrum that has been shifted to smaller scales by a factor β and boosted by a factor α^2 (we ignore the small-scale collisionless damping, with $\Delta^2 \to \Delta^2 \exp[-k^2R^2]$, where $R \simeq 0.25/m$).

The large-scale spectrum, $\Delta^2_{\rm LSS}$ would be just the standard BBKS CDM spectrum with an apparent density

$$\Omega h|_{\rm apparent} = \Omega h \left(\frac{1.45}{1.68}\left[1 + x(m^2\tau)^{2/3}\right]\right)^{1/2}. \tag{8}$$

The shift β is just the ratio of the horizon sizes deduced above

$$\beta = 10.6mh^{-2}\left[1 + x(m^2\tau)^{2/3}\right]^{1/2}. \tag{9}$$

The 'boost factor' α is more subtle, and turns out to be

$$\alpha = y\,[m^2\tau]^{2/3}, \tag{10}$$

where y is a further dimensionless constant which must be determined by fitting an exact integration. To obtain the value of this constant, we compared the power spectrum (7) with the BE results for 17 keV neutrinos decaying at 1, 10, 10^2, 10^3 and 10^4 years. It was found useful to adopt a softening parameter γ such that that

$$\Delta^2(k) = \left([\Delta^2_{\rm LSS}(k)]^\gamma + [\alpha^2\Delta^2_{\rm LSS}(k/\beta)]^\gamma\right)^{1/\gamma}. \tag{11}$$

This smooths the transition region at which the small scale power of the second term becomes dominant over that of the first. The best fit obtained requires the prefactor for α to be $y = 1.14$, with $\gamma = 0.33$.

3 Constraints

3.1 Normalization

We now need to place limits on the model by comparing its predictions with observations of structure in the universe. For this, we require normalization for the theoretical power spectra. This can be expressed as the linear theory rms density contrast when averaged over spheres of radius $8h^{-1}$Mpc, i.e. σ_8. White, Efstathiou and Frenk (1993) use the observed abundance of rich clusters to deduce a hard allowed range of $\sigma_8 = 0.52 - 0.62$. We shall adopt $\sigma_8 = 0.6$, which is perhaps at the higher end of the allowed range; since we are looking to see if extra small-scale power is required, it makes sense to be conservative and adopt the highest reasonable normalization for the large-scale mass spectrum.

To evolve these linear power spectra to the present day ($z = 0$) involves non-linearities which can alter the power spectrum significantly at small scales. In evolving the $\Delta^2(k) \gtrsim 1$ portion of the power spectrum we use the formulae of PD (derived from the work of Hamilton *et al.* 1991, which was based on N-body simulations on the relevant scales) to incorporate the effects of non-linear evolution. Whilst this correction increases the small-scale power for shallow power-law regions it actually removes power in regions with a strong k-dependence.

3.2 High redshift objects

One way of constraining the small-scale ($k \gtrsim 0.2h$ Mpc^{-1}) power is to require that it is sufficient to form the observed abundances of high redshift objects. The most stringent constraints can be derived from recent deep measurements of damped Lyα sytems with H I column densities greater than $\sim 2 \times 10^{20}$ cm^{-2} (Lanzetta *et al.* 1991) These have been used by several authors to investigate rival dark matter models. If the fraction of baryons in the virialized dark matter halos equals the global value $\Omega_{\rm B}$, then these data can be used to infer the total fraction of matter that has collapsed into bound structures at high redshifts (Ma & Bertschinger 1994, Mo & Miralda-Escudé 1994, Subramanian & Padmanabhan 1994). The deepest measurement at $\langle z \rangle \simeq 3.2$ implies $\Omega_{\rm HI} \simeq 0.005$, and hence a collapsed fraction of $\simeq 10\%$ if $\Omega_{\rm B} = 0.05$. Here we apply the Walker *et al.* (1991) constraint to the baryon density, namely $\Omega_{\rm B} h^2 = 0.0125 \pm 0.0025$.

The assumption here will be that the damped Lyα systems are the progenitors of present day spiral galaxies. Evidence for this view has been provided by absorption in lensed quasar systems, showing the systems be of galactic dimensions (Wolfe *et al.* 1992). Furthermore the baryon mass inferred in present day galaxies is comparable to that of the damped Lyα systems at $z \simeq 3$. The photoionizing background prevents systems with circular velocities of less than about 50 km s^{-1} cooling sufficiently to form bound systems. We shall use this conservative velocity limit to estimate the minimum mass of object that the Lyα measurements detect. Virial equilibrium for a halo of mass M and radius r_v demands

$$v_c^2 = \frac{GM}{r_v} \tag{12}$$

For a spherically collapsed object this velocity can be converted directly into a Lagrangian comoving radius containing this mass (White *et al.* 1993)

$$r_0 = \frac{2^{1/2} v_c}{H_0 \Omega^{1/2} (1+z)^{1/2} (1 + 178\Omega^{-0.6})^{1/6}} \tag{13}$$

The values introduced above require $r_0 > 0.15h^{-1}$ Mpc.

To use the measurement to constrain our candidate power spectra we employ the formalism of Press & Schechter (1974) which gives a collapsed fraction Ω_c above some mass scale $M_{\rm coll}$, given a redshift z and a means of computing the $z = 0$ rms density contrast as a function of mass $\sigma(M)$:

$$\Omega_c(> M, z) = 1 - \mathrm{erf}\,[\delta_c(1+z)/\sqrt{2}\sigma(M)]. \tag{14}$$

to obtain $\sigma(M)$, we use a spherical 'top hat' filter of radius R_T Our lower mass limit corresponding to $r_0 = 0.15h^{-1}$ Mpc is $M_{\rm coll} = 10^{9.6}h^{-1}\mathrm{M}_\odot$. $\delta_c = 1.69$ was assumed.

Our power spectra are therefore required to produce an Ω_c which *exceeds* the Ly α collapsed fraction for $z = 3.2$, $M > 10^{9.6}h^{-1}\mathrm{M}_\odot$. $\Omega h = 0.5$ CDM successfully attains a sufficient collapsed fraction at the high redshift end. Models that agree with the APM at large scales do less well. $\Omega h = 0.25$ CDM and MDM with $\Omega_\nu = 0.3$ both fall far short of having enough small scale power. The CDM + relativistic decay model for $\tau = 10$ yr and $m = 1.3$ keV has ample small scale power thanks to the extra bump from the first epoch of matter domination.

How seriously should these constraints be taken? A pure $\Omega h = 0.25$ spectrum does not fail to fit the data by a very large amount, and it is not impossible that such a model could be rescued by tweaking the assumptions in the calculations. Nevertheless, with theoretical predictions falling rapidly, but the data apparently rising, it seems plausible that the inconsistencies may increase greatly with the advent of deeper Lyα measurements in the future. Our work is therefore perhaps best viewed as exploring the potential consequences of such models. We have therefore indicated below the consequences of extrapolating the measured $\Omega_{\rm HI}$ abundance at $z = 3.2$ to $z = 4$ and $z = 5$. The conclusions will be that $\Omega_{\rm HI}$ at these redshifts must fall if there is no small-scale power-spectrum feature, and that $\Omega_{\rm HI}$ cannot rise very much without the whole model being ruled out.

3.3 Galaxy clustering

The above constraints require only some minimum level of power; however, we do not want to exceed this minimum by too large a factor. An upper limit to the power can be provided by comparison of the candidate power spectra with the observed small-scale power spectrum, best determined by angular deprojection of the APM galaxy survey (Baugh & Efstathiou 1993), who give data down to $k \simeq 8h$ Mpc^{-1}. The mass-to-light ratios of clusters strongly encourage us to believe that the small-scale clustering of light must exceed that of mass, if $\Omega = 1$. Although some models have been advocated in which this would not be true (e.g. the paper by Couchman & Carlberg 1992 on $b = 1$ standard CDM), it is reasonable to regard such a situation as observationally unacceptable. We will therefore set a conservative upper limit to the allowed degree of small-scale power by demanding that the theoretical nonlinear power spectrum of the mass at no point exceeds that of the light.

4 Allowed models

4.1 Limits on parameters

The limits on the decaying neutrino models derived by the above methods are summarised in figure 2

The region disallowed by the Lyα structure formation requirement at z=3.2 is indicated by the darker shaded region. In order not to exceed the small-scale

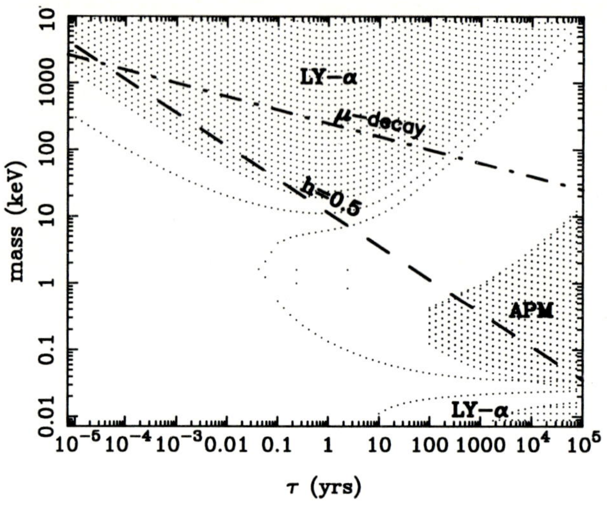

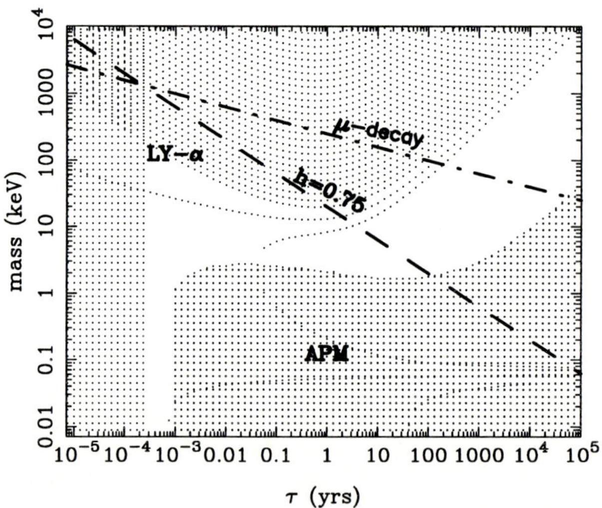

Fig. 2. Constraints on mass and lifetime of hypothetical decaying neutrino. Limits are based on the damped Lyα fraction at redshift $z = 3.2$ (darker shaded region forbidden) with extrapolations for the same fraction at $z = 4$ and 5 (dotted lines). Values which would exceed the APM power spectrum for $k > 1h$ Mpc^{-1} occupy the lighter shaded region. (a) Model constraints for $\Omega h = 0.5$, $\Omega_{\rm B} = 0.050$. The dashed line shows the large-scale structure constraint for $\Omega h = 0.5$, the dot-dashed line the $m^5\tau \simeq const$ law for heavy leptons. (b) Model constraints for $\Omega h = 0.75$, $\Omega_{\rm B} = 0.022$. The dashed line shows the large-scale structure constraint for $\Omega h = 0.75$, the dot-dashed line the $m^5\tau \simeq const$ law for heavy leptons.

APM curve, the parameters must lie somewhere in the plane away from the lighter shaded region. The dashed line from top-left to bottom-right of each plot represents $m^2\tau \simeq 120$ and 500, the values required to reconcile $\Omega h_{\rm apparent} = 0.25$ with $\Omega h_{\rm true} = 0.50$ and 0.75. We can summarize the conclusions from these figures as follows:

$$\Omega h = 0.50 \Rightarrow 0.5 < m < 10 \text{ keV}, \ 1 < \tau < 500 \text{ yr} \tag{15}$$

$$\Omega h = 0.75 \Rightarrow 2.0 < m < 10 \text{ keV}, \ 3 < \tau < 200 \text{ yr}. \tag{16}$$

Are these parameter values physically plausible? The dot-dashed line in figures 2a & 2b shows the equation

$$m^5\tau = 3 \times 10^4 \text{ keV}^5\text{yrs}, \tag{17}$$

which is the form of a naive prediction for the relation between mass and lifetime for a particle which decays via the weak interaction. This takes the usual E^5 scaling of weak-interaction cross sections and scales to the decay of the μ lepton. In neither case does this 'muon-decay' line cross the large-scale structure line in a permitted region; if the model is to be considered plausible, the decay physics involved must be more exotic.

4.2 Early reionization

The existence of significant small scale power in decaying particle CDM can allow structures of mass $10^5 - 10^8 h^{-1}\mathrm{M}_\odot$ to form earlier than in the standard model and so may permit early reionization of the intergalactic medium. We have performed a brief analysis based on the method of Tegmark, Silk and Blanchard (1994). They estimate a parameter $f_{\rm net}$, the net efficiency of ionisation processes from stars, and calculate $f_s(M)$ the collapsed fraction of the universe (that gives rise to the star formation). These they relate to an ionisation fraction χ such that

$$\chi \simeq 3.8 \times 10^5 f_{\rm net} f_s \tag{18}$$

Tegmark *et al.* give a range of $f_{\rm net}$ values they believe permissible, dubbing the top of the range 'optimistic' (in the sense of promoting early reionization), the bottom of the range 'pessimistic' and the median value 'middle-of-the-road'. Once f_{net} is set in this manner the requirement of 100% reionization ($\chi > 1$) becomes a condition on f_s which can be calculated by the same Press-Schechter method as was used to analyze the Lyman-α constraint. In order to get a feel for the sort of redshifts at which reionization could occur in our model we set $m^2\tau \simeq 500$, the value needed to give an apparent $\Omega h = 0.25$ in the case where $\Omega = 1$ and $h = 0.75$. We then vary the mass within the range 0.01 to 100 keV, τ being set at $500/m^2$. f_s is calculated for the resulting power spectra and used to derive the redshift at which reionization is complete.

The lower limit for the Press-Schechter integration relates to the masses of the first galaxies to form and there is a considerable range thought applicable from $10^5 h^{-1}\mathrm{M}_\odot$ (Couchman & Rees 1986) to $10^7 - 10^8 h^{-1}\mathrm{M}_\odot$ (Blanchard *et al.*

1992). Fortunately the power spectra are relatively flat in this range and the collapsed fraction relatively insensitive to $M_{\rm coll}$. Our results relate to the mass scale $10^7 h^{-1} M_\odot$. In rough agreement with the Tegmark *et al.* results typical reionization redshifts are in the range $\sim 10 - 100$ for an optimistic or middle-of the road $f_{\rm net}$. A pessimistic $f_{\rm net}$ permits early reionization over the mass range $\sim 0.5 - 20$ keV. MBR photons will be significantly scattered by the reionized plasma if the optical depth between $z_{\rm ion}$ and $z = 0$ is $\simeq 1$. To obtain this requires $z_{\rm ion} \simeq 50$ (e.g. Padmanabhan 1993) which can occur in the relativistic decay model for $m \sim 1 - 10$ keV – within our allowed range. Such a result may well be of importance in wiping out details of the last scattering surface on angular scales around $\sim 1°$ (e.g. White, Scott & Silk 1994).

Acknowledgements SJM acknowledges the support of a SERC/PPARC research studentship. We thank Dick Hunstead and Rocky Kolb for helpful discussions, and particularly Michael Turner for saving us from error at an early stage of this investigation.

References

Bardeen, J. M., Bond, J.R., Efstathiou, G. 1987, *Astrophys. J.* **321**, 28
Baugh, C.M., Efstathiou, G. 1993, *Mon. Not. R. Astr. Soc.* **265**, 145
Blanchard, A., Valls-Gabaud, D., Mamon, G.A. 1992, *Astron. Astrophys.* **264**, 365
Bond, J.R., Efstathiou, G. 1991, *Phys. Lett. B* **265**, 245
Couchman, H.M.P., Carlberg, R.G. 1992, *Astrophys. J.* **389**, 453
Couchman, H.M.P., Rees, M.J. 1986, *Mon. Not. R. Astr. Soc.* **221**, 53
Davis, M., Lecar, M., Pryor, C., Witten, E. 1981, *Astrophys. J.* **250**, 423
Hamilton, A.J.S., Kumar, P., Lu, E., Matthews, A. 1991, *Astrophys. J. Lett.* **374**, L1
Kolb, E.W., Turner, M.S. 1990, *The Early Universe*, Addison-Wesley, Redwood City
Lanzetta, K., Wolfe, A.M., Turnshek, D.A., Lu, L., McMahon, R.G., Hazard, C. 1991, *Astrophys. J. Suppl.* **77**, 1
Mather, J.C., *et al.* 1994, *Astrophys. J.* **420**, 439
Mo, H.J., Peacock, J.A., Xia, X.Y. 1993, *Mon. Not. R. Astr. Soc.* **260**, 121
Ma, C.P., Bertschinger, E. 1994, *Astrophys. J.* **429**, 22
Mo, H.J., Miralda-Escudé, J. 1994, *Astrophys. J. Lett.* **430**, L25
Padmanabhan, T. 1993, *Structure Formation in the Universe*, Cambridge Univ. Press, pp.237-239
Peacock, J.A., Dodds, S.J. 1994, *Mon. Not. R. Astr. Soc.* **267**, 1020
Press, W.H., Schechter, P. 1974, *Astrophys. J.* **187**, 425
Simpson, J.J. 1985, *Phys. Rev. Lett.* **54**, 1891
Subramanian, K., Padmanabhan, T. 1993, *Mon. Not. R. Astr. Soc.* **265**, 101
Tegmark, M., Silk, J., Blanchard, A. 1994, *Astrophys. J.* **420**, 484
Tormen, G., Moscardini, L., Lucchin, F., Matarrese, S. 1993, *Astrophys. J.* **411**, 16
Walker, T.P., Steigman, G., Schramm, D.N., Olive, K.A., Kang, H.S. 1991, *Astrophys. J.* **376**, 51
White, S.D.M., Efstathiou, G., Frenk, C.S. 1993, *Mon. Not. R. Astr. Soc.* **262**, 1023
White, M., Scott, D., Silk, J. 1994, *Ann. Rev. Astr. Astrophys.* **32**, 319
Wolfe, A.M, Turnshek, D.A., Lanzetta, K.M., Oke, J.B. 1992, *Astrophys. J.* **385**, 151

Galaxies at High Redshift: 1994

Hyron Spinrad

Astronomy Dept., University of California at Berkeley,
Berkeley, CA 94720

1 Summary

In this review I discuss galaxies at $z > 2.0$, mostly radio-loud giants. We can learn much about young galaxies from a combination of radio and optical / IR observations, even for quite faint systems.

I discuss selection effects in locating these distant galaxies and the radio quiet friends of Lyα absorbers. We can now measure the Lyman limit in the most distant radio galaxies, and I discuss those spectra and their repercussions.

Further discussion centers on the chemical abundances and space-densities of the most distant radio galaxies, at epochs corresponding to $z > 2$.

Finally we note that new Keck spectrograms of 3C 265 ($z = 0.81$) shed clear light on the debate about the physical origin of the extended aligned portions of distant radio galaxies; they clearly harbor a "hidden" QSR which is visible through some scattered UV radiation.

2 Introduction: Radio Loud and Quiet Galaxies with $z > 2$

One of our major goals to justify the difficult practical study of faint, distant, galaxies is the expectation that they will constrain galaxy formation. The epoch(s) of formation of the spheroids is likely to be rather large, in some cases $z > 3$. Perhaps the manner of "formation" can be also be glimpsed by observation of galaxy morphologies with the high angular resolution now afforded by the Hubble Space Telescope.

Radio galaxies are still the most fruitful type of galaxy to study at high redshifts, because they are now locatable with reliability, even at very faint light levels. However they are definitely peculiar large systems-with only the nearest sources bearing a resemblance to giant E galaxies at $z < 1$. At larger redshifts the radio galaxies are all somewhat peculiar, but still are information-rich, yielding "early galaxy" clues more readily then the "fuzz" around QSO's (their host galaxy).

On the other hand, we are pleased to note a growing number of Lyα-emitting, radio quiet galaxies are being located at large redshifts because they are physically associated with damped Lyα absorbing (intervening) galaxies.

In this written review I mention some contemporary topics relevant to the $z > 2.0$ galaxies — a really distant subset of AGN, and a few damped Lyα companions. A few topics have been updated and / or abbreviated since the verbal presentation at Ringberg Castle in September 1994.

3 The Status of Very Distant Galaxies in Mid-1994

About 79 galaxies that are strong or fairly strong radio emitters and have secure redshifts in excess of $z > 2.0$ are now known.

This large number is the result of 6 years of hard work by several well known optical / radio groups with industrious P.I. and graduate students. The radio galaxy redshifts now extend to $z = 4.25$ (Lacy *et al.* 1994; Spinrad *et al.* 1995). Note the z - distribution; it is still easier to locate and observe emission-line galaxies with $z \sim 2.0$ than at large redshift, provided the near-UV spectral region between 3700 - 4000 Åis well covered (some spectrographs are still blue -UV- 'blind'). Figure 1 shows the histogram of z - distributions for radio galaxies known to me in mid-1994.

The selection criteria for locating these radio galaxies is worth some discussion. Most of the radio samples utilized recently for our distant galaxy quest have been at $\sim$ "1 Jy strength"; e.g. $S_{408\mathrm{MHz}} \leq 1$ Jy (equivalent to $S_{1.4\mathrm{GHz}} = 0.3$ Jy) for typical steep-spectrum sources. The 3CR (strong source) sample allows us to observe galaxies only to $z = 2.47$ (one galaxy with $z > 2.0$) because of the high flux density limit (9 Jy) and the steepness of the luminous end of the radio luminosity function (RLF). The much weaker LBDS or Westerbork / VLA mJy sources tend to have weak optical emission lines, balancing our assumption that many of them could be at truly high redshifts. They will not usually enter our positive result discussion unless Lyα is visible and strong ($z > 1.8$), and this is rare.

The radio spectral-idex bias, first suggested by Miley and by Gopal-Krishna, and exploited successfully by Chambers, Miley, & van Breugel (1990; CMB) in the late 1980's has been a useful way of leaning toward luminous and distant radio galaxies. Ultra-steep-spectrum radio sources with $\alpha > 1$ ($S_\nu \sim \nu^{-\alpha}$) are normally associated with galaxies; those with $\alpha > 1.2$ are rare and are the "sources of choice" in recent surveys which hope to bias their selection criteria toward the location of high redshift galaxies.The most recent discussion of the success of this "α - selection" is in Röttgering (1993). Following CMB and McCarthy & van Breugel (1989), we note three effects that contribute to the (α, z) correlation in the radio and optical studies of distant radio galaxies. These are: the K-correction for the radio spectra that are curved (a non-constant α with ν), secondly a true change in the RLF at higher z, causing the utltrasteep sources to be preferentially included in low frequency surveys and third, the redshift-spectral index correlation could reflect the existence of a set of powerful sources

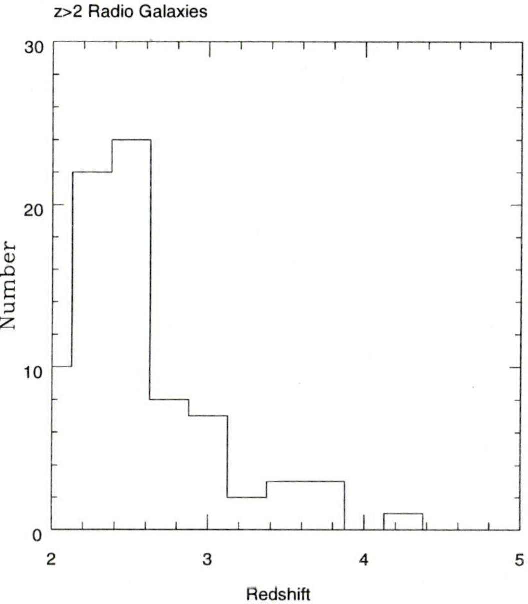

Fig. 1. A histogram representation of the distribution of redshift measures for $z >$ 2.00 radio galaxies known to the author is mid-1994. Note the general decrease in the number of AGN beyond z = 2.9. Also the first bin (2.0 - 2.2) may be artificially depressed by observational bias against successfully observing Lyα emission below $\lambda =$ 4000Å with certain spectrographs. This redshift distribution is compared to models of the evolving RLF later in the paper.

that would have intrinsically steeper radio spectra. In our present samples it is still unclear as to which of those 3 effects is dominant . Clearly it will be advantageous for our science if not all investigators select only steep-spectrum radio sources for intensive study.

Fortunately in 1994 we have more than just radio-loud galaxies available for study at $z > 2$. Historically the first non-radio galaxies found at such large redshifts were the rare galaxy companions to radio-loud QSR's, discriminated by their moderately strong Lyα emission. Narrowband imaging by Djorgovski and colleagues (1985, 1987). Hu & McMahon (1993) and Steidel *et al.* (1991) have only located a few such Lyα companions, with $<z>\sim 3.1$. Some may be slightly "active" also, shown by the emission line ratios and line widths in the spectrum of the PKS 1614+058 companion.

Perhaps more important are the companions to intervening damped Lyα absorption galaxies at z >2. Historically searches for these distant, and perhaps more-normal galaxies, have begun with a damped Lyα line from an extensive gas column-like that seen looking through the disk of a spiral galaxy. Lyα imaging then locates galaxies at the common absorption redshift that have star formation or an AGN, perhaps $z_e \approx z_{damped}$, usually to within $500\,\mathrm{km\,s^{-1}}$. The survey limitations are the necessary spatial proximity of the damped absorber to a QSO sight line, and for the other galaxies in the group, no complete loss of Lyα emission through resonant scattering and dust-absorption.

The most exciting specific Lyα emitters of this type are galaxies studied by Møller & Warren (MW) (1993), by Lowenthal *et al.* (1991), and by Francis *et al.* (1995). MW's three faint galaxies with Lyα emission all have resolved lines with some spatial extent in at least one. Most of these companions do show detectable C IV / He II when re-observed to adequate S/N; the big question is, are these radio-quiet Lyα emitters some sort of AGN [as the presence of C IV emission might suggest] or just star-formers with little dust or neutral H in the path toward earth?

Steidel & Hamilton (1992) have proposed locating $z > 3$ galaxies utilizing the generic property of the UV spectra of galaxies, even those with active on-going star formation. All systems should show a large discontinuity at the 912 (rest) Lyman limit. This ought to be a very robust finding procedure; by deep imaging at the U and B bands the Lyman limit can be placed strategically between the photometric bands ($z \approx 3.4$] so that $(U - B)$ is numerically large. At the same time any QSO with an intervening optically thick system at the Lyman limit will go dark in the band below 912 (U in the Q0000-263 case). Then 'nearby' foreground interloping galaxies should be detectable. In my opinion an extension of this procedure may be the best method to locate clustered galaxies at $z > 3$ (c.f. Giavalisco *et al.* 1994).

4 Some Speculative Topics in Radio Galaxy Research

4.1 Comments on the Lyman Edge and Lyman Limit Systems (LLS)

The Lyman limit has now been spectroscopically measured for two distant radio galaxies; both 4C 41.17 ($z = 3.80$) and 8C 1435+63 ($z = 4.25$) have detectable UV continua longward of the limit, given substantial observing efforts. The break amplitudes are large (c.f. Spinrad, Dey, and Graham 1995), corresponding to flux ratios of 8. This effect was anticipated whether the intrinsic ultraviolet seen is directly from hot stars or is attenuated by circumnuclear or even circumgalactic neutral gas. Opaque circumgalactic clouds would be expected to leave very marked holes in the strong Lyα emission line peak of the radio galaxy; indeed Hippelein & Meisenheimer (1993) suggest a spatially restricted gap in the 4C 41.17 Lyα profile. But that probably would not cover nearly enough continuum source to cause the Ly limit "system" we measure for that galaxy.

Within the constraint of our tiny radio galaxy sample we note a difference from the more- familiar QSO Lyman limit systems caused by intervening H clouds ($N_H > 10^{17}$). The QSO spectra usually show a "free-redshift-space" of $\delta z = 0.1 - 0.5$ units from the emission line redshift to the LLS z-value. Thus the QSO-nucleus is visible from earth at $\lambda_o = 911$, while the 2 radio galaxies are "obscured" or intrinsically very faint at that limit wavelength. Why the difference? The only safe statement is that the QSO's do not have circumnuclear neutral gas in large quantities.

The radio galaxies are visible in the deep UV by either hot starlight or more probably scattered radiation from a buried QSR. But should not the scattered ultraviolet "reflection" resemble the direct beam supposed for the typical QSO/QSR itself? One reply could be: perhaps the gaseous Lyman limit is prevented in QSO's environment by the higher UV radiation field in the beamed direction only. No neutral gas remains. The radio galaxies, in this "unification" scenario, illustrate the system at a non-beamed angle and thus can show LLS absorption from the remaining neutral gas out of the nuclear and extranuclear ionization cone.

Of course the unification concept may not be correct or complete; perhaps the difference between galaxies and QSR's at λ_0 912 is an intrinsic one. We need to eventually quantitatively measure the higher members of the Lyman series in absorption to differentiate starlight from QSO- scattering. Of course spectral measures longward of Lyα should eventually accomplish this goal also.

In conclusion of this subsection, we note the distant (early) Universe is probably not flooded with ionizing photons from radio galaxies and other non-QSO-like AGN. 4C 41.17 and 8C 1435+63 have not added much to the metagalactic radiation field.

We look forward to the future (difficult) measures of the Lyman edge in star-forming galaxies-like those observed by Steidel & Hamilton (1992). Do they allow any ionizing radiation to escape?

Perhaps no further ionizing radiation (besides that from known optical and X-ray-bright QSO's) is needed to satisfy the extant Gunn-Peterson limits.

4.2 Speculations on the Chemical Abundances at $z > 3$

Steidel (1990) and Steidel & Sargent (1989) have shown that the number of C IV absorption line systems suggest a noticeable decline in the C/H abundance ratio at large z ($z > 3.0$) compared to $z \sim 1.5$. Do we anticipate any similar compositional change in the massive galaxies that we observe as radio -loud "monsters" ?

Unfortunately there is only one strong "metallic" line available to compare with hydrogen or helium over the common redshift range, $1.7 < z < 4.25$. That is, of course, the 1549 C IV doublet; it is often a well-detected emission line in the spectra of high redshift radio galaxies (c.f. Spinrad *et al.* 1985; McCarthy 1993).

If we compare the spectra of well observed radio galaxies over the above z range to examine the ratios of Lyα and He II 1640 to the C IV 1549 line we do note a slight ,but noisy trend toward a decrease in the strength of C IV emission compared to Lyα of H and / or the He II line. Besides the S/N problem, Lyα is subject to blue-wing absorption from any cold neutral gas and dust in the sight line; probably the comparison to 1640 He II is safer, but that weaker line is much more difficult to flux-measure than Lyα. There is less ionization difference with the C IV/He II ratio also.

We note that distant QSO's do not indicate any clear-cut decline in C IV strength with z. Of course their broad-line region originates in a tiny and perhaps non-representative volume near the AGN, while radio galaxy emission lines are far more spacially extensive and could reflect true abundances of the host galaxy's ISM.

In any case, the extent of the carbon (metal) deficiency we nay detect in distant radio galaxies is of order ~ 2 in C/H or C/He. This is a very modest metal-deficiency compared to that observed for the oldest halo stars in our Milky Way or the Lyα clouds at $z > 2.3$ (Cowie *et al.* 1995).

4.3 The Comoving Space Densities of Radio Galaxies Beyond z = 2.0

A major question in the early history of the Universe is whether the AGN phenomenon had an abrupt beginning — or phrased in reverse, is there a cutoff of radio-loud galaxies (for example) at the highest measurable redshift? How soon after structure formation can the galaxies assemble central black holes?

Optimistically we hope to use the radio galaxy evolutionary pace as a 'clock' to guide our guesses about the time-scales of other, normal massive stellar systems.

The evolution of the Radio Luminosity Function (RLF) is a basic problem. Fortunately we can lean upon and extend the seminal work of Dunlop & Peacock (1990) who suggested that the co-moving space density of radio AGN declines substantially after the dramatic rise in density from small redshifts z to $z \sim 2$. They are predicted to be reduced in density by about a factor 5 at $z = 4$ from $z = 2.0$. Dunlop and Peacock's work was somewhat hindered by a lack of spectroscopic redshifts for the faintest radio galaxies.

We have chosen two data sets on radio galaxies to improve the statistical comparison with confidence. First we compiled a list of radio galaxies with reliable redshifts from the restricted subset of the MG catalogue described by Spinrad *et al.* (1993). These 216 small, moderate steep radio sources are bias against very low z's (angular size limited) but should not be size biased between $z = 2$ and $z = 4$.

Through the courtesy of Dr. John Peacock we compare in Table 1 the Dunlop-Peacock model predictions of the anticipated number of radio galaxies in coarse redshift bins to those observed so far in the optical-radio research using the MG set. Our list of $z- >2$ galaxies has only 7 entries, but it conforms well to the uniform density model ($2 < z < 4$) without any high-z cutoff. There is no rapid decline in the co-moving space density of radio galaxies beyond z = 2 from this small-sample of data, at least. The RLF at $z > 3$ seems much like that at $z = 2$.

Perhaps a statistically stronger radio galaxy sample may be assembled by simply listing all known radio galaxies with $z > 2.0$. Thanks to Drs. Miley, Röttgering, McCarthy, Djorgovski, and my own students we now have 79 such distant radio galaxies with secure redshifts.

Figure 1 shows their z-distribution declines steadily from $z = 2.3$ onwards; it is assumed the "deficit" at the $z = 2.0$ bin is aphysical — probably due to the location of the strong Lyα line below 4100Å at $z < 2.3$. Several spectrographs

Table 1. Comparison of 1994 MG Redshift Data with the Dunlop & Peacock Model. Ω =1, RLF and Density constant above $z > 2.0$

z $>$	Predicted Fraction	Predicted Number	Observed Number
2.0	0.130	9	7
2.5	0.055	4	4
3.0	0.025	2	1
3.5	0.012	0.8	1
4.0	0.0057	0.4	0

that have been utilized for the red spectra of faint galaxies have quite poor response at near-UV wavelengths.

Ignoring the redshift space at $z < 2.3$ then, we chose to compare this "unrestricted" sample to theoretical distribution models like these of Isobel Hook (1994). We normalized at $z = 2.6$; the parameters are Ω (not critical) and the co-moving density of galaxies. There is also a slight dependence on the radio spectra index; we assume $\alpha = 0.8$.

A decent fit to our observed z distribution is found with only a mild decline in the radio galaxy density; a factor of two decline per unit z interval from $z = 2.6$ onward is reasonable, but a hard "cut-off" in the density of loud galaxies at $z >$ 3 is not required by our overall comparison.

4.4 Scattered Nuclear Light in an Aligned Radio Galaxy

In closing with a more physical discussion, I think it is a permissible to discuss a topic that can be attacked with evidence at an intermediate redshift. What is the source of the extended continuum emission in aligned radio galaxies (McCarthy *et al.* 1987; Chambers *et al.* 1987)? Is it radiation scattered by dust (or electrons) originating in a hidden (buried) anistropically-radiating central QSR, or are the extended regions of the radio galaxies bright in the UV-blue because they have young stars, recently formed in a post-shock ISM? These are the two most-popular current hypothesis.

To try to decide between these 2 very different scenarios, Arjun Dey and I have obtained high S/N slit spectra of the radio galaxy 3C 265 ($z = 0.811$) using the low-resolution spectrograph on the Keck 10-m telescope. Fortunately, along our major-axis slit there lies an extra-nuclear knot, some 35 kpc from the galaxy nucleus. Superposed on a weak continuum, the knot spectrum shows a moderately strong and wide Mg II emission line, whose FWHM $\sim$ 4000 km s^{-1}! This feature is distinctly QSO-like! There is also the weak broad Mg II at the galaxy nucleus measured on these same long-slit spectra. The clinching argument in favor of a scattering model for the aligned galaxy extensions is the constancy of the Mg II line width over the face of 3C 265; independent (more-or-less)of position and line flux. The Mg II line carries a similar profile over the radio galaxy, center to "edge". The scattering medium consistent with the observed line width would be dust or relatively cold ($< 10^5$K) electrons. There is little evidence for starlight on these spectrograms, so the morphology of 3C 265 is most-consistent with the "buried Quasar" model for radio galaxies, in accord with the unified hypothesis to explain QSR's and radio galaxies with one type of active nucleus, viewed at different orientations.

My prejudicial view of the radio galaxy morphology and physical emission mechanisms in 1994/5 lead to a "mixed model". That model still contains stars, many of them quite old even as seen near $z = 1$. But starlight dominates in a symmetric and relaxed-looking galaxy only at emitted wavelengths beyond the visible ($\lambda_o \sim 1\mu$m , typically). At shorter wavelengths other processes, like the anisotropic scattering from a buried QSR nucleus is often dominant. And at the deep UV region $\lambda_o < 1216$Å, the presence or absence of gas and some dust can critically modify the emergent spectrum of a distant active galaxy.

It is our responsibility to clarify these points by innovative observations of distant galaxies in the remaining years of the 20th century.

I thank the U.S. National Science Foundation for their continued financial support; stimulating conversations with Drs. Dey, Dickinson, Djorgovski, van Breugel, McCarthy, Röttgering, Peacock, Dunlop, Hook, Miley, and Steidel are acknowleged, also.

References

Chambers, K., *et al.* 1987, *Nature* **329**, 609

Chambers, K., Miley, G., & van Breugel, W. 1990, *Astrophys. J.* **363**, 21

Cowie. L., *et al.* 1995, *Astron. J.*, in press

Djorgovski, S., *et al.* 1985, *Astrophys. J. Lett.* **299**, L1

Djorgovski, S., *et al.* 1987, *Astron. J.* **93**, 1318

Dunlop, J., & Peacock, J. 1990, *Mon. Not. R. Astr. Soc.* **247**, 19

Francis, P., *et al.* 1995, *Astrophys. J.*, in press

Giavalisco, M. *et al.* 1994, *Astrophys. J. Lett.* **425**, L5

Hippelein, H., & Meisenheimer, K. 1993, *Nature* **362**, 224

Hook, I.M. 1994, *Ph.D.Thesis*, Cambridge University

Hu, E., & McMahon, R. 1993, in *First Light in the Universe*, 9th IAP Astrophys. meeting, ed. Rocca-Volmerage *et al.* p.87

Lacy, M., *et al.* 1994, *Mon. Not. R. Astr. Soc.* **271**, 504

Lowenthal, J., *et al.* 1991, *Astrophys. J. Lett.* **377**, L73

McCarthy, P., *et al.* 1987, *Astrophys. J. Lett.* **321**, L1

McCarthy, P. 1993, *Ann. Rev. Astr. Astrophys.* **31**, 639

McCarthy, P., & van Breugel, W. 1989, in *Epoch of Galaxy Formation*, eds. Frenk *et al.* , Kluwer Dordrecht, p.57

Møller, P. & Warren, S.J. 1993, *Astron. Astrophys.* **270**, 43

Röttgering, H. 1993 *Ph.D. Thesis*, Leiden University

Spinrad, H., *et al.* 1985, *Astrophys. J. Lett.* **299**, L7

Spinrad, H., *et al.* 1993, in *Observational Cosmology*, A.S.P. Vol. **51**, eds. Chincarini, Iovino, Maccacaro, & Maccagni, p.585

Spinrad, H., Dey, A., & Graham, J. 1995, *Astrophys. J. Lett.* **438**, L51

Steidel, C. & Sargent, W.L.W. 1989, *Astrophys. J. Lett.* **343**, L33

Steidel, C. 1990, *Astrophys. J. Suppl.* **74**, 37

Steidel, C. Sargent, W.L.W, & Dickinson, M. 1991, *Astron. J.* **101**, 1187

Steidel, C., & Hamilton, D. 1992 *Astron. J.* **104**, 941

High-Redshift Milli-Jansky Radio Galaxies

James S. Dunlop[1], *John A. Peacock*[2] *and Rogier A. Windhorst*[3]

[1] Institute of Astronomy, Dept. of Physics and Astronomy, University of Edinburgh, Blackford Hill, Edinburgh EH9 3HJ, UK.
[2] Royal Observatory, Blackford Hill, Edinburgh EH9 3HJ, UK.
[3] Department of Physics and Astronomy, Arizona State University, Tempe, AZ 85287-1504, USA.

1 Summary

We present the first results of a programme of deep infrared imaging of mJy radio galaxies selected from the Leiden Berkeley Deep Survey (LBDS; $S_{1.4} > 0.58$ mJy, Windhorst *et al.* 1985). In combination with deep optical g, r, i photometry, these new data have enabled us to derive 'reliable' redshift estimates for a complete sub-sample of these galaxies. While spectroscopic confirmation is clearly required for a subset of these sources, our estimated values indicate that (i) the redshift cutoff displayed by powerful radio galaxies (Dunlop & Peacock 1990) also applies to these much less powerful sources, and (ii) weak radio galaxies at $z > 2$, unlike their more radio-powerful counterparts, have colours and magnitudes consistent with passively evolving ellipticals, indicative of a high redshift for elliptical galaxy formation ($z_f > 5$).

2 Background

The study of radio galaxies selected at mJy flux levels has the potential to resolve two important issues in observational cosmology provided redshifts can be determined or reliably estimated for complete samples of such sources. First, the deep flux limit combined with the shape of the radio luminosity function means that the redshift distribution of such samples provides a much more powerful and unambiguous test of the existence of the high-redshift cutoff for radio sources (Dunlop & Peacock 1990) than can be provided by further studies of brighter radio samples. Second, as a consequence of selection from bright radio surveys, the detailed study of galaxies at $z > 2$ has to date been confined to objects of extreme radio power (*e.g.* 4C 41.17, Chambers *et al.* 1990; B2 0902+34, Lilly 1988), and it has now become clear that the ultraviolet-infrared properties of such sources are strongly contaminated by processes connected to the AGN (Eales & Rawlings 1993, Dunlop & Peacock 1993) . Being 100–1000 times less radio luminous than these extreme sources, mJy radio galaxies at comparable redshifts should provide much more representative probes of the formation and evolution of elliptical galaxies in general.

3 Redshift Estimation

Unfortunately the relative inactivity which makes mJy galaxies of interest also makes it extremely difficult to determine their redshifts (because emission line luminosity scales with radio power, Rawlings & Saunders 1991). Thus, for the study of statistically useful samples it is necessary to find a means of estimating redshifts for the majority of sources. Our K-band photometry has enabled us to do this by two independent routes. First we have estimated redshifts directly from the established $K - z$ relation for powerful radio galaxies (Eales *et al.* 1993) after correcting this relation for the contribution made to the K-band light by the blue component in these more active objects with known redshifts (Dunlop & Peacock 1993). Second we have estimated redshifts from spectral fitting to our combined g, r, i, K data (Lilly 1989), a method which essentially constrains the location of the 4000Å break. Each method yields a range of feasible redshift estimates, and by considering the overlap of the two confidence regions we can obtain what do indeed appear to be reliable redshift estimates for the vast majority of the galaxies in the sample (Fig. 1).

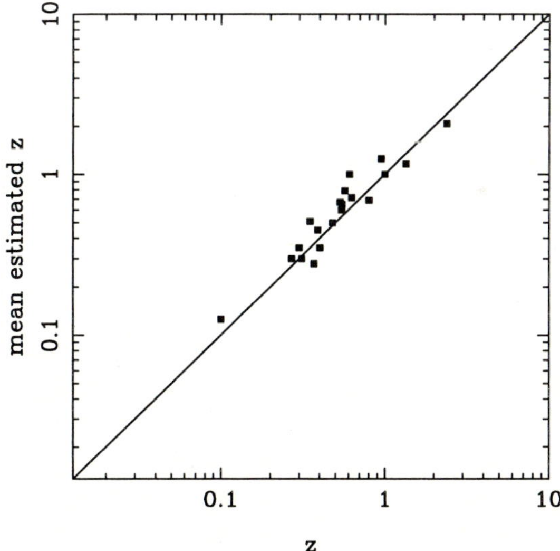

Fig. 1. Confirmation of the reliability of our two-pronged method of redshift estimation. The best estimate of the redshift (as taken from the centre of the overlap region of the two independent redshift-estimate confidence regions) is plotted against actual redshift for the 20 galaxies in the Herc region of the LBDS sample which do have spectroscopic redshifts. Little confirmation is currently available at very high redshifts, but the demonstrated success of this method out to $z \simeq 2$ gives confidence that the faintest galaxies in this sample really do lie at $z > 2$.

4 The Redshift Cutoff for Radio Sources

We can use the results of our previous investigation of the high-redshift evolution of powerful radio sources (based on complete samples with $S_{2.7GHz} > 100$ mJy, Dunlop & Peacock 1990) to predict the expected redshift distribution of galaxies in the LBDS mJy sample under different evolutionary scenarios. For example if one assumes that the radio luminosity function remains constant at $z > 2$ we predict that 15 out of the 70 galaxies in the LBDS Herc sample should lie at $z > 3$, whereas if one assumes a power-independent cutoff this number drops to 7. It is therefore interesting that the actual number of galaxies in this sample for which we have derived redshift estimates of greater than $z \simeq 3$ is actually 8. Spectroscopic confirmation of at least a subset of these high redshift estimates is clearly required, but our existing data are certainly consistent with a picture in which mJy radio galaxies suffer a similar decline in their comoving density at $z > 2$ to that which afflicts their much more powerful counterparts (Dunlop & Peacock 1990).

5 The Formation and Evolution of Elliptical Galaxies

As shown in Fig. 2, several of the mJy galaxies with redshift estimates as high as $z \simeq 4$ have very red optical-infrared colours ($R - K \simeq 5$). These red colours are certainly not due to unusually large near-infrared luminosities. Indeed these sources describe a $K - z$ relation which, at $z > 2$ is approximately 0.5 magnitudes fainter than that displayed by the most powerful radio galaxies, providing further support for the expectation that the light from high-redshift mJy galaxies is relatively uncontaminated by emission lines and/or a reddened quasar nucleus. We can thus be reasonably confident that the red colours of the mJy galaxies are genuinely indicative of their stellar content, in which case they indicate the presence of an evolved stellar population with a pronounced 4000Å break.

As illustrated in Fig. 2 such apparently evolved stellar populations are difficult to reconcile with a formation redshift $z_f < 5$, even if one attempts to maximise the rate of reddening by invoking a short-lived formation startburst (0.1 Gyr). The exact age of these galaxies is of course rather model dependent, but what seems clear is that the most distant galaxies in these mJy samples are markedly redder than their well-studied more radio-luminous counterparts. Indeed, the relatively blue colours of powerful radio galaxies such as 4C 41.17 ($z = 3.8$, $R - K = 2.7$, Graham 1994) and B2 0902+34 ($z = 3.4$, $R - K \simeq 1.9$, Eisenhardt & Dickinson 1992) have led to claims that they may genuinely primeval objects observed during their first major starburst. This contrast suggests that either extreme radio power is linked to the process of galaxy formation (*i.e.* powerful high-redshift radio galaxies really are younger than the mJy galaxies), or that by moving to mJy flux levels we are managing, for the first time, to see the true colours of the stellar populations in high-redshift elliptical galaxies, relatively uncontaminated by processes associated with the nuclear activity.

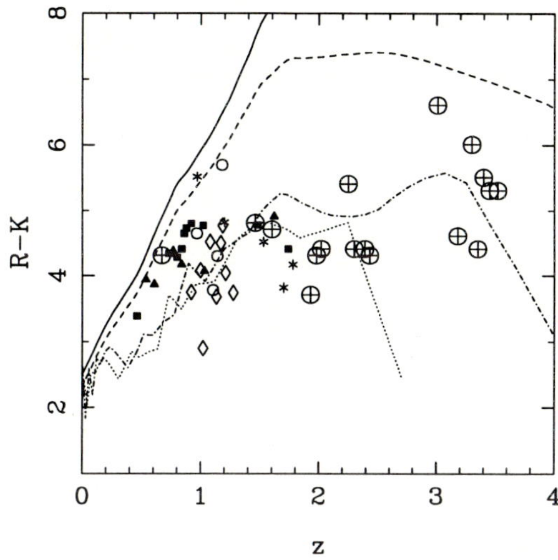

Fig. 2. A comparison of the expected evolution of R−K colour of a giant elliptical galaxy under different evolutionary scenarios with the R−K colours of radio galaxies drawn from a number of different samples. The solid line shows the effect of simply redshifting the spectrum of a present-day UV-cold elliptical galaxy while the dashed line shows the same for a UV-hot elliptical galaxy such as M87 (Guideroni & Rocca-Volmerange 1987). These two lines thus bracket the locus of unevolving ellipticals and indicate that at $z > 2$ an unevolving elliptical should have R−K > 7. The dot-dash line shows the predicted colour evolution for a 0.1 Gyr burst occurring at $z_f = 5$ (assuming $\Omega_0 = 1$; $H_0 = 50$ km s^{-1} Mpc^{-1}) and the dotted line shows the same for $z_f = 3$. Solid symbols are the Parkes galaxies, open the symbols the 3CR galaxies, and asterisks the 1Jy galaxies discussed by Dunlop & Peacock (1990). The large symbols are the high-redshift mJy radio galaxies from the LBDS sample which at present provide the best estimate of the red envelope at $z > 2$.

References

Chambers, K.C., Miley, G.K. & van Breugel, W. 1990, *Astrophys. J.* **363**, 21

Dunlop, J.S. & Peacock, J.A. 1990, *Mon. Not. R. Astr. Soc.* **247**, 19

Dunlop, J.S. & Peacock, J.A. 1993, *Mon. Not. R. Astr. Soc.* **263**, 936

Eales, S.A. & Rawlings, S. 1993, *Astrophys. J.* **411**, 67

Eales, S.A., *et al.* 1993, *Astrophys. J.* **409**, 578

Eisenhardt, P. & Dickinson, M. 1992, *Astrophys. J.* **399**, L47

Graham, J.R., Matthews, K., Soifer, B.T., Nelson, J.E., Harrison, W., Jernigan, J.G., Lin, S., Neugebauer, G., Smith, G., Ziomkowski, C. 1994, *Astrophys. J.* **420**, L5

Guiderdoni, B. & Rocca-Volmerange, B. 1987, *Astron. Astrophys.* **186**, 1

Lilly, S.J. 1988, *Astrophys. J.* **333**, 161

Lilly, S.J. 1989, *Astrophys. J.* **340**, 77

Rawlings, S. & Saunders, R. 1991, *Nature* **349**, 138

Windhorst, R.A. *et al.* 1985, *Astrophys. J.* **289**, 494

Spectroscopy and Imaging of a Forming Galaxy Causing Damped Ly α Absorption at $z = 2.81$*

Palle Møller[1,2] *and Stephen J. Warren*[3]

[1] STScI, 3700 San Martin Drive, Baltimore, MD 21218, USA
[2] On assignment from the Space Science Department of ESA
[3] Blackett Laboratory, Imperial College of Science Technology and Medicine, Prince Consort Rd, London SW7 2BZ, UK

1 Searching for Progenitors of Normal Galaxies

While the subject of this workshop, "Galaxies in the Young Universe" is certainly both timely and of high interest, it is also, from an observers perspective, somewhat frustrating because there is not yet, despite extensive searches, a single object known of redshift $z > 2$ which one can point to and assert that it is a normal galaxy.

Consequently we know next to nothing about the sizes, morphologies, or spatial distribution of normal galaxies when the universe was young. The detection and detailed study of just one normal galaxy of redshift $z > 2$ would therefore be a very valuable contribution to our understanding of when and how galaxies formed. However there is no prospect of progress in this field by simply obtaining spectra of fainter magnitude-limited samples of galaxies.

1.1 Ly α Imaging of Damped Ly α Galaxies

An alternative detection strategy is to identify a galaxy firstly in absorption in the spectrum of a higher redshift quasar, and then to try to image the absorber, usually through a narrow band filter tuned to Ly α. Galaxies found in this way are just what we are looking for, because the sight–line to a quasar is a random skewer through the observable volume of the universe. The extensive searches for the galaxies responsible for the damped Ly α ($N_{\mathrm{HI}} > 10^{20}\mathrm{cm}^{-2}$) absorption lines are summarised by Møller & Warren (1993a). With the exception of our own observations (below) there is no case of a detection confirmed by both imaging and spectroscopy.

There are, nevertheless, three reported detections of nearby *companions* to damped systems (Lowenthal *et al.* 1991, Macchetto *et al.* 1993, Francis *et al.* 1995). High redshift objects found as companions do not, unfortunately, make up a representative sample of normal galaxies. They are most likely AGNs, and are hence not what we seek.

* Based on observations collected at the European Southern Observatory, La Silla, Chile

The newest of the companions listed above were reported at this meeting. The field contains at least two Ly α emitting galaxies at $z = 2.38$ (Francis *et al.* 1995). A deep K band image of the field reveals what appears to be a massive high redshift cluster. This field could become an important laboratory for the study of cluster galaxies at high redshifts, but it still does not reveal the presumably "typical" galaxy actually causing the absorption.

2 The Damped Ly α Absorber in Q0528-250

2.1 Summary of Results from Previous Observations

The lack of success at high redshifts led us to devise a new technique. We decided to attempt to use the ionising flux from the quasar to detect the Ly α "silver lining" of the absorbing galaxy (see e.g. Møller & Warren, 1993b, Fig 1a,b). In a combined ten hours Ly α narrow-band image of Q0528-250, obtained at the ESO 3.6m in December 1991, we isolated 3 candidates at $z = 2.81$. The average seeing of these images was 1.7 arcsec. The three sources (S1, S2, and S3) are extremely faint, and are lying 1.2, 12, and 21 arc seconds from the quasar, respectively. These projected separations correspond to distances of 5, 45, 78 kpc ($H_o = 100, q_o = 0.5$). Full details are reported in Møller & Warren (1993a).

2.2 New Observations

In Dec 1992 we obtained NTT/EMMI spectra of the sources S1 and S2, as well as deep B band images. In Dec 1993 we obtained a higher resolution (average seeing around 0.9 arcsec) NTT/SUSI narrow band image, and also obtained NTT/EMMI spectra of sources S2 and S3.

The 4 hours SUSI narrow band image is shown in Figure 1. The two panels show the image with and without the quasar psf subtracted, and with some smoothing applied as detailed in the caption. The EMMI spectra of the three sources are shown in Figure 2. The B band image is not shown here but S2 and S3 are both detected in B. S1 is too close to the quasar and hence cannot be observed in broad bands from the ground. A total of 30 orbits of HST time has been allocated to observations of this field in Cycle 5. With those observations we should be able to detect (or put interesting upper limits on) continuums flux also from S1.

2.3 Summary of Results from Our New Observations

The results from our 1992/93 observations can be summarised as:

1. The Ly α line emission from all three sources has now been confirmed spectroscopically, they all have redshifts which match that of the absorber to within less than 200 km/s.

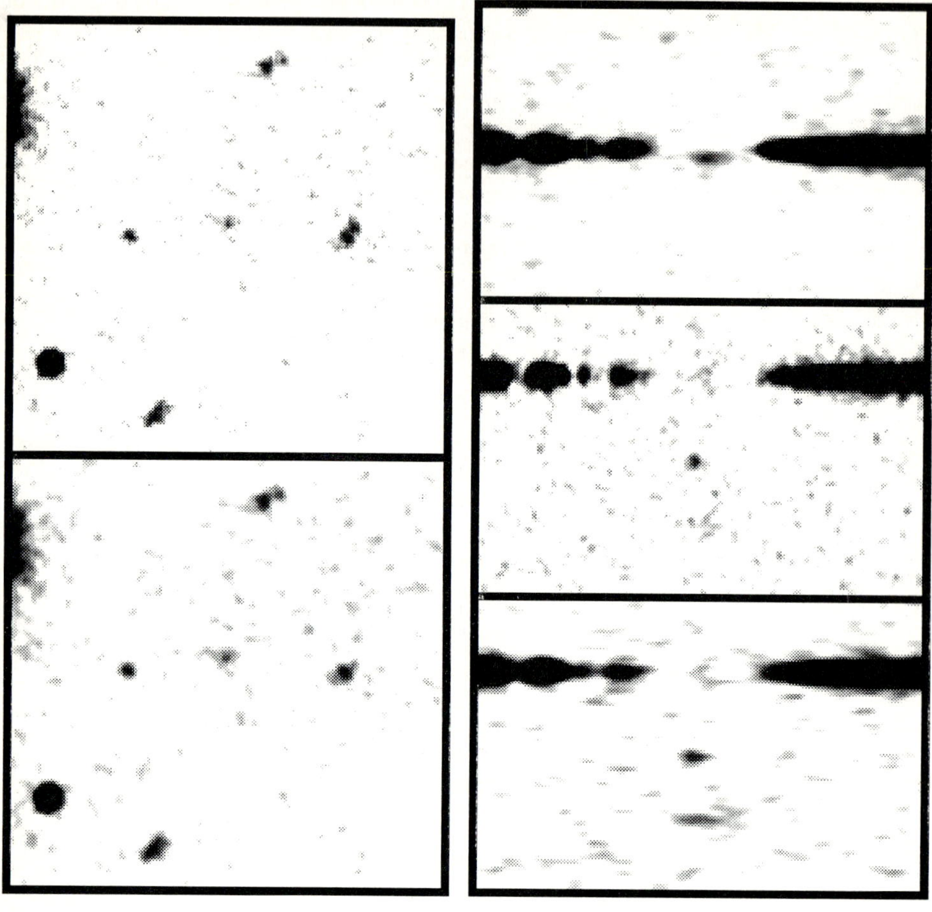

Fig. 1. *left panel* Image of S1, S2, and S3 obtained at the 3.5m NTT using SUSI. The image was obtained through a narrow band filter centred on Ly α redshifted to z=2.81. The upper panel still contains the flux from the quasar. In the lower panel a point spread function was subtracted at the position of the quasar and the image was smoothed. The object just above and left of the centre in both panels is S2, the object left of S2 is S3 and the right object is S1. Note that S1 appears to be double in the upper panel, where the upper object is the quasar. N is up and E is to the left.
Fig. 2. *right panel* Two dimensional spectroscopic frames, of dimensions 140Å × 40 arcsec, showing the Ly α emission lines from S1, S2, and S3. In each frame the dark band is the spectrum of the quasar (blue is to the left, red is to the right), and the prominent gap is the damped absorption line. (*top*) Frame containing source S1 (1.2 arcsec offset under quasar), smoothed by convolution with an elliptical Gaussian profile of σ 2 by 1 pixel, (*middle*) frame containing sources S2 (upper, 12 arcsec offset) and S3 (lower, 22 arcsec offset), optimised for S2, smoothed with a circular Gaussian of σ one pixel, (*bottom*) frame containing sources S2 and S3, optimised for S3, smoothed with an elliptical Gaussian of σ 3 by 1 pixel. S3 is extremely broad, and only becomes visible to the eye when the strong smoothing in the bottom panel has been applied.

2. We have resolved line shapes of all three sources. They all have different FWHM ranging from 200-1200 km/s, the broadest line appears to be more flat–topped than a gaussian profile and is consistent with the shape expected for a line resulting from a narrow line broadened by radiative transfer inside an H I cloud of high column density.
3. The B band continuum magnitudes of S2 and S3 are rather faint, $m_B \approx 26.5$, and they imply Ly α rest frame equivalent widths of roughly 65 Å. This has not been corrected for the possible effect of dust (Charlot & Fall 1993).
4. The narrow band image does not show any sign of the sought for Ly α "silver lining" ring or arc structure which we predicted in the case that the emission is caused by the underlying quasar. The Ly α emission is hence more likely caused by star formation.

3 Discussion

A detailed report of the observations and analysis of the data shown here will be given elsewhere (Warren & Møller 1995), and we shall here only summarise the most important conclusions.

The central question is: "What is the source of the Ly α emission". To answer this we note that

1. for S2 and S3 we have now detected a stellar continuum under the emission lines
2. the large width of the lines from S1 and S3, and their profiles, are consistent with being caused by resonant scattering of Ly α emission from a star forming region embedded in a slab of neutral hydrogen
3. the morphology of the Ly α emission from S1 does not support the expectation that the Ly α emission is caused by the nearby quasar.

All of our new observations hence point toward star formation as the source of Ly α in all three sources.

Calculating the star formation rate of the whole group (assuming Case B recombination and the prescription of Kennicutt, 1983) we find that the total star formation rate is no more than a few solar masses per year. On the assumption that the broadening of the lines is due to resonant scattering, we can calculate the column density of the HI between the star forming region and us. This, combined with the known Ly α flux which escapes the cloud, gives us an upper limit on the dust to gas ratio which is much smaller than in present day galaxies, and which is in good agreement with the conclusion reached by Meyer & York (1987) from an absorption line study of the absorption system belonging to this galaxy.

The picture that emerges is that of a galaxy which, at a redshift of 2.8, is not yet fully assembled, but consists of small separate subunits which are forming stars at a fairly moderate rate. We estimate that dynamical friction will cause the individual subunits to merge in less than 10^9 yrs, at which time the star formation rate may increase.

The multiple structure of this object, and the apparent alignment of its subunits, is in good agreement with the filamentary structure found in many simulations of galaxy formation (see in particular Katz, Hernquist & Weinberg, 1992; Katz & White, 1993; Evrard, Summers & Davis, 1994; Navarro, Frenk & White, 1994).

4 Conclusion

The tentative conclusion from our study of this system is that the sources S1, S2 and S3 are the (so far) most likely candidate progenitors of todays normal galaxies.

We speculate that it is likely that we here have "caught a galaxy in the act" of forming. If this is a typical case, we can conclude that galaxies in their initial state of star formation are fairly moderate, which could explain why previous Ly α searches have been unsuccessful.

References

Charlot, S., Fall, S.M. 1993, *Astrophys. J.* **415**, 580

Evrard, A.E., Summers, F.J., Davis, M. 1994, *Astrophys. J.* **422**, 11

Francis, P.J., Woodgate, B.E., Warren, S.J., Møller, P., Lowenthal, J.D., Williams, T. 1995, *Astrophys. J.*, in press

Katz, N., Hernquist, L., Weinberg, D.H. 1992, *Astrophys. J. Lett.* **399**, L109

Katz, N., White, S.D.M. 1993, *Astrophys. J.* **412**, 455

Kennicutt, R.C. 1983, *Astrophys. J.* **272**, 54

Lowenthal, J.D., Hogan, C.J., Green, R.F., Caulet, A., Woodgate, B. E., Brown, L., Foltz, C.B. 1991, *Astrophys. J. Lett.* **377**, L73

Macchetto, F., Lipari, S., Giavalisco, M., Turnshek, D.A., Sparks, W. B. 1993, *Astrophys. J.* **404**, 511

Meyer, D.M. York, D.G. 1987, *Astrophys. J. Lett.* **319**, L45

Møller, P., Warren, S.J. 1993a, *Astron. Astrophys.* **270**, 43

Møller, P., Warren, S.J. 1993b, in *Observational Cosmology*, Chincarini, G., Iovino, A., Maccacaro, T., Maccagni, D. eds., A.S.P. Conference series **51**, 598

Navarro, J.F., Frenk, C.S., White, S.D.M. 1994, *Mon. Not. R. Astron. Soc.* **275**, 56

Warren, S.J., Møller, P. 1995, *Astron. Astrophys.*, submitted

Ly α Absorption in 4C 41.17

Hans Hippelein, Klaus Meisenheimer and Hermann-Josef Röser

Max–Planck–Institut für Astronomie, Königstuhl 17,
D-69117 Heidelberg, Germany

1 Introduction

From the rich absorption line spectra of QSOs it becomes obvious that a considerable part of the Ly α radiation from extended emission line regions (EELR) around distant radio galaxies may be absorbed by Lyman forest clouds right in front of the galaxy. Whereas in the case of QSOs the clouds close to the galaxy may be highly ionized by the strong UV radiation from the quasar, radio galaxies are not bright enough to influence their surrounding and the clouds are expected to have an ionization factor of $N_{\rm HI}/N_{\rm HII} \sim 10^{-4}$. Thus, it is possible to get "direct" images of Lyman forest clouds and give an answer to the discussion about the typical size of these clouds.

Comparing the Fabry-Perot images obtained by scanning the profile of the Ly α line emitted by 4C 41.17, at $z = 3.8$, we detected an extended depression in the line profiles at the western edge of the EELR (Hippelein & Meisenheimer 1993). This first direct observation of a Ly α absorption cloud at high redshift proved that forest clouds indeed have sizes of several tens of kpc (Chaffee *et al.* 1986; Dinshaw *et al.* 1995).

From its column density and linear extension, and adopting $b \sim 35\,\mathrm{km\,s^{-1}}$, the mass of this cloud resulted to be somewhat high for a typical Lyman forest cloud. In order to find out if the observed line depression was due to a superposition of several smaller clouds with similar velocity, and to look for more and fainter line absorbers we repeated the observation at the MMT with both, high spatial and spectral resolution.

2 Observational Results and Discussion

Fig. 1 presents a digitized spectrum obtained at the MMT telescope along the major axis of 4C 41.17. It is the sum of 12 spectra each of 1 hour integration, collected during two photometric nights with seeing $\leq 1\farcs0$. The spectral resolution was 1.3 Å and the slitwidth $1\farcs1$. The tracings represent line profiles at every pixel position along the slit, separated by $0\farcs6$, each. The trace through the brightness maximum of the EELR is indicated as a thicker line. There is a prominent line depression at $v_{\rm rel} \sim -480\,{\rm km\,s^{-1}}$, at the same wavelength where an absorption feature was found by Hippelein & Meisenheimer (1993). Two other marginal features are seen at $v_{\rm rel} \sim -750\,{\rm km\,s^{-1}}$ and on top of the emission line, the first one also indicated in the spectrum published by Chambers *et al.* (1990).

The strong line depression runs from E to W all along the EELR, at almost constant radial velocity, width, and depth. This regularity stays in contrast to the broad emission line, which changes width and velocity along the slit, and has a complicated profile, with two peaks at $v_{\rm rel} = 0$ and $-600\,{\rm km\,sec^{-1}}$. A scenario explaining the depression by a superposition of two line components at these velocities is extremely unlikely considering their strongly varying profiles. Thus, the line depression must be due to absorption by neutral hydrogen in front of 4C 41.17.

In Hippelein & Meisenheimer (1993) we reported line absorption only around an area 2$''$.5 west from the EELR maximum (4th trace from below in Fig. 1). Indeed, the old data show a line absorption feature around the bright part of the EELR as well, but at that time it was not considered significant since it falls on the steep, blue slope of the emission line profile.

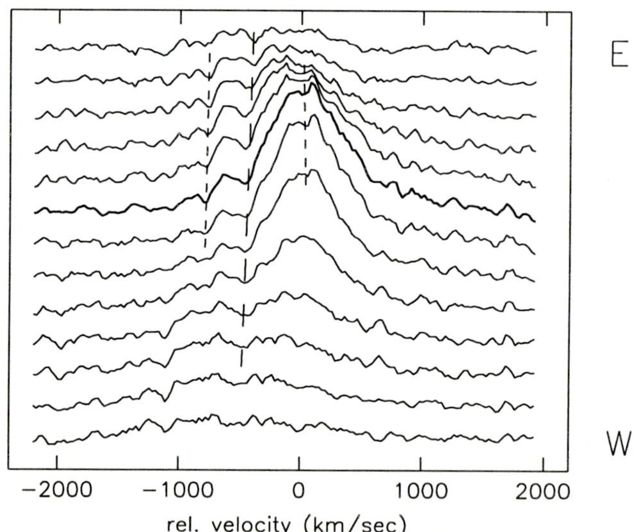

Fig. 1. Line tracings of the spectrum of 4C 41.17. The emission line maximum is indicated by a thicker line.

What is the nature of this H I cloud? The absorption feature is relatively broad, $b \sim 150\,{\rm km\,s^{-1}}$, and shallow, ~40 % deep. Its rest equivalent width is ~0.4 Å, corresponding to a column depth of $N_{\rm HI} \sim 10^{14}\,{\rm cm^{-2}}$. The cloud has an extension of at least 80×50 kpc, the extension perpendicular to the spectrograph slit being derived in our first study. Assuming $N_{\rm HI}\,/\,N_{\rm HII} \sim 10^{-4}$ gives a total gas mass of $\sim 3 \times 10^{7}\,{\rm M_{\odot}}$.

Having estimated the physical properties of the absorption cloud we now will discuss its location along the line of sight. Though its mass is rather large for a normal forest cloud, the cloud is certainly not massive enough to be associated with the disk of the galaxy embedded in the giant Ly α halo. It is also too small to be associated with the dust cloud recently detected by Dunlop *et al.* (1994) and by Chini & Krügel (1994), which has an estimated mass of $M_{\rm dust} \sim 3 \times 10^8\,{\rm M}_\odot$.

Additional information about absorbing clouds can be gained from the comparison of the morphology of 4C 41.17 as seen in different emission lines. It happens that the redshifted Hβ and the [OIII] lines from 4C 41.17 fall right into the K-window at 2.330 μm and 2.400 μm, respectively. Therefore, in March 1994, we took two narrow-band images with filters peaked at these two lines, and with $\Delta\lambda/\lambda = 0.02$ with the NIR camera MAGIC mounted at the 3.5 m telescope on Calar Alto. For comparison, another image was taken with the K' filter, which is essentially free from emission lines of the observed object. Integration times were inbetween 1.5 and 2.0 hours for each filter, yielding a $5\,\sigma$ limit of 20.6^{mag} for the K' image.

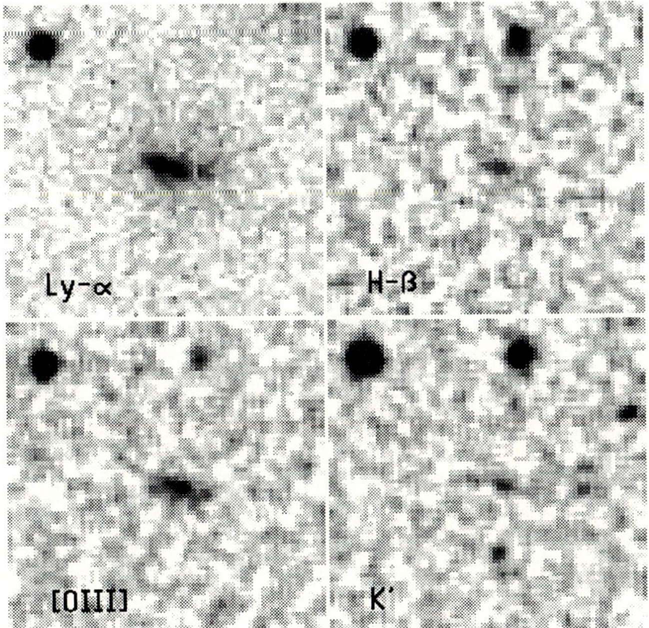

Fig. 2. Images (30″×30″) of 4C 41.17 in Ly α, $H\beta$, [OIII] and in K'.

The three NIR images are shown in Fig. 2 together with the Ly α frame. The [O III] image disagrees with that shown by Graham *et al.* (1994), who from Keck observations claimed that it has a different shape than the Ly α image; their Hβ + [O III] line image was, however, derived from the difference between a K

and a K' image, only the first of which includes the two lines. Our results show, in contrast, an identical morphology of all images within the error. In particular, the position of the brightness dip at the western edge matches pefectly in the Ly α and [O III] images.

This dip seems to be the place of the dust cloud mentioned above, where also the center of the galaxy is assumed (Miley *et al.* 1992). If it were more extended we should have seen it as a significant difference between the emission line images.

The line ratio from CCD photometry yields $\mathrm{I}(\mathrm{Ly}\alpha)/\mathrm{I}(\mathrm{H}\beta)=7.4$, close to the value obtained by Eales & Rawlings (1994) from infrared spectra. This is equivalent to an absorption measure of only $A_V \sim 0.35$, which again argues against large amounts of dust associated with the gas cloud.

There are two possible scenarios for the Ly α absorber. Either it is a large forest cloud at cosmological redshift distance of $-450\,\mathrm{km\,s^{-1}}$; or it belongs to the (lower ionized) boundary of the halo of 4C 41.17 at about the same velocity as the central galaxy. In the latter case, the bright EELR would be falling towards the galaxy at a velocity of roughly $450\,\mathrm{km\,s^{-1}}$, and since we can only see the front part of the deep EELR, it would appear redshifted against the absorption layer (Meisenheimer *et al.* 1994). The uniformity of the absortion line and the indication of a second absorption feature at $-750\,\mathrm{km\,s^{-1}}$ favour the first of these two interpretation.

All the red galaxies listed in Graham *et al.* (1994) are detected on our K' frame as well. None of them, however, shows an emission line counterpart in the [O III], Hβ or Ly α frames, which does not support the speculation that they are galaxies at the redshift of 4C 41.17.

Acknowledgements. We thank the MMT director for the allocation of visitors time. We are grateful to Fred Chaffee and Craig Foltz for supervising the observations.

References

Chaffee, F.H., Foltz, C.R., Bechtold, J., Weymann, R.J. 1986, *Astrophys. J.* **301**, 116
Chambers, K., Miley, G., van Breugel, W. 1990, *Astrophys. J.* **363** 21
Chini, R., Krügel, E. 1994, *Astron. Astrophys.* **288**, L33
Dinshaw, N., Foltz, C.B., Impey, C.D., Weyman, R.J., Morris, S.L. 1995, *Nature* **373**, 223
Dunlop, J., Hughes, D., Rawlings, S., Eales, S., Ward, M. 1994, *Nature* **370**, 347
Eales, S., Rawlings, S. 1993, *Astrophys. J.* **411**, 67
Graham, J., Matthews, K., Soifer, B., Nelson, J., Harrison, W., Jernigan, J., Lin, S., Neugebauer, G., Smith, G., Ziomkowski, C. 1994, *Astrophys. J.* **420**, L5
Hippelein, H., Meisenheimer, K. 1993, *Nature* **362**, 224
Meisenheimer, K., Hippelein, H. 1992, *Astron. Astrophys.* **264**, 455
Meisenheimer, K., Hippelein, H., Neeser, M. 1994, in *The Physics of Active Galaxies*, ASP Conference Series Vol. **54**, 397
Miley, G., Chambers, K., van Breugel, W., Macchetto, F. 1992, *Astrophys. J.* **401**, L69

The Revolution in Studies of Distant Radio Galaxies

Huub Röttgering[1,2]

[1] Mullard Radio Astronomy Observatory, Cavendish Laboratory, Madingley Road, Cambridge CB3 0HE
[2] Institute of Astronomy, Madingley Road, Cambridge CB3 0HA

1 Introduction

In this contribution I will be discussing and defending the claim that studies of distant radio galaxies are undergoing a dramatic revolution. Two main arguments supporting this claim will be put forward.

Firstly, the advances in instrumentation allow the study of distant radio galaxies in unprecedented detail. The high resolution images from the Hubble Space Telescope give morphological information on scales about a factor of ten smaller than is possible with conventional ground-based optical telescopes. Advances in the instrumentation of ground-based telescopes allow studies of the cold ($\ll 10^4$ K) material in these objects. High resolution spectroscopy of the extended Lyα emission regions allows the neutral H I gas in the proximity of the radio galaxy to be probed. In nearby galaxies, molecular gas such as CO is an important probe of star formation. The improvement in receiver technology, in conjunction with mm/sub-mm telescopes, makes it worthwhile attempting to detect this gas phase in distant radio galaxies.

Secondly, at the moment several new radio surveys are being produced. These surveys contain $10 - 100$ times more sources than previously published surveys, allowing studies of radio source populations that were hitherto not possible. Two such surveys, the 7C survey at 151 MHz and the WENSS survey at 325 MHz, will be discussed later.

The hope is that the detailed observations of radio galaxies will help to answer the numerous questions concerning their properties and evolution. Most distant radio galaxies ($z \gtrsim 1$) show a clear alignment between the optical axis and the axis defined by the double radio source (McCarthy *et al.* 1987, Chambers *et al.* 1987). This indicates that the radio source has a profound influence on the properties of the host galaxy. It is therefore essential to understand the cause of this phenomenon before we can properly address other questions such as 'what are the properties of gas and stars in these systems, what is their age, and are they in clusters?'. Finally, we would like to understand their relation to "normal" galaxies and to primeval galaxies.

Since these high-redshift radio galaxies contain the most powerful radio sources known, it is perhaps not surprising to find that the radio source has a profound influence on the properties of the host galaxies. The new radio catalogues will allow faint samples of radio galaxies to be defined. Although it will be a large amount of work, in principle it is possible to obtain from these new surveys significant samples of distant galaxies that have radio powers a factor 10 – 100 times less than the radio galaxies in the samples presently available. This should allow us to distinguish between the properties of the galaxy that are due simply to large look-back times and the properties that arise directly from the presence of a powerful radio source.

2 HST Observation of Radio Galaxies at $z \sim 1$

At the moment we are carrying out a programme to survey a representative sample of 24 galaxies from the 3CR radio catalogue, spanning a redshift range $0.65 \leq z \leq 1.8$ (PI M. Longair). Images of 3C 368 ($z = 1.13$), 3C 324 ($z = 1.21$) and 3C 265 ($z = 0.81$) are shown in a letter by Longair *et al.* (1995). During this workshop Mark Dickinson showed his images of 3C 324 ($z = 1.21$). Here I will discuss briefly thc HST WFPC2 and VLA images we obtained for 3C 368 (see Fig. 1). The radio image was taken with the VLA in A-array configuration at a frequency of 8.4 GHz, giving a resolution of 0.15 arcsec, comparable to the 0.1 arcsec resolution of the HST. The WFPC2 images are taken with the filters F702W ($\lambda_0 = 6900$ Å) and F791W ($\lambda_0 = 7830$Å). We estimate that 25 – 40 % of the emission in these filters is due to line emission, mainly [O II] 3727. The HST images show a remarkable 'cigar-shaped' emission region oriented along the radio axis. This morphology is a spectacular confirmation of the alignment between the optical and radio emission from $z > 0.6$ radio galaxies (e.g. McCarthy 1993 and references therein). The bright unresolved knot close to the centre is the foreground M-dwarf identified by Hammer *et al.* (1991).

An optical jet-like feature links the supposed centre of the galaxy (just north of the M-dwarf) to the emission regions in the northern parts of the image. At the centre of the southern emission region there is a clear intensity minimum, reminiscent of a bow shock associated with the southern radio lobe, as discussed by Meisenheimer & Hippelein (1992). The source has extended optical polarisation (Scarrott *et al.* 1990), indicating that a significant fraction of the aligned emission is scattered light from a hidden quasar. However, the absence of a clear, cone-like morphology suggests that this is not the only mechanism causing the alignment between the optical and radio emission. Additional mechanisms include (i) jet induced star formation (Begelman & Cioffi 1989; Rees 1989) and (ii) excess of material along the radio axis (Eales 1992; West 1994). Further details can be found in Longair *et al.* (1995).

The main conclusion is that the new HST images show a wealth of fine structure down to 0.1 arcsec scales. It is not possible to explain the observed structures with any single theory.

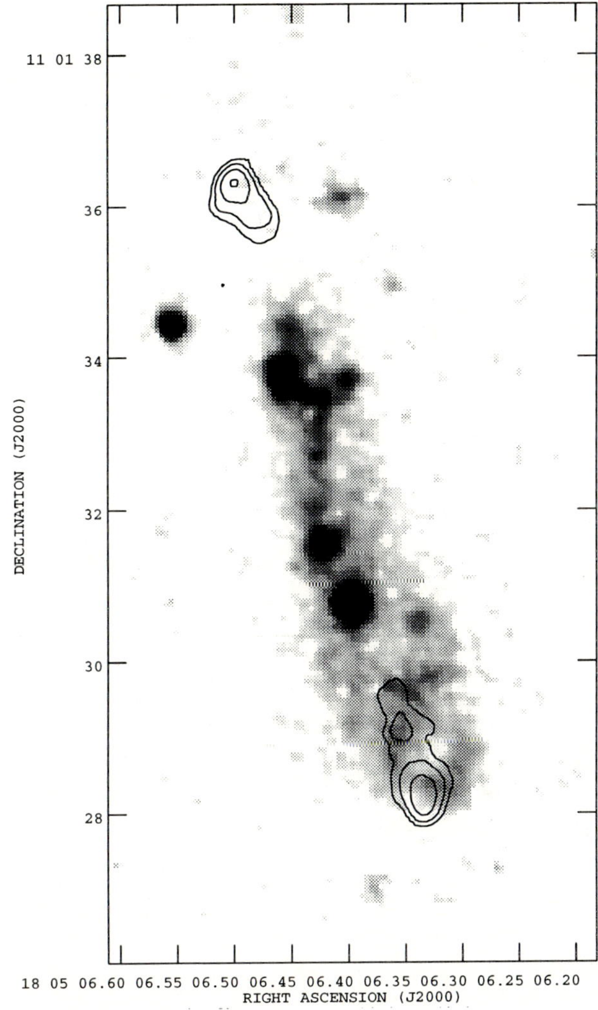

Fig. 1. 3C 368 The image is the sum of both filters of the WFPC2 observations. The VLA data is superimposed, the contour levels being at $(2, 8, 16, 64) \times 1.18 \times 10^{-4}$ Jy beam^{-1}.

3 Cold Gas in Distant Radio Galaxies

At least three different gas phases are observed, or are expected to be abundantly present, in distant radio galaxies. The emission line gas at temperatures of $\sim 10^4$ K has been studied extensively through the observations of lines such Lyα and [O II] 3727. There seems to be an intimate relation between the radio source and the presence of this gas. A second, much hotter (10^{7-8} K) component is the X-ray emitting gas and is more difficult to detect. The ROSAT results so far suggest that at least some of the distant radio galaxies are surrounded by dense halos of hot gas (Crawford & Fabian 1993). The third component is cool gas, typically at temperatures of order 100 K. Since, in some models, distant radio galaxies are young and forming objects — the long sought after protogalaxies — this would be the material from which stars are being formed at a high rate. In what follows we will discuss observations of H I and CO gas in distant radio galaxies.

3.1 Neutral Hydrogen

For the past two decades the study of quasar absorption lines has been one of the most important techniques for gaining information about the physical conditions and element abundances in the early universe (e.g. Sargent 1988; Wolfe 1991 and references therein). Unfortunately, the continuum emission from distant radio galaxies is too weak ($R \sim 22-24$) for absorption studies with 4 m-class telescopes. However, their Lyα emission is often spectacular (e.g. Chambers *et al.* 1990; McCarthy *et al.* 1991): the line luminosities can be as high as 5×10^{44} erg s^{-1} and their spatial extents sometimes exceed 100 kpc. We therefore embarked on a project to obtain spectra of high redshift radio galaxies at a significantly higher spectral resolution (1.5 Å) than previous investigations, and aimed specifically at studying any absorption that might be seen against the extended Lyα emission. The first detection of extended absorption (Hippelein & Meisenheimer 1993) was a system with a relatively low column density associated with 4C 41.17 ($z = 3.8$).

The high resolution AAT spectra (1.5 Å) of the Lyα region of the $z = 2.9$ radio galaxy 0943−242 revealed a complex emission line profile which is dominated by a black absorption trough centred at 250 km s^{-1} blueward of the emission peak (see Röttgering *et al.* 1995 and Röttgering 1995). This absorption covers the entire Lyα emission region which has a spatial scale of 1.7 arcsec. Its linear size is thus at least 13 kpc, making this the first direct measurement of the spatial scale of an absorber with a column density of $\sim 10^{19}$ cm^{-2}.

From the H I column density distribution of QSO absorption line systems (e.g. Petitjean *et al.* 1993) the number of absorbers with column densities $> 10^{19}$ cm^{-2} expected within the width of the Lyα emission (50 Å, corresponding to $\Delta z \sim 0.04$) is 0.008. This indicates that the detection of such a strong absorber in 0943−242 is remarkable.

Two relevant questions regarding such extended absorption lines are: (i) Do the other emission lines such as C IV show signs of absorption related to the main absorber in 0943−242? and (ii) How often does the Lyα emission of radio galaxies show such strong absorption?

We were able to obtain high resolution spectra of 0943−242 in the region of C IV and He II in two hours of poor seeing and moderate weather conditions. Although we clearly need better signal-to-noise, already a number of exciting features seem to be present in the data: (i) there is a strong suggestion that the C IV emission is heavily absorbed at the redshift of the damped absorber, indicating that the absorber is metal-enriched, and (ii) there is a hint from the spectrum that the C IV emission may have a broad component with $\Delta V \sim 3000\,\mathrm{km\,s^{-1}}$.

To investigate how common such strongly absorbed Lyα emission is, we carried out high resolution spectroscopy on a sample of eight other high redshift radio galaxies using the EMMI spectrograph at the NTT and the ISIS spectrograph at the WHT. As an example, we show in Fig. 2 the high resolution spectra of 2202+128 ($z = 2.7$) and 1558−003 ($z = 2.5$). These sources are drawn from a sample of distant radio galaxies found during an ESO key programme (Miley *et al.* 1992; Röttgering *et al.* 1994). The spectra were obtained in June 1994 at the WHT and have a resolution of 1.7 Å. Details of the observations and further analysis will be given by Van Ojik *et al.* (1995). The first conclusions from the observation are that 2202+128 shows similar strong absorption to 0943−242, and that 1558−003 does not show any sign of absorption. This turned out to be typical for our sample. Four out of the eight objects show strong absorption and the other four show none. Given the QSO absorption line statistics discussed above, this implies that strong associated absorption is more common in radio galaxies than in quasars.

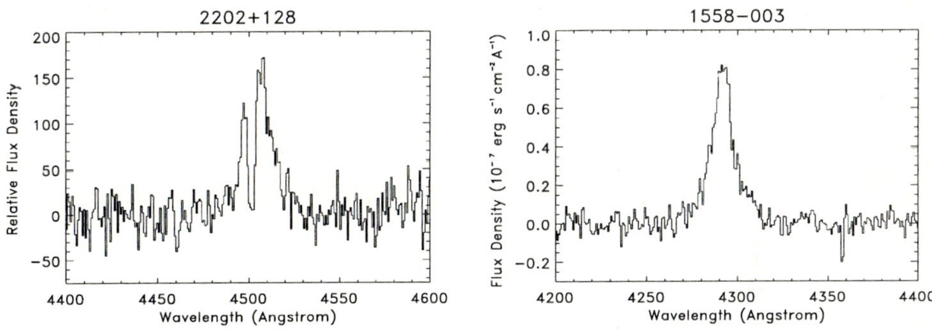

Fig. 2. Part of the high resolution ISIS spectra (1.7 Å) of the Lyα regions of the radio galaxies 2202+128 ($z = 2.7$) and 1558−003 ($z = 2.5$).

Some models of radio-loud quasars and powerful radio galaxies propose that they are intrinsically similar objects observed from different angles (see Antonucci 1993, and references therein). The bright nucleus is surrounded by an obscuring torus of dust and gas, resulting in twin beams of continuum emission being emitted perpendicular to the torus. In broad outline, quasars are objects viewed along the beam, whilst radio galaxies are seen perpendicular to the beams. In this picture, the beam ionises the surrounding medium, resulting in the extended emission line region (EELR) around quasars and radio galaxies that can have sizes up to 100 kpc (Baum *et al.* 1988; Heckman *et al.* 1991). These EELR tend to be brightest and show their highest ionisation state along the radio axis. In this model the neutral gas will be preferentially located outside the beams. The suggestion that strong associated absorption might be more common in radio galaxies than in radio quasars is consistent with the existence of ionising beams: the quasar ionises clouds that are located between it and us, resulting in a deficiency of associated absorbers in quasar spectra compared with radio galaxies.

The tentative detection of a broad component in the C IV emission in the spectra of 0943−242 could be consistent with this model. The broad emission line component would originate in a region well inside the torus and be scattered into our line of sight by dust or electrons outside the (obscuring) torus.

3.2 Carbon Monoxide

Since the CO molecule is a good tracer of star formation in galaxies, studying this molecule at cosmological distances is important for understanding the evolution of young galaxies. Some theories for the radio-optical alignments in distant radio galaxies assume that the powerful radio jet induces massive amounts of star formation. Whether this is indeed the case is not yet clear (e.g. McCarthy *et al.* 1993).

We have started a programme to search for CO emission from distant radio galaxies using the JCMT and IRAM telescopes. From our spectra we have clear indications that we are indeed detecting CO lines from some of these distant objects. If confirmed, these lines indicate that the density of CO gas is in the range $10^4 - 10^5$ cm^{-3} and that the temperature is of order 50 – 100 K. Using galactic conversion factors, the total H_2 masses are then $\sim 10^{11} M_\odot$, comparable to the mass obtained for F10214+4724.

If correct, these detections would support the idea that distant radio galaxies are undergoing vigorous starbursts, suggesting that the radio jet does indeed trigger large amounts of star formation.

4 The 7C and the WENSS Survey

At the moment four extensive surveys of the radio sky are being undertaken. The VLA is carrying out two surveys at 1.4 GHz, one with a resolution of about one arcmin (Condon, this Volume) and one with a resolution of 4 arcsec (Becker *et al.* 1994). These two surveys will produce of order a few million sources.

The surveys I will briefly discuss here are the Westerbork Northern Sky Survey (WENSS) and the 7C survey. The WENSS survey is a large sky survey, carried out with the Westerbork Synthesis Radio Telescope at 325 and 608 MHz (PIs de Bruyn and Miley). The main parameters of the survey are given in Table 1 (see also de Bruyn *et al.* 1994). At the time of writing, 75 % of the survey area has been observed and close to 50 % of the data has been reduced.

The 7C survey is being made with the Cambridge Low Frequency Telescope (CLFST) at 151 MHz. It covers most of the sky north of Dec. 20°, potentially providing positions and flux densities for around 10^5 sources. Further details of the survey are given in Table 1. Selected regions of the survey have been published (see Visser *et al.* 1995, and references therein).

Table 1. Parameters of the 7C and WENSS radio surveys

	7C	— WENSS —	
Frequency	151 MHz	325 MHz	608 MHz
Region	$\delta > 20°$	$\delta > 30°$	$\delta > 30°$, $\lvert b \rvert > 30°$
Limiting flux density	100 – 200mJy	15 mJy I,Q,U,V	15 mJy I,Q,U,V
Number of sources	70,000	300,000	60,000
Resolution	$77'' \times 77''$cosecδ $100'' \times 100''$cosecδ	$55'' \times 55''$cosecδ	$30'' \times 30''$cosecδ
Position Accuracy (stronger sources)	$3'' \times 3''$cosecδ	$2'' \times 2''$cosecδ	$1'' \times 1''$cosecδ

In Fig. 3 we show how the sensitivities of these surveys compare with each other and with older surveys. The lines correspond to sources with spectral indices $\alpha = -1.0$. Almost all the sources from WENSS will be present in the VLA survey and a significant fraction (~ 20 %) will be present in the 7C survey. This will allow the construction of radio colour-colour diagrams that will enable selection of large numbers of sources with specific radio spectral features over a wide range of flux densities.

Such colour-colour diagrams will be an important tool in helping to distinguish between the various types of source. These types include: H II regions and pulsars in our own galaxies; normal and starburst galaxies; radio halos and head-tailed sources in clusters of galaxies; various types of radio galaxies and quasars ranging from the very compact objects such as CSS (Compact Steep Spectrum) and GPS (Gigahertz Peaked Spectrum) sources to the the giant radio sources that have sizes larger than a few Mpc.

Since this contribution is for a large part concerned with distant radio galaxies, I would like to note that these new surveys will play a crucial role in future research of these objects. A very successful way of finding distant radio galaxies ($z > 2$) is to study samples of bright radio sources ($S_{\sim 100\,\mathrm{MHz}} > 1$ Jy) with the steepest radio spectra (e.g. Spinrad, this Volume). The new low frequency surveys allow samples of radio sources with very steep spectra to be defined at flux levels of over a magnitude fainter than previously was possible. There is the interesting prospect that these sources have luminosities similar to those of the 1 Jy selected samples but are at appreciably higher redshifts. Alternatively, they could be at similar or lower redshifts than the present samples and form an excellent sample for comparison with the brighter distant radio galaxies, allowing us to study the relative impact of the radio source on the host galaxies.

At the beginning of this contribution I said that I would like to discuss and defend the claim that studies of distant radio galaxies are undergoing a dramatic revolution. I hope that the discussion on both detailed observations of selected radio galaxies and on the large radio survey has helped to convince the reader that this might indeed be the case.

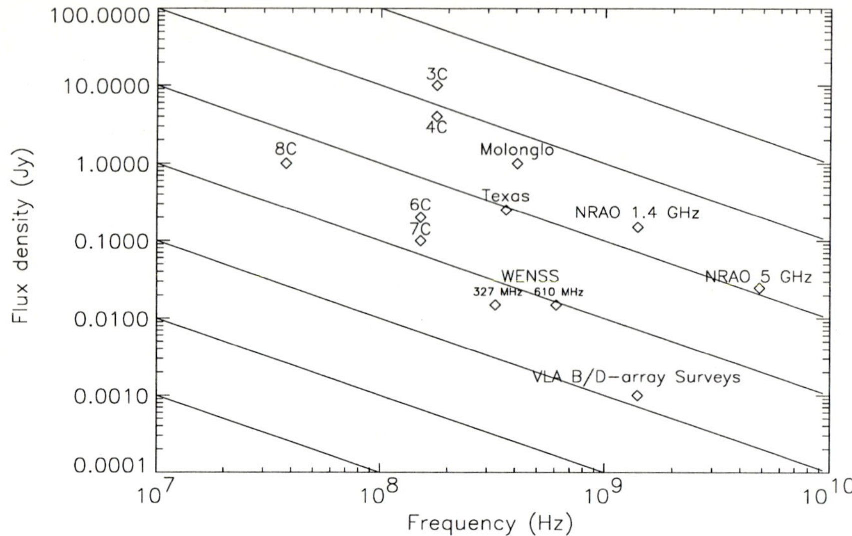

Fig. 3. The flux density limits of various catalogues as a function of frequency. The lines correspond to sources with spectral indices $\alpha = -1.0$ ($S_\nu \propto \nu^\alpha$).

Acknowledgements. I would like to thank my collaborators, Philip Best, Ger de Bruyn, Dick Hunstead, Malcolm Longair, George Miley, Rob van Ojik, Liz Waldram, Paul van der Werf and Mark Wieringa for continuous discussion.

References

Antonucci, R. 1993, *Ann. Rev. Astr. Astrophys.* **31**, 473

Baum, S., Heckman, T., Bridle, A., van Breugel, W., Miley, G. 1988, *Astrophys. J. Suppl.* **68**, 643

Becker, R.H., White, R.L., Edwards, A.L. 1994, in *Astronomical Data Analysis Systems and Software III*, ASP conference series, Vol.61, eds D.R. Crabtree, R.J. Hanisch & J. Barnes, p.165

Begelman, M.C. & Cioffi, D.F. 1989, *Astrophys. J. Lett.* **345**, L21

Chambers, K.C., Miley, G.K., van Breugel, W. 1987, *Nature* **329**, 604

Chambers, K.C., Miley, G.K., van Breugel, W.J.M. 1990, *Astrophys. J.* **363**, 21

Crawford, C., Fabian, A. 1993, *Mon. Not. R. Astr. Soc.* **260**, L15

de Bruyn, G., Miley, G., Tang, Y., Bremer, M., Brouw, W., Rengelink, R., Bremer, M., Röttgering, H. 1994, *Astron NFRA newsletter, Extra IAU Issue*, August, 4

Eales, S.A. 1992, *Astrophys. J.* **397**, 49

Hammer, F., Le Févre, O.L., Proust, D. 1991, *Astrophys. J.* **374**, 91

Heckman, T.M., Lehnert, M.D., van Breugel, W., Miley, G.K. 1991, *Astrophys. J.* **370**, 78

Hippelein, H., Meisenheimer, K. 1993, *Natute* **362**, 224

Longair, M.S., Best, P.N., Röttgering, H.J.A. 1995, *Mon. Not. R. Astr. Soc.* **275**, L47

McCarthy, P., van Breugel, W., Kapahi, V., Subrahmanya, C. 1991, *Astron. J.* **102**, 522

McCarthy, P., van Breugel, W., Spinrad, H., Djorgovski, S. 1987, *Astrophys. J. Lett.* **321**, L29

McCarthy, P.J. 1993, *Ann. Rev. Astron. Astrophys.* **31**, 639

Meisenheimer, K., Hippelein, H. 1992, *Astron. Astrophys.* **264**, 455

Miley, G., Röttgering, H., Chambers, K., Hunstead, R., Macchetto, F., Roland, J., Schillizi. R., van Ojik, R. 1992, *ESO Messenger* **68**, 12

Petitjean, P., Webb, J.K., Rauch, M., Carswell, R.F., Lanzetta, K. 1993, *Mon. Not. R. Astr. Soc.* **262**, 499

Rees, M.J. 1989, *Mon. Not. R. Astr. Soc.*, **239**, 1P

Röttgering, H.J.A., Lacy, M., Miley, G., Chambers, K., Saunders R. 1994, *Astron. Astrophys.* **108**, 79

Röttgering, H. 1995, in *Examining the Big Bang and Diffuse Background Radiation*, IAU Symposium **168** ed. M. Kafatos, Kluwer, (in press)

Röttgering, H., Hunstead, R., Miley, G.K., van Ojik, R., Wieringa, M.H. 1995, *Mon. Not. R. Astr. Soc.* (in press)

Sargent, W.L.W. 1988, in *QSO Absorption lines: Probing The Universe*, p.1, Blades *et al.* eds, Cambridge: University Press

Scarrott, S.M., Rolph, C.D., Tadhunter, C.N. 1990, *Mon. Not. R. Astr. Soc.* **243**, 5P

Visser, A.E., Riley, J.M., Röttgering, H.J., Waldram, E.M. 1995, *Astron. Astrophys. Suppl* **110**, 419

West, M.J., 1994, Mon. Not. R. Astron. Soc. **268**, 79

Wolfe, A.M. 1991, in Shaver, P.A., Wampler, E.J., Wolfe, A.M., eds, *Proceedings of the ESO Mini-Workshop on Quasar Absorption Lines*, Vol.9, p.97

Detection of $10h^{-1}$ Mpc Quasar/Absorber Correlation at High Redshift

Palle Møller

STScI, 3700 San Martin Drive, Baltimore, MD 21218, USA,
on assignment from the Space Science Department of ESA

1 Introduction

Determining when the first structures in the early universe were formed, and how they evolved to form present day galaxies and galaxy clusters, is one of the central problems of todays observational cosmology. Between the smooth universe of the Microwave Background Era mapped by COBE and the highly structured local universe mapped by the low redshift galaxy surveys, we presently have access to significant samples of only two types of objects: Quasars and quasar absorption systems.

On the assumption that both quasars and high redshift absorption systems are related to galaxies, we should be able to study the evolution of the galaxy clustering out to very high redshifts. In particular we would, in the simplest case, predict quasar/quasar, quasar/absorber, and absorber/absorber clustering on the same length scale.

One encouraging new result in that respect is that of Andreani & Cristiani (1992) who reported that quasars at redshifts out to $z = 2.15$ cluster on a length scale of $6 - 10h^{-1}$ Mpc. Following the above argument, we should therefore expect to see quasar/absorber clustering on the same scale, at similar redshifts.

A decade ago a series of papers by the ESO group, headed by Peter Shaver, (see e.g. Shaver & Robertson, 1983 for a summary) were addressing the question of a possible correlation between quasars and absorbers on scales of up to $\approx 1h^{-1}$ Mpc, and it was indeed shown that there is some tendency for high redshift quasars with a close line of sight neighbour at lower redshift, to exhibit absorption at a redshift close to that of the foreground neighbour. This type of quasar/absorber pairing was what the authors named "associated absorption". The question now is whether it is possible, with the observational data collected since then, to find the predicted large scale correlation.

I shall here present results which show that the expected quasar/absorber correlation (i.e. associated absorption in the above sense) is indeed seen in a complete sample, and I shall discuss how this can be used to determine the size, and covering factor for C IV absorption (CF_{CIV}), of high redshift structures.

2 Method

2.1 Fighting Biases and Incompleteness

It is well known that most quasar samples are strongly biased and incomplete. To circumvent problems arising from this, I chose the following method for the correlation test.

1. Define a sample of quasar spectra which is complete and unbiased with respect to the distribution of intervening C IV absorption systems (the C IV sample). "Intervening" is for the current purpose defined as "not being within 5000 km $s^{-1}$of the emission redshift of the underlying quasar".
2. Pick any sample of quasars which may be both biased and incomplete.
3. Extract a cylindrical volume (radius R) around the sight–line towards each quasar for which there is a spectrum in the C IV sample. Pick all the quasars from the quasar sample which happen to be inside the volume so defined.
4. For each sight–line cross correlate the list of C IV absorption systems in that sight–line with all the quasars found in the corresponding volume.
5. Combine all the individual cross correlations.

The great advantage of this method is that biases and incompletenesses of the quasar catalogue does not matter, because the test which is being done is: "For each known quasar inside a given volume, find the excess probability that a C IV systems is seen at the same redshift, as compared to any other redshift". In an incomplete sample there will be lots of undiscovered quasars in the same volume, but they will obey the same correlation, and their exclusion will hence not bias the result. Therefore only the C IV sample needs to be complete.

2.2 Sample Selection

For a preliminary test of the method described in section 2.1 above, I chose two C IV samples from the literature, and cross correlated both with the Véron–Cetty and Véron #5 quasar catalogue (1991) (hereafter V&V). One C IV sample was based only upon the C IV doublet line list from Sargent, Boksenberg & Steidel (1988) (hereafter SBS), the other sample was the SBS sample augmented by other data, also found in the literature (the SBS+ sample). Great care was exercised to ensure that those samples were both complete and unbiased. A detailed discussion of the two C IV samples can be found in Møller (1995).

3 Results

The cross correlation analysis, as described above, of the two C IV samples against V&V gave results which were identical to within the errors, and I shall therefore restrict the discussion here to the SBS+ sample. The full analysis can be found in Møller (1995). All numbers given below are calculated using $H_0 = 100h$ km s^{-1} Mpc^{-1} and $q_0 = 0.5$.

To be able to combine all the individual cross correlations, which were performed over a wide range of redshifts, the analysis was done in velocity space. In Figure 1 I show the cross correlations of the strong C IV lines (for "strong" I adopt the common definition of rest equivalent width of the 1548 Å C IV line ≥ 0.3 Å) versus V&V. The correlation is shown for a radius of the volume of $R = 10h^{-1}$ Mpc (a and c), and also for a volume with a radius of $R = 50h^{-1}$ Mpc where the central $R = 10h^{-1}$ Mpc volume has been excluded (b and d). The upper panels (a and b) show the correlation by number count (each absorber counts one), the lower panels (c and d) show the correlation weighted by rest equivalent width. The dotted curves in Figure 1 are the "survey functions", defined as the total redshift interval sampled in each bin, normalised to the observed absorber density.

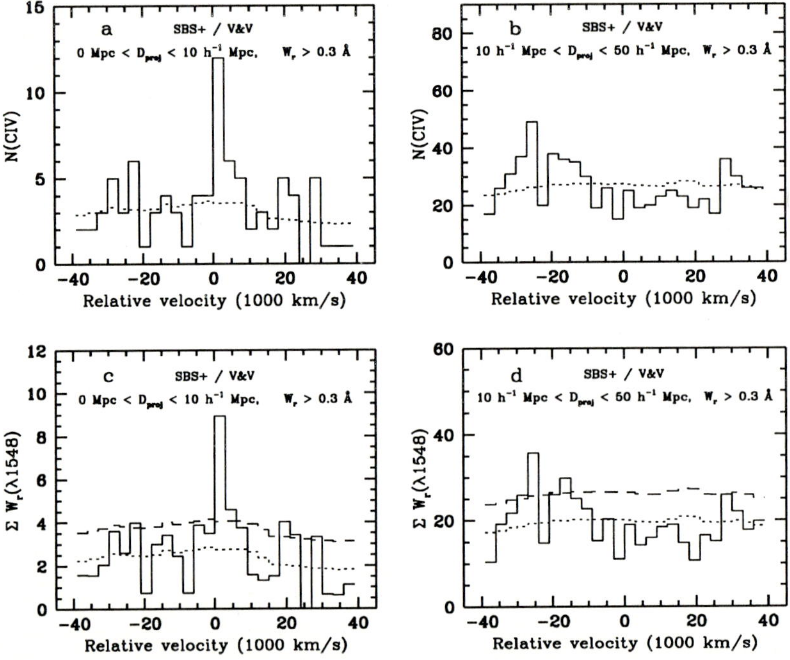

Fig. 1. Cross correlation of the SBS+ strong doublet C IV sample with the Véron–Cetty and Véron #5 quasar catalogue. (a) and (b) show the correlation of the number of C IV absorbers with line of sight neighbour quasars at projected distances inside $10h^{-1}$ Mpc radius, and between $10h^{-1}$ Mpc and $50h^{-1}$ Mpc radius respectively (full line). (c) and (d) show the corresponding correlations of the summed equivalent width of the 1548 Å C IV lines. In all four plots the dotted curve is the expectation value. In (c) and (d) the dashed curve is the 1σ upper limit as determined from the observed distribution.

The most striking feature in Figure 1a is the peak in the bin centred on +1500 km s^{-1}. From the normalised survey function we expect to find 3.54 absorbers in this bin if no correlation is present, but we observe 12. From Figure 1c it is seen that for the equivalent width weighted correlation the peak is 4.8σ above the null hypothesis expectation. No corresponding peak is found at distances in excess of $10h^{-1}$ Mpc (Figure 1b,d). For a detailed discussion of the significance of the cross correlation peak, see Møller (1995).

Several interesting conclusions can immediately be drawn from the results presented in this section.

1. First of all a strong peak has been found in the absorber/quasar correlation for strong C IV absorbers, and we can conclude that quasars and strong C IV absorbers do in fact occupy the same regions of space.
2. The size of the structures is of the order $10h^{-1}$ Mpc.
3. The correlation peak was found to be centred at a relative velocity of +1500 km s^{-1}. This is in agreement with reports of blueshift of high ionisation quasar emission lines of a similar amount (Espey etal., 1989 and references therein), and constitutes a completely independent way of getting a statistical determination of the mean emission line blue shift.

Finally, it is of interest to establish what the C IV absorption covering factor, CF_{CIV}, of the identified $10h^{-1}$ Mpc structures is. A simple estimate can be obtained from the fact that we observe 12 pairs in 42 sight–lines, ≈29%. However, some of the detections, as well as some of the sight–lines, are multiple, and it is not clear that those two effects will cancel. If we define P_{poiss} as the probability of finding one or more strong C IV absorber in a velocity bin with one or more foreground quasars within $D_{proj} < 10h^{-1}$ Mpc, and weed out the two databases to reflect this, we are left with only 9 pairs in 33 sight–lines, still giving a probability of $27^{+12}_{-9}\%$. The 1σ Poisson errors are true errors because all degeneracy has been deleted from the samples. The actual covering factor depends upon the size and shape of the structures, as well as the measured P_{poiss}. It is, however, impossible to construct a $CF_{CIV} = 1$ model consistent with the low value of P_{poiss}, and at the same time produce the large size of the structures. It hence follows that CF_{CIV} is less than unity, a result which is in agreement with reports by Møller and Kjærgaard (1992) and by Jakobsen and Perryman (1992), both reporting cases of very close sight–lines with no associated absorption.

References

Andreani, P., and Cristiani, S. 1992, *Astrophys. J.* **398**, L13

Espey, B.R., Carswell, R.F., Bailey, J.A., Smith, M.G., Ward, M.J. 1989, *Astrophys. J.* **342**, 666

Jakobsen, P., and Perryman, M.A.C. 1992, *Astrophys. J.* **392**, 432

Møller, P. 1995, *Astron. Astrophys.*, in press

Møller, P., Kjærgaard, P. 1992, *Astron. Astrophys.* **258**, 23

Shaver, P.A., and Robertson, J.G. 1983, *Nature* **303**, 155

Sargent, W.L.W., Boksenberg, A., and Steidel, C.C. 1988, *Astrophys. J. Suppl.* **68**, 539

Véron-Cetty M.-P., Véron P. 1991, A Catalogue of Quasars and Active Nuclei (5th Edition), Scientific Report No. 10, European Southern Observatory, Garching

Large-Scale Structure at z ~ 2.5

G.M. Williger[1], *C. Hazard*[2,3], *J.A. Baldwin*[4] *& R.G. McMahon*[3]

[1] MPI für Astronomie, Königstuhl 17, D-69117 Heidelberg, Germany
[2] Dept of Physics & Astronomy, Univ. of Pittsburgh, Pittsburgh PA 15260, USA
[3] Institute of Astronomy, Madingley Road, Cambridge CB3 0HA, England
[4] Cerro Tololo Inter-American Observatory, Casilla 603, La Serena, Chile

1 Summary

We report results from a program of 2Å resolution spectroscopy of intervening C IV absorbers in a ~ 1 deg^2 field near the South Galactic Pole. We have made a complete, statistically unbiased survey of C IV systems covering $1.5 < z < 2.8$ toward 12 lines of sight to rest equivalent width detection threshold $W = 0.15$ Å. We find 22 C IV systems. The number and redshift distribution of the intervening absorbers is consistent with what would be expected from all-sky C IV absorber statistics. However, there is a signal of structure on the $15 - 35h^{-1}$ Mpc scale ($H_0 \equiv 100$ km s^{-1} Mpc-1) as determined by the 2-point correlation function. We reject the null hypothesis that C IV systems are spatially Poisson distributed on scales $5 - 50h^{-1}$ Mpc at the > 99.9% significance level. The structure likely reflects the distance between two groups of absorbers subtending $\sim 13 \times 5 \times 21h^{-3}$ and $\sim 7 \times 1 \times 15h^{-3}$ Mpc^3 at $z \sim 2.3$ and $z \sim 2.5$ respectively.

2 Introduction

QSO absorption lines provide the means to study large-scale structure far beyond the $z \lesssim 0.5$ limits of current galaxy surveys. They are a natural extension of pencil beam galaxy surveys (Broadhurst *et al.* 1990) which claim to find structure at $z < 0.5$, and which have sparked interest in "foam" models for large scale structure (e.g. Coles 1990, van de Weygaert 1991, SubbaRao & Szalay 1992). However, the surface density of bright, high-z QSOs and the abundance of metal absorbers are generally too low to provide enough systems to outline structure in three dimensions at high redshift and to discriminate between some of the various models (Kaiser & Peacock 1991). In this paper we describe observations in a field with a particularly high surface density of QSOs, offering many closely spaced lines of sight.

The most investigated precedent for this work is the common C IV absorption in the vicinity of Tol 1037-2703 and Tol 1038-2712 (Jakobsen & Perryman 1992). The 17 C IV systems at $1.8 \lesssim z \lesssim 2.2$ extend over nearly a degree, outlining a sheet-like structure $5 - 10 \times 40 - 50h^{-2}$Mpc^2 seen nearly edge-on ($H_0 = 100h$

km s^{-1} Mpc^{-1}, $q_0 = 0.5$ and $\Lambda = 0$ are assumed throughout). A similar investigation was made toward PKS 0237-233. Foltz *et al.* (1993) selected it *a priori* on the evidence of a C IV cluster found in an all-sky survey by Sargent, Boksenberg & Steidel (1988, hereafter SBS). There appears to be a spatial overdensity of C IV systems over $1.58 \lesssim z \lesssim 1.67$ along the lines of sight to 6 high z QSOs with $12 - 16$ C IV systems spanning up to $65h^{-1}$Mpc, but most of the absorption is seen only toward two QSOs.

The new QSO field which forms the basis for the present work was found by Hazard *et al.* (1995) from an objective prism survey near the South Galactic Pole. Within an ~ 1 deg^2 region are 25 confirmed QSOs with redshifts $1.5 \lesssim z \lesssim 3.4$. These QSOs are more closely spaced, brighter and at higher redshift than the QSO fields studied by Jakobsen & Perryman (1992) and Foltz *et al.* (1993). Therefore, it is possible to search a relatively longer baseline for C IV systems at higher spectral and spatial resolution than in previous studies. The results can then be compared with global C IV properties determined from all-sky surveys. A study of this region will indicate whether or not such C IV absorber associations as those in the Tol 1037 – 2703/1038 – 2712 and PKS 0237 – 233 fields are common. We present results for 12 QSOs for which we have coverage between Lyα and C IV emission with sufficient signal-to-noise to make a complete survey of C IV systems. The field is plotted in Fig. 1.

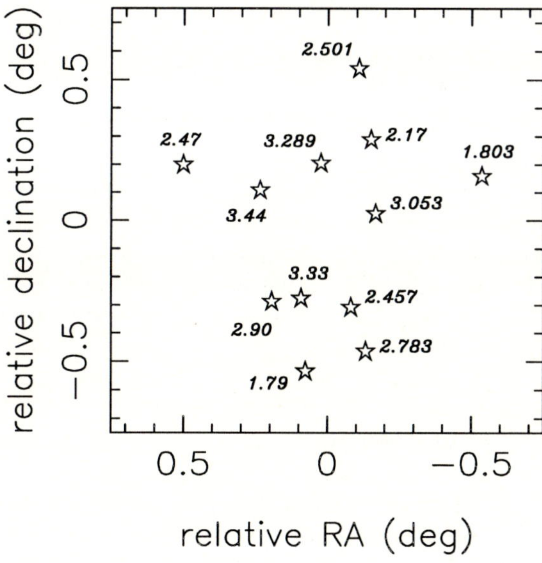

Fig. 1. The SGP QSO field. Stars indicate QSOs, with emission redshifts listed adjacent. The field is centered at $\alpha = 00^{\rm h}42^{\rm m}00^{\rm s}$, $\delta = -26°40'00''$ (1950).

3 Observations and Reductions

Data were taken with the CTIO 4m Argus multifiber spectrograph using a Reticon CCD over 25-27 Sep 1992 and 14-17 Oct 1993. Coordinates for fiber positioning were measured with Automatic Plate Measuring Machine at the IOA, Cambridge. Two 632 line/mm gratings were used, obtaining data spanning 3835 – 5933Å. All confirmed QSOs in the Hazard *et al.* field with $z > 1.5$ and falling in the unvignetted Argus field of view, plus QSO candidates, were observed subject to fiber positioner limitations. The data were extracted optimally using a routine adapted from that used in Rauch *et al.* (1992). A variance for each pixel was determined based on photon counting statistics from the object, sky and readout noise. Final spectra for each object were formed by adding the sky-subtracted, extinction corrected, flux calibrated spectra from each frame, using inverse variance weighting. The resolution is ~ 2Å.

4 C IV Absorption Systems

Absorption lines were searched for using the statistically based procedure described in Young *et al.* (1979). Standard IRAF routines were used interactively to deblend complexes. It is estimated that there are typical variations of 10–20% in equivalent widths due to continuum uncertainties. All wavelengths used were transformed to the vacuum heliocentric frame.

The C IV systems were selected in the same way that was used by SBS and Steidel (1990) in all-sky surveys, so that their large sample could be used as a comparison data base. All line pairs between Lyα and C IV emission were noted which had both components significant at the $\geq 3\sigma$ level with wavelength and equivalent width ratios consistent with those of the C IV $\lambda\lambda$1550 doublet. If the corresponding Lyα line fell within the spectral coverage, it was examined for wavelength and equivalent width consistency as well. Complexes with component separation $\Delta v \leq 150$km s^{-1} were counted as one system. "Associated absorption systems" within 5000 km s^{-1} of the emission redshift (e.g. Foltz *et al.* 1986) are excluded from further analysis. A total of 22 C IV systems with $\langle z_{abs} \rangle = 2.34$ were included in the sample (Fig. 2).

5 Results and Discussion

The total numbers and redshift distribution of our SGP C IV absorber sample are not significantly different from those found in all-sky samples. Steidel's (1990) sample ES2, $d\mathcal{N}/dz \propto (1+z)^{\gamma}$, $\gamma = -1.26$ for $z > 2$, $\gamma = 0$ for $z < 2$ predicts that we should have found 13.2 C IV absorbers in our survey, indicating a $\sim 2\sigma$ overdensity. Using the KS test, the 22 observed C IV systems are consistent with the broken power law redshift distribution found by Steidel for $\gamma = -1.26 \pm 0.56$; the probability of the data arising from the distribution for $-1.82 < \gamma < -0.70$ is $\geq 17\%$.

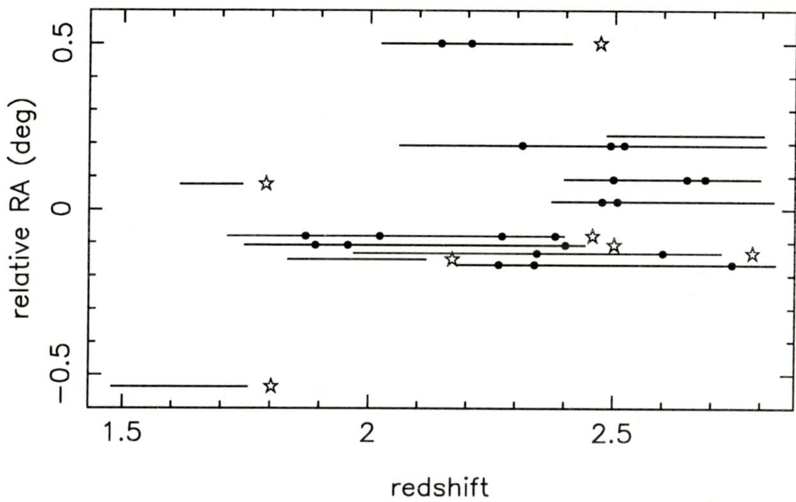

Fig. 2. The SGP field in RA-z space, centered at $\alpha = 00^h42^m00^s$ (1950). Stars represent QSOs, dots C IV systems in the sample and lines redshift sensitivity limits.

To test for clustering, we calculate the three dimensional two point correlation function as in Davis & Peebles (1983), in which the observed data are cross-correlated with a randomly generated dataset to provide the normalization for the distribution of C IV - C IV pairs from the observed data. The C IV redshift density function $d\mathcal{N}/dz \propto (1+z)^\gamma$ for each line of sight was calculated using the broken power law from Steidel (1990) with $\gamma = -1.26, 0$ for $z > 2$, $z < 2$ respectively. The density functions for each line of sight were then normalized from the sum for all lines of sight. A cumulative distribution was then made, producing a one-to-one correspondence between numbers from 0 to 1, and a redshift, RA and declination where the spectra were complete to $W_0 = 0.15$ Å for the C IV λ1550 line. Thus, choosing a number between 0 and 1 corresponds to one of the lines of sight in the survey, and a redshift along that line of sight to which the survey is complete to C IV λ1550 at W=0.15 Å. In this way 22 numbers were chosen to create a comparison dataset, and cross-correlated with the data; this was repeated 10^4 times. In the nomenclature of Davis & Peebles, the correlation function is calculated as $\xi = \frac{DD}{DR} - 1$. The 1σ uncertainty in the DR bin counts was calculated as the first moment about the mean for each bin. The pair separation r_{12} for systems at redshifts z_1, z_2 and angular separation α are calculated by the law of cosines

$$r_{12} = \frac{(r_1^2 + r_2^2 - 2r_1 r_2 \cos\alpha)^{1/2}}{1 + z_{12}} \tag{1}$$

where $z_{12} = (z_1 + z_2)/2$, and r is the proper distance in a $q_0 = 0.5$, $\Lambda = 0$

cosmology

$$r = \frac{c}{H_0 q_0^2}\{q_0 z + [q_0 - 1][(1 + 2q_0 z)^{1/2} - 1]\} \tag{2}$$

as presented by Misner, Thorne & Wheeler (1973) and applied by Crotts (1985).

The results are shown in Fig. 3. There is a feature at $15 - 35h^{-1}$ Mpc, significant at the $\sim 4\sigma$ level (adding the Poisson errors in quadrature). The result is similar when each of the twelve lines of sight is removed from the sample in turn: there is always a signal over the same comoving separation at the $\gtrsim 3\sigma$ level. This feature may be interpreted by noting that over half the sample can be divided into two small groups: seven systems at $2.26 < z < 2.40$, and five at $2.47 < z < 2.52$. The line of sight comoving distance between the group members corresponds to $12 - 37h^{-1}$ Mpc, while the groups themselves subtend $\sim 13 \times 5 \times 21h^{-3}$ and $\sim 7 \times 1 \times 15h^{-3}$ Mpc^3 respectively. Removing three systems from each group reduces the significance of the feature, while removing six systems from outside the groups increases it. We note that the complex at $2.47 < z < 2.52$ could be associated with the QSOs at $z = 2.46, 2.47, 2.50$.

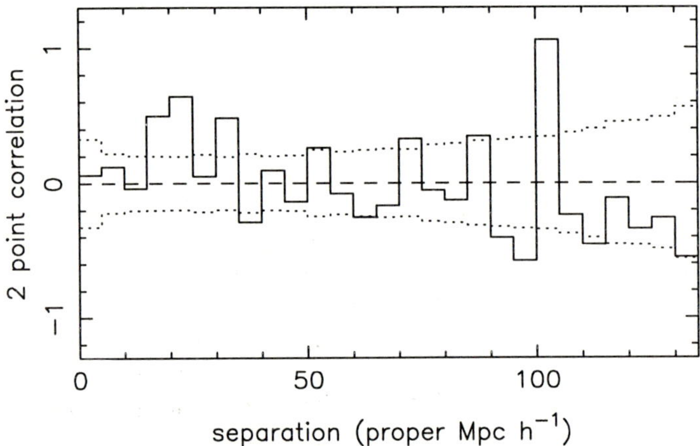

Fig. 3. The two point correlation function for the 22 C IV absorbers. The dotted line shows the 1σ uncertainty limits. At $z = 2.5$, 1° on the sky is $14h^{-1}$ Mpc, and $\Delta z = 0.1$ is $13h^{-1}$ Mpc.

The only other feature of note is an overdensity at $100 - 105h^{-1}$ Mpc, which arises from the relation of a pair of absorbers at $1.87 < z < 1.89$ with the group at $2.47 < z < 2.52$. This feature is not significant as removal of one of the $1.87 < z < 1.89$ pair virtually eliminates the it, and in any case the 1σ error bars for the two-point correlation function tend to be underestimated by Poisson estimates at these large scales (Hamilton 1993).

The bins are not independent, so we make a more direct test of the significance of the $15 - 35h^{-1}$ Mpc feature. Keeping as a statistic the maximum number of pairs in any bin, in any contiguous pair of bins and likewise up to any eight contiguous bins, we used the same 10,000 Monte Carlo simulations. The data were matched or exceeded by the simulations 246 times for a single bin, 40 times for two bins, then 5, 2, 38, 116, 299, 391 times for three to eight bins. Results are similar for bin sizes of $10h^{-1}$ and $15h^{-1}$ Mpc. To test for purely spatial correlations, we excluded pairs arising along the same line of sight as in Crotts (1985); this reduced the available number of C IV - C IV pairs in the data by only 10% and did not make a significant difference in the results. We conclude that it is very unlikely for a large number of C IV - C IV pairs arising from a Poisson-distributed sample at scales $> 2h^{-1}$ Mpc to occur in the range $15 - 35h^{-1}$ Mpc, as in the dataset.

SBS and Steidel (1990) found significant correlation at the $200 < \Delta v < 600$ km s^{-1} scale ($\sim 0.3 - 1.0h^{-1}$ Mpc at $z = 2.3$), and marginal correlation from $10^3 - 10^4$ km s^{-1} ($\sim 2 - 17h^{-1}$ Mpc at $z = 2.3$). Our survey is less sensitive at small scales, due to the small number of lines of sight and the minimum comoving distance between sightlines $\sim 2h^{-1}$ Mpc. Heisler, Hogan & White (1989) found the all-sky excess power at large scales attributable to PKS 0237−233. Conservatively, that absorber group spans $1.5773 < z < 1.6731$ toward PKS 0237−233 and Q 0233−2430, subtending $\sim 22 \times 26h^{-2}$ Mpc^2, larger than either association toward the SGP.

We reject the null hypothesis at $> 99.9\%$ significance that C IV absorbers are distributed randomly on scales $5 - 50h^{-1}$ Mpc. We appear to be finding $10 - 20h^{-1}$ Mpc structures separated by gaps of similar sizes. The beamwidth of $\sim 15h^{-1}$ Mpc for this survey is at the small end of the scale length where the two point correlation function indicates structure. We plan to widen the survey to find out whether the apparent groups of C IV absorbers form "sheets and walls", as is seen in local galaxy surveys, or more separated structures.

References

Broadhurst, T.J., Ellis, R.S., Koo, D.C. & Szalay, A.S. 1990, *Nature* **343**, 726

Coles, P. 1990, *Nature* **346**, 446

Crotts, A.P.S. 1985, *Astrophys. J.* **298**, 732

Davis, M. & Peebles, P.J.E. 1983, *Astrophys. J.* **267**, 465

Foltz, C.B., Hewett, P.C., Chaffee, F.H. & Hogan, C.J. 1993, *Astron. J.* **105**, 22

Foltz, C.B., Weymann, R.J., Peterson, B.M., Sun, L., Malkan, M.A. & Chaffee, F.H. 1986, *Astrophys. J.* **307**, 504

Hamilton, A.J.S. 1993, *Astrophys. J.* **417**, 19

Hazard, C., *et al.* 1995, in preparation

Heisler, J., Hogan, C.J. & White, S.D.M. 1989, *Astrophys. J.* **347**, 52

Jakobsen, P. & Perryman, M. 1992, *Astrophys. J.* **392**, 432

Kaiser, N. & Peacock, J.A. 1991, *Astrophys. J.* **379**, 482

Misner, C.W., Thorne, K.S. & Wheeler, J.A. 1973, *Gravitation* (San Francisco: Freeman)

Rauch, M., Carswell, R.F., Chaffee, F.H., Foltz C.B., Webb J.K., Weymann R.J., Bechtold J. and Green R.F. 1992, *Astrophys. J.* **390**, 387

Sargent, W.L.W., Boksenberg, A. & Steidel, C.C. 1988, *Astrophys. J. Suppl.* **68**, 539 (SBS)

Steidel, C.C., 1990, *Astrophys. J. Suppl.* **72**, 1
SubbaRao, M.U. & Szalay, A.S. 1992, *Astrophys. J.* **391**, 483
van de Weygaert, R. 1991, *Mon. Not. R. Astr. Soc.* **249**, 159
Young, P.J., Sargent, W.L.W., Boksenberg, A., Carswell, R.F. & Whelan, J.A.J. 1979, *Astrophys. J.* **229**, 891

Spectroscopy of 600 Faint Field Galaxies at CFHT: Luminosity Function to z = 1 and Properties of Blue Emission-Line Galaxies at z < 0.3

Francois Hammer[1], *Laurence Tresse*[1], *Simon Lilly*[2]
Olivier Le Févre[1], *David Crampton*[3], *Claudia Rola*[1]

[1] DAEC, Observatoire de Meudon, 92195 Meudon Principal, France
[2] Astronomy Dpt., University of Toronto, Toronto, Canada M5S 1A7
[3] Dominion Astrophysical Observatory, NRCC, Victoria, Canada

1 Summary

Our French-Canadian collaboration has obtained the spectra of 85% of 943 objects with $I < 22$ mag, 591 of which are galaxies, 6 are quasars and 200 are stars. Galaxy redshifts range from $z = 0.02$ to $z = 1.3$, with a $<z>= 0.56$. This survey provides the first opportunity to study in detail normal galaxies at a time when the Universe was less than half its present age. While the luminosity distribution of red galaxies is apparently similar to the local one, blue galaxies brighter than L^* show evolution of their luminosity since $z = 0.5$. The luminosity function of blue galaxies also presents a steepening at lower redshift and luminosity range ($z < 0.3$, $L < 0.1\ L^*$).
Further analysis shows that roughly 40% of these blue galaxies at $z < 0.3$ have emission-line properties intermediate between Seyfert 2 galaxies and LINERs.

2 Introduction

Deep counts of galaxies have revealed the presence of an overwhelming population of blue galaxies (Tyson 1988), corresponding to a number excess of 3 to 5 times more galaxies at $B > 23$ mag than one can expect from the local luminosity function. These objects apparently lie at moderate redshift ($<z>= 0.3$ for $B = 23-24$ mag) as shown by spectroscopy of B-selected galaxies (Colless *et al.* 1990). Beyond $z = 0.2$ the B filter samples the range below the 4000 Å break in the rest frame of the galaxies, which is sensitive to short-lived stars. Therefore, star forming galaxies are likely crowding B-selected samples. Moreover at large $z (z \sim 0.7)$ the B-band K-corrections for different galaxy spectral types can span 4 magnitudes and the observed B-band is redshifted into the ultraviolet for which the ultraviolet properties of galaxy types are poorly known. We believe that the B-selected samples are not well suited for studying the population of normal galaxies at large redshifts. More recently $\sim$ 50 galaxies have been selected at redder wavelengths ($I_{AB} < 22.5$, $I_{AB} = I + 0.48$, (Tresse *et al.* 1993; Lilly 1993), in order to estimate the older stellar light from faint galaxies which is perhaps better related to their stellar mass. We have formed the current collaboration (the Canada France Redshift Survey, or CFRS) with the aim of gathering a much larger sample redshifts for 1000 faint objects selected at I band. This sample is designed to study the luminosity function in detail at various redshift and color ranges, and has been completed in 30 nights at CFHT within two years.

3 Observations

Our deep I-photometry is complete up to $I_{AB} = 23.5$ mag, from which we select 943 objects with $I_{AB} < 22.5$ mag. No attempt has been made to separate stars from galaxies. The multiplexing capability of the MOS at CFHT was optimized for our project, and spectra for up to 80 objects were obtained at the same time using the grism with a spectral resolution of 40 Å. We define a 'set' of observations for each object to be eight, one hour spectra for that object. We have obtained at least one 'set' (8 hour total integration time) for 943 objects. More than one 'set' of spectra were obtained for 187 objects, for a total of 1150 object 'sets'.

Each spectrum has been extracted, combined and then analysed independently by three of us. Comparison of the three analyses has led to a final redshift and the corresponding note of reliability (6 different notes, see Figure 1). Multiple 'sets' of observations for an object were independently analyzed and then intercompared to give an estimate of the reliability of our redshift note system (Figure 1).

In addition to the deep I band imaging, we have also imaged these fields in the B, V and K bands. Of the 797 objects for which we identified the spectra (85% success rate), 200 are stars (26%), 6 quasars (0.6%) and 591 confirmed galaxies (74%). Most of the 137 objects for which spectra were not obtained appear to be galaxies having low surface brightnesses and/or objects belonging to the faintest luminosity range.

4 Redshift Distribution and Galaxy Luminosity Function

The luminosity function (LF) for the galaxies has been derived after calculating the B absolute magnitude of each galaxy from the V and I band magnitude, using K-corrections derived from a grid of 14 galaxy SEDs (Bruzual & Charlot 1990). All points of the global LF for the galaxies in our survey are above the local LF for present-day galaxies (Loveday *et al.* 1992). However, the meaning of our global LF is not obvious, as it is a mix of galaxies with different luminosities and at different redshifts.

Figure 3 presents the LF of galaxies in two redshift ranges ($0.05 < z < 0.5$ and $0.5 < z < 1$). We have divided the sample in two categories, according to whether the V-I colour of each galaxy was bluer or redder than the median colour defined by the average of elliptical and Sa colours. The LF for galaxies as red as non-evolved E/S0 galaxies shows little difference for the $z < 0.5$ sample

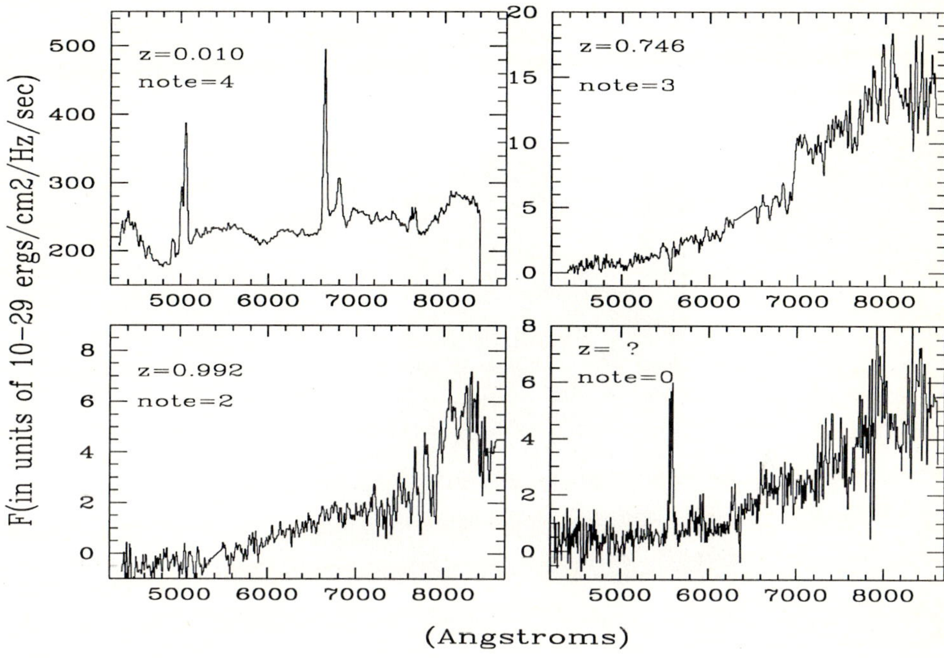

Fig. 1. Four spectra (8 hour integration) from the CFRS data basis. Redshift and notes are indicated ; note = 4,3 and 2 correspond to reliabilities higher than 99.5%, 98%, and 82% respectively; note = 9 are distinguishing the single emission line objects; note = 1 redshift are only tentative, and note = 0 an absence of determination, these two last categories accounting for the incompletness

relative to the $z > 0.5$ sample, consistent with the idea that stellar populations in the most luminous early type galaxies are not much evolved since the epoch when the Universe was less than half its present age. The apparent decrease of the number density of red galaxies towards the faint luminosity range might be real and comparable to the one found locally (Loveday *et al.* 1992), or might be due to the increasing incompletness in the faintest luminosity range which may included intrinsically fainter early-type galaxies. On the other hand, galaxies bluer than non-evolved, present-day spirals show a clear evolution from $z < 0.5$ to $z > 0.8$, which can be interpreted either by 1 mag of brightening or an increase of their number density by a factor 2.5 on a scale time of 3 Gyr.

At moderate redshift ($0.05 < z < 0.5$) we also find that intrinsically faint blue galaxies ($L < 0.1L^*$) were more numerous in the (recent) past than now, which confirms previous studies of faint B-selected samples (Cowie *et al.* 1991). This is apparently due to steepening of the LF in the lower luminosity range, which suggests that these faint blue galaxies are the low redshift counterpart of the faint blue population which crowds the blue counts.

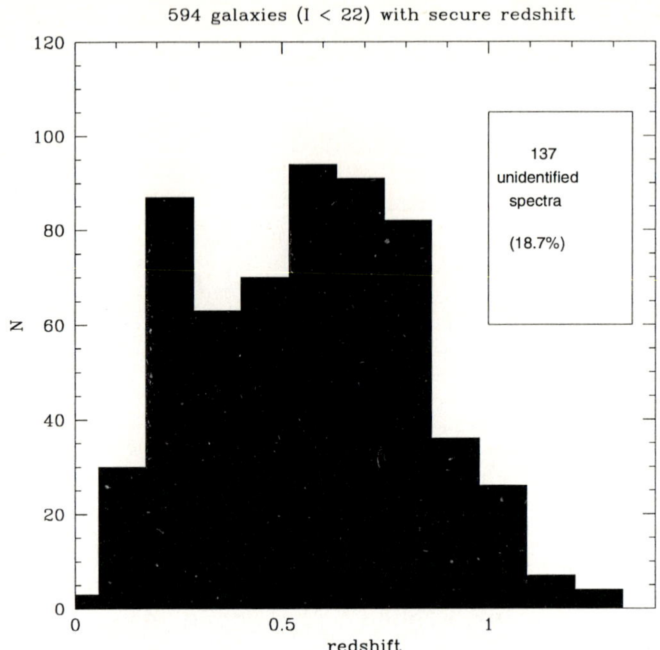

Fig. 2. Redshift distribution of the $I_{AB} < 22.5$ galaxies. Average z is 0.56.

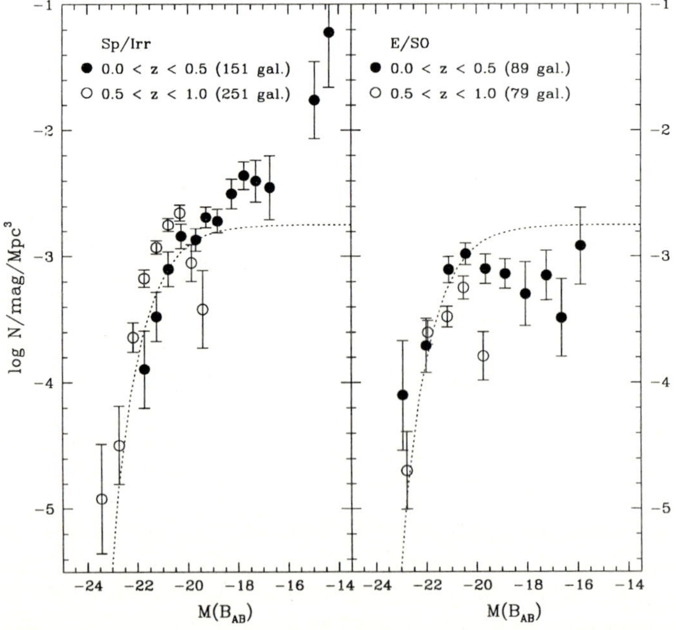

Fig. 3. Luminosity function for red galaxies (defined to be as red as non-evolved E/S0) and blue galaxies in two redshift ranges. Dotted line represents the local luminosity function from Loveday *et al.* (1992).

The redshift distribution for galaxies ranges from $z \sim 0$ to $z = 1.3$ (Figure 2), but one can expect an increasing incompleteness beyond $z = 1$, where strong spectral features in galaxy spectra go beyond our spectroscopic window. Also, the decreasing CCD efficiency below 4500 Å may affect the identification of the 4000 Å break of red absorbing-spectrum galaxies at low redshift.

5 Emission-Line Ratio Diagrams for Galaxies up to z =0.3

Several properties in the CFRS (see above and Tresse *et al.* 1995) have allowed a spectral classification of the 138 galaxies at $z \leq 0.3$: 74 spectra present H_α and H_β in emission, and forbidden lines ([O II] λ3727, [O III] λ4959, [O III] λ5007, [S II] λ6725); 43 spectra have H_α in emission and H_β in absorption; 21 spectra have H_α and H_β in absorption.

The two diagnostic diagrams shown in figure 4, lead to classify narrow emission-line spectra according to their principal ionization source, consequently to better understand the nature of the corresponding galaxies. In such diagrams, objects photoionized by active nucleus (Seyfert 2 galaxies, LINERs) are well separated from those photoionized by massive stars (H II region-like objects, starburst galaxies (SB)).

Since the CFRS fields are at high latitude ($b_{II} \geq 50^o$), our spectra are not affected by dust in our Galaxy, and the reddening has been calculated in using the Balmer decrement H_α/H_β equal to 2.86 (case B, T = 10 000 K, n = 100 cm^{-3}, Veilleux & Osterbrock 1987) and the Seaton's law (Seaton 1979). The [O III] $\lambda5007/H_\beta$ versus [S II] $\lambda6725/H_\alpha$ diagram has the advantage of being independant of the reddening due to intrinsic dust in galaxies.

The corrected emission-line ratios have been plotted in both diagrams: a significant fraction of them belong to the AGN area. The CFRS sample is a good representation of the field galaxies in the Universe, and allows statistics. We found that at least 20% of all the field galaxies up to $z = 0.3$ have emission-line ratios similar to AGN (Veilleux & Osterbrock 1987).

Our spectral resolution ($\sim$ 40 Å) does not allow to examine properly the stellar absorption under the Balmer lines. It is probably the reason why for 18 spectra amongst the 74 emission-line spectra, $H\beta$ does not appear above the continuum despite a strong $H\alpha$ in emission. Combinated spectra of data being in the H II/SB area, and being in the AGN area are shown in the figure 5. The host galaxie for AGN is seen by the absorption lines like CaH&K, and H_δ. However we believe unlikely that strong Balmer absorption lines can significantly affect the location of objects in the diagnostic diagrams (especially because that it should not affect the ratio [S II] $\lambda6725/H_\alpha$).

To go further, models have been built with the photoionization code PHOTO (Stasińska 1990) for H II regions (Kurucz model atmospheres, Kurucz 1992, T = 60 000 K, n = 10 cm^{-3}) to draw the upper limits for objects photoionized by massive stars (typically OB stars) in these diagrams. Again, objects are outside these upper limits. These models have been also confronted with local samples of H II regions, starburst galaxies, Seyfert 2 galaxies and LINERs.

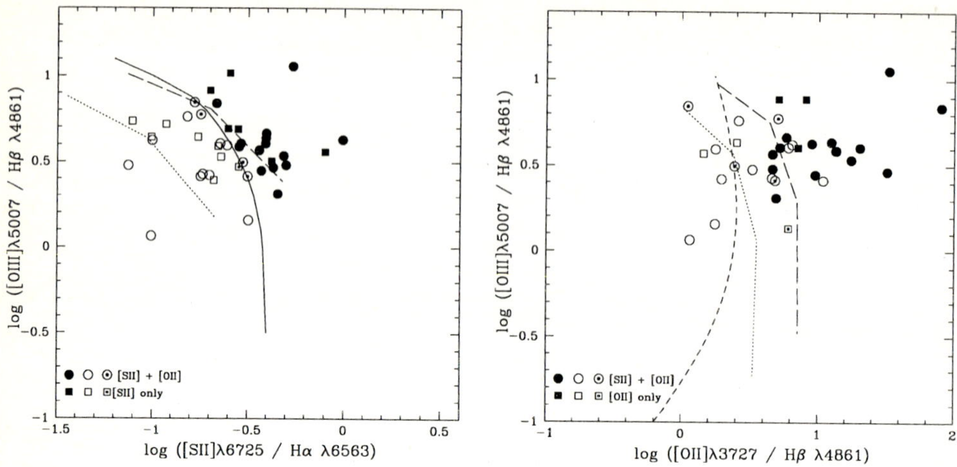

Fig. 4. Diagnostic diagrams for spectral classification (Veilleux & Osterbrock 1987). Round symbols represent spectra with both [S II] λ6725 and [O II] λ3727 in the observed spectral range. Square symbols represent spectra with only [O II] λ3727 or only [S II] λ6725 inside the spectral range. The solid line represents the separation between AGN (right) and H II/SB galaxies (left). The short-dashed line represents the main sequence of H II region-like objects. Black symbols are data in the AGN area from the diagram on the left, and reported in the one on the right. With the same procedure, white symbols are data in the H II/SB area, and the composite symbols are data in the intermediate area. Long-dashed lines and dot lines are the modelised upper limits for photoionization by OB stars (T = 60 000 K, n = 100 cm^{-3}) with respectively $Z = 0.25\ Z\odot$ and $Z = 0.1\ Z\odot$.

These AGN have an ionization parameter in between the Seyfert 2 galaxies and LINERs with [O III] λ5007/$H_\beta \approx 0.6$, closer to low-ionization Seyfert 2 galaxies. The rest-frame [O II] λ3727 equivalent width of all the emission-line spectra span from 10 Å to 80 Å, with no distinction between AGN, H II region-like objects, or starburst galaxies. These AGN represent a significant fraction ($\approx$ 40%) of the bluest galaxies with irregular and spiral spectral type at $z < 0.3$, and are faint galaxies ($-20.5 < M(B_{AB}) < -16.5$). At this stage, our analysis provides evidence that nuclear activity may be associated with the rapidly evolving blue galaxy population.

The low-luminosity AGN fraction ($\approx$ 2%), found locally in magnitude-limited spectroscopic sample (Huchra & Burg 1992), is much smaller than the 20% found in this analysis. Morphological and high spectral resolution observations will be done to confirm this result.

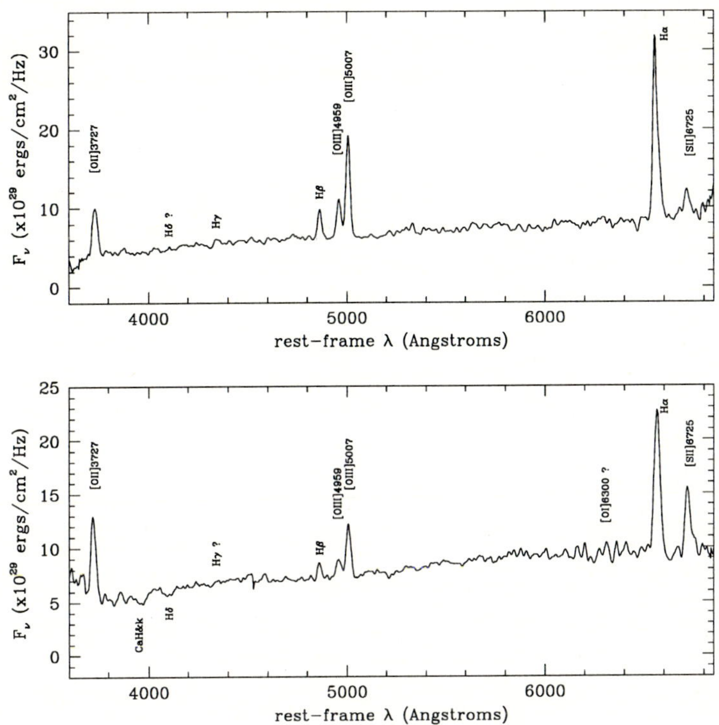

Fig. 5. Combined spectra for spectra with line ratios similar to H II region-like objects or starburst galaxies (top), and similar to AGN (bottom).

6 Conclusion

The CFRS project has succeeded in gathering one of the largest data bases of very faint galaxies, including two-thirds (594) of the available spectra of all galaxies with $I < 22$ in our photometered fields, plus deep imagery in the visible and near infrared. The survey provides the first opportunity to analyse properties of "normal" galaxies up to $z = 1$, which corresponds to a time when the Universe was younger than half of its present age. Because of the large number of objects we can compare galaxy properties in several redshift and colour bins, to derive observational constraints on their evolution. This analysis is expected to be reliable, as it compares objects selected in an identical way, and is not affected by uncertainties in the local luminosity function. First results show that the luminosity function of galaxies as red as present-day E/S0 galaxies does not evolve much from $z = 0.2$ to 1, while at $z > 0.5$, blue L^* galaxies were either more abundant or were brighter than present day L^* blue galaxies. We also notice the presence of a potentially important population of fainter blue galaxies at moderate redshift ($z < 0.3$) with emission-line properties apparently driven by AGN. We intend to further use our data basis to pursue several projects as follows:

1. Spectrophotometric analyses of blue and red galaxies, combined with morphological studies based on HST data expected this year (Cycle 4) and next year (Cycle 5).
2. Multiwavelength analyses of galaxies, based on our recent identification, in imagery and spectroscopy, of a complete sample of μJy radiosources, in one of our fields that is coincident with the Fomalont *et al.* (1991) field. This study will be completed by observations in the far IR (ISO) and at sub-mm wavelengths.

Acknowledgements. We are very grateful to the directors of CFHT as well as to the two CFHT time allocation committees for supporting our observational program during three years. L.T. thanks the Max-Planck Society for travelling support.

References

Bruzual, G., Charlot, S. 1990, *Astrophys. J.* **405** , 538
Colless, M., Ellis, R., Taylor, K., Hook, R. 1990, *Mon. Not. R. Astr. Soc.* **235**, 827.
Cowie, L., Songaila, A., Hu, E. 1991, *Nature* **354** , 460.
Huchra, J., Burg, R. 1992, *Astrophys. J.* **393**, 90.
Kurucz, R. 1992, in *The Stellar Population of Galaxies*, eds. Barbuy & Renzini, Kluwer Academic Publishers, p.225
Lilly, S. 1993, *Astrophys. J.* **411**, 501
Loveday, J., Peterson, B., Efstathiou, G., Maddox, S. 1992, *Astrophys. J.* **390** , 338
Seaton, M. J. 1979, *Mon. Not. R. Astr. Soc.* **187**, 73
Stasińska, G. 1990, *Astron. Astrophys. Suppl.* **83**, 501
Tresse, L., Hammer, F., Le Févre, O., Proust, D. 1993, *Astron. Astrophys.* **277**, 53
Tresse, L., Rola, C., Hammer, F., Stasińska, G. 1995, in *Wide field spectroscopy and the distant Universe*, 35th Herstmonceux Conference, eds S. Maddox & A. Aragon-Salamanca, World Sci. Publ. Co., Singapore, in press.
Tyson, J. 1988, *Astron. J.* **96**, 1
Veilleux, S., Osterbrock, D. 1987, *Astrophys. J. Suppl.* **63**, 295

Deep Near-Infrared Imaging with the Keck Telescope

Matthew A. Bershady[1,2]

[1] Penn State University, State College PA 16802, USA
[2] Hubble Fellow

1 Introduction

We present first results from a study of faint field galaxies in deep J and K band images from the Keck Telescope in several fields at high galactic latitudes (Bershady *et al.* 1995b). The total area of the survey is small (~3e-4 deg^2), but reaches depths of K=24 and J=24.5 for the most compact objects. Compared to other surveys (Cowie *et al.* 1994, Djorgovski *et al.* 1995) at comparable depths, we find that the K band differential counts neither flatten nor steepen beyond K=22, but continue to rise with a slope of log(A) near 0.3. Based on new, empirical models of galaxy counts, we find the *slope* of the faint end of the K band counts is not sensitive to q_0. Our galaxy count analysis is only the first step in probing the galaxy distribution at high redshift.

Deep near-infrared images offer a special regime for the study of distant galaxies for a number of reasons, not the least of which is the fact that κ-corrections are uniform and negative to z~3. Depths achievable in a few hours on 10m-class telescopes are capable of seeing L* galaxies at redshifts well in excess of 2 (assuming no evolution). Perhaps the most appealing aspect of studying distant galaxies in the near-infrared is that we can make predictions about their appearance (colors, sizes, surface-brightness, shapes) based on our extensive knowledge of the optical properties of local galaxies. In contrast, we know little about the ultra-violet properties of galaxies at low redshift. Hence it is very difficult to model reliably the observed galaxy distributions in deep, *optical* images.

To make strictly empirical models of distant galaxies in the near-infrared, we are required to make but one assumption (aside from a choice of cosmology), namely that there is no evolution. We expect this assumption to be wrong if we probe to sufficient look-back times, but it is still well-defined to measure the *difference* between predicted and observed distributions. A difference, if found, will likely have a non-unique interpretation, while the difference itself is an enduring measurement. In this spirit, we introduce an empirical model based on a relatively local galaxy survey (Bershady *et al.* 1995a, Bershady 1995) that we will use in future studies of our deep near-infrared images.

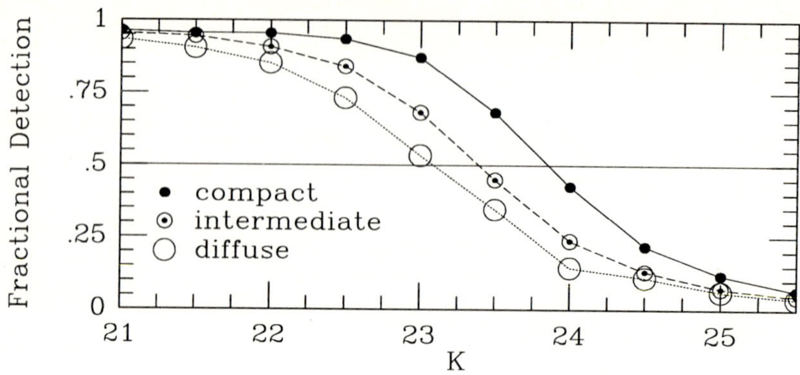

Fig. 1. Detection completeness as a function of image size for our deepest K band image (17,500 sec, 0.55 arcsec FWHM seeing) using FOCAS with detection parameters discussed in Bershady *et al.* 1995b.

2 Survey Limits: Completeness and Image Size

Before discussing galaxy counts, we first emphasize that detection limits in imaging surveys depend critically on both source magnitude and size (i.e. size and surface-brightness). While this may be known in principle, the magnitude of the effect is important to quantify for the particular application of counting sources in deep near-infrared images taken in exquisite seeing. We have performed extensive Monte Carlo simulations using real object images spanning the observed range of galaxy sizes, inserted back into our images at random locations. The fraction of detections as a function of image size and simulated apparent magnitude are shown in Fig. 1. Image classes "compact", "intermediate" and "diffuse", as defined by *non*-isophotal sizes, roughly divide the sample of all objects detected to K=23 into thirds. Note there is nearly a one magnitude range in the 50% detection limit for the different image classes, where the completeness function is steepest. Without knowing *a priori* the distribution of galaxy sizes as a function of magnitude, adopting any single completeness function to correct the observed counts would invariably produce a bias near the detection limit.

3 K-Band Galaxy Counts

In Fig. 2, we have plotted a compilation of galaxy counts from 10< K <23 as well as several models. For our own data (Bershady *et al.* 1995b), corrections for completeness and reliability have been done separately as a function of image size, and the corrected counts then summed. Our photometry is based on aperture magnitudes, corrected to total also as a function of image size. All other data is presented as published except we have applied aperture corrections to

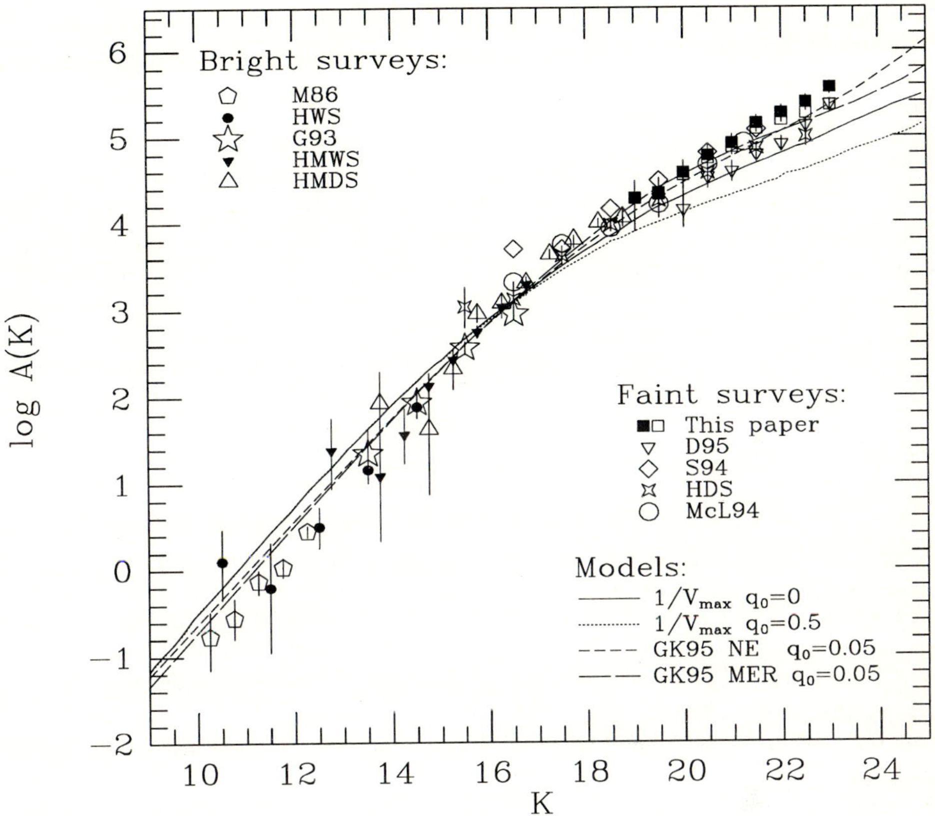

Fig. 2. K Band differential galaxy counts (number mag^{-1} $degree^{-2}$) from discrete-source surveys. Bright surveys: M86 (Mob), G93 (Glz), HWS, HMWS, and HMDS adopted from Gardner *et al.* (1993) and references therein. Faint surveys: McL94 (McLeod *et al.* 1995), HDS (Cowie *et al.* 1994 and Gardner *et al.* 1993), S94 (Soifer *et al.* 1994), D95 (Djorgovski *et al.* 1995). Boxes represent our counts (Bershady *et al.* 1995b); open boxes correspond to our counts prior to corrections for completeness. Models: GK95 (Gronwall & Koo 1995) NE (no evolution), and MER (mild evolution with reddening), and $1/V_{max}$ models (see text and Bershady *et al.* 1995b).

the data of Djorgovski *et al.* (1995) to make their photometry consistent with our own and other surveys.

At any given magnitude, there is considerable range ($\sim$ a factor of 2) in *normalization* between surveys, but the *slopes* are comparable except at the faintest magnitudes. The 'scatter' may be due to the confluence of two effects: Small numbers of objects at the bright magnitudes, and small field sizes (field-to-field fluctuations) at faint magnitudes. Over the last 1.5 magnitudes of our survey, there is roughly a 50% difference in the counts between our two fields, although the *slopes* are the same. The normalization of our counts at the bright end is in close agreement to McLeod *et al.* (1995); both surveys include fields within 20 arcminutes of each other in SA 57. Had we not applied aperture corrections to Djorgovski *et al.'s* counts, our slope would be in agreement with theirs; both surveys find steeper slopes at the faintest magnitudes than found previously (Cowie *et al.* 1994).

By design, the models of Gronwall & Koo (1995) fit the data to K=22.5 prior to Keck observations (Bershady *et al.* 1995b, Djorgovski *et al.* 1995, Soifer *et al.* 1994), and are a bench-mark for comparison. The $1/V_{max}$ models (Bershady *et al.* 1995b) underpredict the observed Keck counts and slope for K>20, but yield close to the same slope as the other models *even for* q_0=*0.5.* The $1/V_{max}$ models are strictly empirical, based on a real data set (Bershady *et al.* 1995a, Bershady 1995), which is scaled, object by object, by the relative accessible volumes in the input survey and the output simulation. Because the input survey has U through K band photometry, κ-corrections in the K band are determined directly for each simulated object z$\sim$7 (not exceeded at these depths). The input sample, however, contains only L>0.0006L* galaxies in the K band because it is a finite, magnitude-limited sample (roughly B<20.5). Lower luminosity galaxies are expected to be present in significant numbers by $K\sim$20, but the precise contribution depends on the currently unknown faint-end of the K band luminosity function. Inclusion of lower-luminosity galaxies in the $1/V_{max}$ models (e.g. from larger, deeper local surveys), will only raise and steepen the predicted counts. Therefore the amount of evolution needed to fit the counts may be quite mild, particularly in an open universe. Similar conclusions have been reached using independent no-evolution models (Djorgovski *et al.* 1995, Gronwall & Koo 1995).

Unlike other no-evolution models (Cowie *et al.* 1994; Djorgovski *et al.* 1995; Gardner *et al.* 1993), for q_0=0.5 our $1/V_{max}$ model predicts the counts do *not* role over around K=22, but continue to rise beyond K=25. This reflects a steeper faint-end slope of the K band luminosity function than previously observed, as we will present elsewhere. As a result, the counts at faint magnitudes are increasingly dominated by low-L galaxies at relatively low redshifts. This in turn makes the *slope* of the faint end of the K band counts insensitive to q_0 as well as the cosmological constant.

The work presented here involves a collaboration with J. Lowenthal and D. Koo (Lick Observatory). MAB was funded by NASA through grant HF-1028.02-92A from STScI (operated by AURA, Inc. under contract NAS5-26555).

References

Bershady, M.A., Hereld, M., Kron, R.G., Koo, D.C., Munn, J.A., Majewski, S.R. 1995a, *Astron. J.* **107**, 87

Bershady, M.A. 1995, *Astron. J.* **108**, 870

Bershady, M.A., Lowenthal, J.D., Koo, D.C. 1995, to be submitted to *Astrophys. J. Letters*

Cowie, L.L, Gardner, J.P., Hu, E.M., Songalia, A., Hodapp, K.-W., Wainscoat, R.J. 1994, *Astrophys. J.* **434**, 114

Djorgovski, S., *et al.* 1995, *Astrophys. J. Lett.* **438**, L13

Gardner, J.P, Cowie, L.L. Wainscoat, R.J. 1993, *Astrophys. J. Lett.* **415**, L9

Gronwall, C., & Koo, D.C. 1995, *Astrophys. J. Lett.* **440**, L1

McLeod, B.A., Bernstein, G.M., Rieke, M.J., Tollestrup, E.V., Fazio, G.G. 1995, *Astrophys. J. Suppl.* **96**, 117

Soifer, B.T. *et al.* 1994, *Astrophys. J. Lett.* **420**, L1

Galaxy Evolution in the Infrared

John Peacock

Royal Observatory, Blackford Hill, Edinburgh EH9 3HJ, UK.

1 Introduction

After several sessions devoted to saturation bombing by Keck and HST, what I have to say may seem somewhat old-fashioned, but it is a project which began in 1987 at a time when 58×62 pixels in an infrared detector was a revolutionary number. This was used to construct the Edinburgh K-band survey described by Glazebrook *et al.* (1994), covering $552\,\mathrm{arcmin}^2$ to $K \simeq 17.5$. That paper presented our results for the K-band star and galaxy counts; the present paper is concerned with a K-selected redshift survey and the resulting colour-redshift and luminosity function analyses.

A study of galaxy evolution in the near infrared is of great interest. Historically the main evidence for bulk evolution in samples of field galaxies derives from optical surveys, culminating in the 'faint blue galaxy problem': number-magnitude counts over $15 < b_{\mathrm{B}J} < 28$ are much steeper than predicted by a non-evolving model (see Ellis 1990 for a review). The K band is however in many ways preferable to the optical. The optical work samples the rest-frame UV at $z \gtrsim 0.3$, so that the optical luminosity depends sensitively on the rate of star formation. The observed morphological mix in the optical will change greatly with redshift, complicating the interpretation. In contrast galaxy colours and K-corrections in the near-infrared are dominated by old stars and are uniform across Hubble types (Aaronson 1978), thus yielding a constant morphological mix.

2 Data

Our redshift catalogue consists of 124 galaxies. Our spectroscopic runs were used to ensure that the redshift sample was as statistically representative as possible. The $(R-K, K)$ colour-magnitude diagram was inspected, and targets were chosen randomly to approximate a uniform sampling of $R-K$ at given K, down to $K = 17.25$ (see Fig. 1).

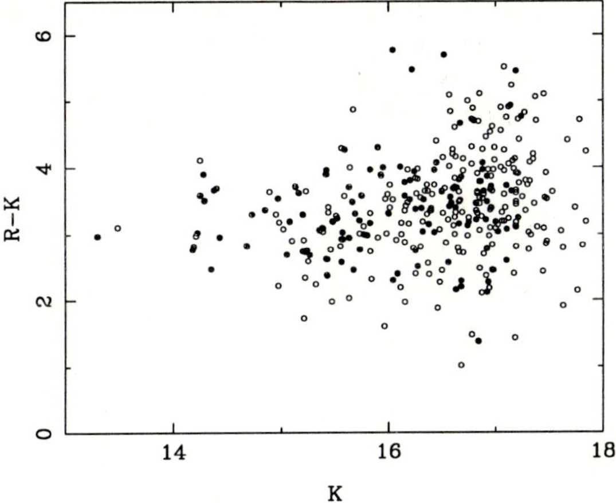

Fig. 1. The $R - K$ vs K colour-magnitude plane for our survey. The open circles show all the data, with spectroscopic sample members being indicated by solid points. Brightwards of the spectroscopic selection at $K = 17.25$, the sampling of colour at fixed K is very close to uniform.

2.1 Aperture corrections

The majority of our data were measured through a 4-arcsecond diameter aperture; an aperture of about 6 arcsec was used for most of the Hawaii work (see Gardner 1992). In their recent paper on the K-band luminosity function, Mobasher *et al.* (1993) corrected all their data to a standard isophotal aperture based on the optical light profiles of their galaxies. How much difference do these various choices make?

We might expect the growth in metric luminosity to be well parameterised following the approach of Gunn & Oke (1975) as:

$$L(< r) \propto r^{\alpha}.$$

For the sample of Mobasher *et al.* (1993), typical effective values are $\alpha \simeq 0.4$, and this is consistent with the Hawaii data (Gardner 1992). Our maximum redshift is 0.8, and all but 2 have $z > 0.06$; the range of proper diameters corresponding to our 4 arcsecond apertures is thus 3.2 to 16.5 h^{-1} kpc. Over this range, the $r^{0.4}$ prediction is an excellent parameterisation of our growth curves.

2.2 Faint red galaxies

Generally, the reddest galaxies are approximately consistent with the model red envelope to within observational error. This was not the case when we used 4-arcsecond aperture colours, which contained many galaxies with $z < 0.3$ much redder than the envelope, particularly at low redshift. This is an effect of colour gradients: at low redshifts, the 4-arcsecond apertures sample the galaxy nuclei only – and these are very red in some cases. Significant optical-infrared colour gradients in ellipticals were previously noted by Peletier *et al.* (1989). They find up to 0.6 mag of reddening in $V - K$ for a factor 10 in radius, and our results seem to be consistent with these more extreme values.

3 Luminosity function analysis

We now proceed to derive the K-band galaxy luminosity function from our data. We shall be particularly interested in the comparison between our results and those of Mobasher *et al.* (1993) and the Hawaii survey (Cowie & Songaila 1993; Cowie *et al.* 1995). In fact, our results turn out not to agree very well with either of these pieces of work.

Our raw data are shown in Fig. 2, in the form of redshift versus absolute magnitude. Throughout, we scale results to the usual dimensionless Hubble parameter: $h \equiv H_0/100 \text{ km s}^{-1}\text{Mpc}^{-1}$. Unless otherwise stated, we assume a cosmological model with $\Omega = 1$ and zero cosmological constant.

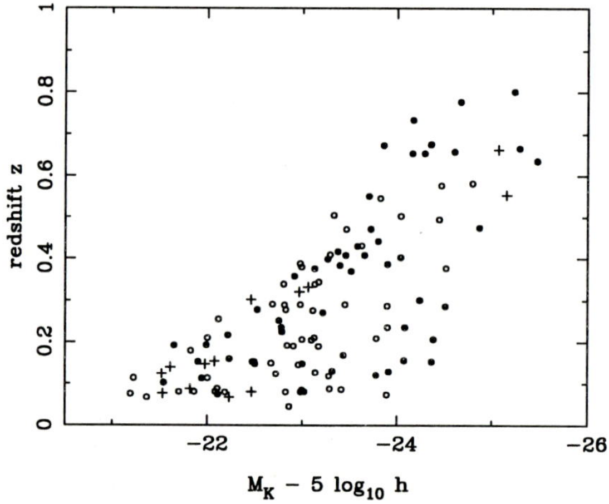

Fig. 2. The redshift-magnitude data, translated to the redshift-absolute magnitude plane. The different symbols correspond to three equal classes of restframe colour: filled circles denote E/Sa; open circles Sb; crosses Sc/Im. Note the fainter characteristic luminosity of the last class.

3.1 Counts and incompleteness corrections

We now need to know the effective K-dependent incompleteness, and this may be deduced by comparing the number of galaxies in our spectroscopic sample as a function of magnitude with that expected from the overall number counts. A convenient analytical fit for these is

$$\frac{dN}{dK} \,/\, \mathrm{deg}^{-2} = \frac{10^{0.75(K-12.1)}}{\left[1+10^{0.35(K-17.2)}\right]^{1.5}},$$

which is a statistically acceptable best fit to the data from Paper I plus the Hawaii counts from Gardner, Cowie & Wainscoat (1993), and the data of Jenkins & Reid (1991) – all corrected to 4-arcsec apertures.

3.2 K-corrections

In order to obtain absolute magnitudes, we require a knowledge of the luminosity distance $D_{\text{ß}L}$, the K-band K-correction $K(z)$, and the aperture correction $A(z)$:

$$M(z) = m - 5\log_{10}[D_{\text{ß}L}/10\ \mathrm{pc}] - K(z) + A(z).$$

For simplicity, we shall throughout quote absolute magnitudes assuming $h = 1$ for the Hubble parameter.

One of the advantages of the infrared waveband is that the K-corrections are very similar for all classes of galaxy, reflecting the dominance of giants in this waveband. The widely different amounts of star formation in different Hubble types only affects the spectra at wavelengths somewhat shorter than 1μm. We use theoretical K-corrections taken from the evolutionary synthesis models of Bruzual & Charlot (1993; BC); there is little model dependence of the K-corrections. The following is a good fit to the 5-Gyr data for $z \lesssim 1.5$:

$$K(z) = \frac{-2.58z + 6.67z^2 - 5.73z^3 - 0.42z^4}{1 - 2.36z + 3.82z^2 - 3.53z^3 + 3.35z^4},$$

and we use this as our standard K-correction.

3.3 Luminosity function estimates

The simplest estimator of the luminosity function is to bin up the data in redshift slices as a function of absolute magnitude. The estimator for the density in a given bin is then the traditional

$$\hat{\phi} = \sum_i \hat{\phi}_i = \sum_i \frac{w_i}{V(z_{\mathrm{max}}) - V(z_{\mathrm{min}})}$$

(Felten 1976), where z_{max} is the smaller of the maximum redshift within which a given object could have been seen, and the upper limit of the redshift band under consideration; z_{min} is the lower limit of the band. The result is shown in Figure 3a, for various redshift bins.

An alternative way of presenting the same data has been favoured by the Hawaii group, which is to use the cumulative luminosity density. The obvious estimator for this is

$$\hat{\rho}(>L) = \sum_{L_i>L} L_i\, \hat{\phi}_i,$$

and the results are shown in Fig. 3b.

In both cases the message is the same, although the cumulative estimator appears (perhaps misleadingly) less noisy. While the two low-z slices are very similar, it is clear that the characteristic luminosity is higher in the $0.4 < z < 0.8$ slice, by at least 0.5 mag. It also seems as though the overall luminosity density is very nearly constant.

We now quantify these visual impressions by model fitting. It is convenient to describe the galaxy luminosity function via a Schechter function fit at each redshift

$$d\phi = 0.921\,\phi^*(L/L^*)^{\alpha+1}\exp[-L/L^*]\,dM.$$

The optimal way of fitting such models to moderate discrete datasets such as ours is to use maximum likelihood. In the absence of clustering, one would define likelihood by

$$\mathcal{L} = \prod_i \frac{d^2p}{dM\,dz}(M_i, z_i),$$

and extra constraints such as operating over a redshift band can be applied by restricting the product to the relevant objects and normalizing the model probability distribution to the required region of (M, z) space.

The presence of clustering renders the vertical normalization of the luminosity function uncertain. We can allow for this by working in an infinitesimal redshift band, since only the probability distribution for M at given z is involved and amplitude scalings normalize away:

$$\mathcal{L} = \prod_i \frac{dp}{dM}(M_i \mid z_i).$$

This expression can be immediately generalized to a finite redshift range by continuing to use the conditional probability of M at given z – but this must now be normalized individually for each z_i of interest.

This method gives a value for the characteristic luminosity in a redshift band, $L^*(z)$; the normalization $\phi^*(z)$ can then be determined from the overall numbers of objects (although it is still subject to clustering fluctuations). The errors quoted below assume that luminosity density can be measured exactly, so that the fractional error on ϕ^* is the same as that on L^*. The results of the analysis are given in the following table, assuming $\Omega = 1$ and a Schechter-function slope of $\alpha = -1$ (letting this float yielded a best-fitting value of $\alpha = -1.04 \pm 0.31$).

These numbers confirm earlier visual impressions. There indeed appears to be some evidence for luminosity evolution in the sense that M_K^* was brighter in the past. The no-evolution hypothesis is ruled out at about the 4 per cent significance level, considering the variation of M_K^* alone. On the other hand, there is no evidence for evolution for $z < 0.6$. Furthermore, there is evidence that the overall normalization of the luminosity function is a declining function of redshift.

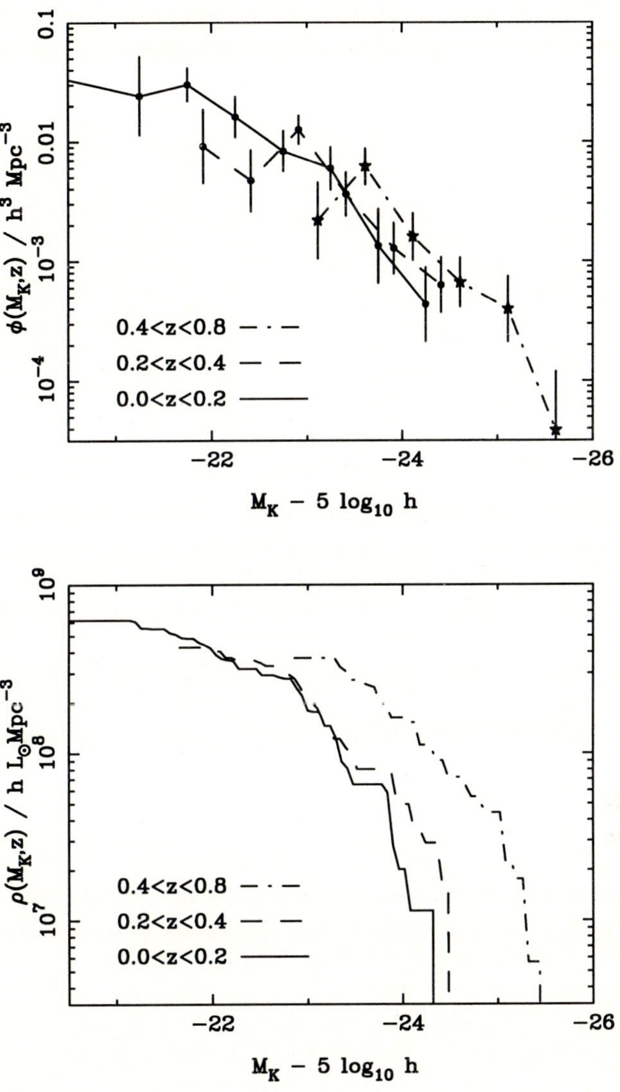

Fig. 3. The luminosity function results, expressed in two ways: (a) the binned luminosity function; (b) the cumulative luminosity density, assuming $M_K(\odot) = 3.4$. There is little evidence for evolution out to $z = 0.4$. At $0.4 < z < 0.8$, however, the characteristic luminosity is 0.5 – 1 mag. brighter, with some suggestion that the characteristic density has declined.

z	$M^*_K(z)$	$\phi^*(z)/h^3\,\mathrm{Mpc}^{-3}$
0.0 – 0.2	-22.72 ± 0.23	0.029 ± 0.007
0.2 – 0.4	-22.85 ± 0.17	0.020 ± 0.003
0.4 – 0.6	-23.23 ± 0.23	0.013 ± 0.003
0.6 – 0.8	-23.68 ± 0.30	0.009 ± 0.002
0.0 – 0.4	-22.75 ± 0.13	0.026 ± 0.003
0.4 – 0.8	-23.41 ± 0.24	0.011 ± 0.003
0.0 – 0.8	-23.01 ± 0.11	0.019 ± 0.002

3.4 Comparison to other results

These numbers are very different from the results of Mobasher *et al.* (1993), who (for $h = 1$) obtained $M^* = -23.6 \pm 0.3$ and $\phi^* = 0.0046 \pm 0.0011$. How can it be that we have obtained a characteristic luminosity a magnitude fainter and a normalization over 5 times higher? Mobasher *et al.* used isophotal magnitudes, rather than a fixed metric aperture. If we consider their objects with $z \simeq 0.1$ (the most luminous objects which thus dominate the determination of M^*), their median aperture is approximately 40 h^{-1} kpc diameter, which immediately makes their magnitudes 0.3 mag. brighter than ours, if we adhere to the power-law aperture correction. Also, the K-corrections used are different: they adopt $K_K(z) = -0.7z + 3.9z^2$, a much weaker dependence than our $K_K(z) \simeq -2.58z$. At $z = 0.1$, the difference in K-correction is 0.22, so that this plus the aperture difference accounts for 0.52 mag. of the 0.8 mag. difference in M^*; the remaining difference is not statistically significant. As for the difference in ϕ^*, this may well be partly due to density fluctuations, but it is also possible that the Mobasher *et al.* sample is systematically incomplete: since their data were based on infrared measurements of blue-selected galaxies, the population of very red nearby galaxies will not be sampled adequately.

Similarly, our results diverge quite markedly from those of the Hawaii group (Cowie *et al.* 1995). They obtain a total $M^*_K \simeq -23.4$, which is claimed not to evolve, and a normalization which changes approximately as $(1+z)^2$. In their lowest redshift bin, $0 < z < 0.2$, Cowie *et al.* have a extremely low normalisation for their luminosity function, equivalent to $\phi^* \simeq 0.006h^3\,\mathrm{Mpc}^{-3}$. Most workers estimate the local optical ϕ^* to lie in the range $0.01 - 0.03h^3\,\mathrm{Mpc}^{-3}$ (e.g. Loveday *et al.*'s (1992) luminosity function analysis); these values are approximately 2 – 5 times larger.

4 Implications for faint counts

The obvious question now is how the models which fit the data at $K \lesssim 17$ and $z \lesssim 1$ will fare when extrapolated to fainter K magnitudes and higher redshifts. The interesting thing here is how well the no-evolution model fits the data, contrary to previous claims, which were made on the basis of analyses which ignored substantial aperture corrections. Since we only rule out the no-evolution model at a moderate level of significance, this is something that should still be taken seriously. Moving to pure $L^* \propto 1 + z$ luminosity evolution with the same local normalization now significantly exceeds the faint counts. However, conserving luminosity density by scaling $\phi^* \propto (1+z)^{-1}$ at high redshifts restores the good fit at most magnitudes. The predicted numbers are too small at $K > 21$, but the typical luminosities of galaxies at that level are fainter than we have been able to probe in our luminosity function determination.

One simple possibility is that the luminosity function has an extra dwarf component which makes it steeper at the faint end; several authors have argued for such a component both in cluster luminosity functions (e.g. Driver *et al.* 1994) and in the field (e.g. Gronwall & Koo 1995). Driver et al. obtain the following parameters for the dwarf luminosity function: $M^*_{\rm dwarf} = M^*_{\rm normal} + 3.5$; $\phi^*_{\rm dwarf} = 2\phi^*_{\rm normal}$; $\alpha_{\rm dwarf} = -1.8$. Adding such a (non-evolving) low-luminosity contribution to our antimerging normal luminosity function fits the faint counts with no adjustment of parameters.

Can these possibilities be constrained by fainter redshift data? Cowie *et al.* (1994) have described redshift surveys as faint as $K(6'') = 20$, and shown that the median redshift continues to follow their no-evolution prediction down to $K(6'') = 18 - 19$ ($z_{\rm med} = 0.65 \pm 0.15$), but to diverge at $K(6'') = 19 - 20$ ($z_{\rm med} = 0.5 \pm 0.2$). Our predictions for these median redshifts are 0.70 & 0.97 respectively for the no-evolution model, 0.87 & 1.17 respectively for the luminosity/density evolution model, and 0.84 & 1.02 respectively for the luminosity/density evolution model with extra dwarfs. All these models disagree with Cowie *et al.*'s faint K = 19–20 bin, but more faint redshift data are really required to constrain this portion of the luminosity function adequately.

5 Conclusions

How do we relate these results to optical studies of the galaxy population? We are unable to say much about the 'faint blue galaxies' that dominate the faint optical counts: given their colours, we would expect them to have $M_K \sim -21.5$, and at $K < 17.25$ we would not be able to see them beyond $z > 0.15$. Rather they would not manifest themselves until K = 19–20, and so this suggests an obvious connection between the faint excess in the blue counts and the possible dwarf population discussed above. Such an idea is given further support by the results of Glazebrook *et al.* (1995), based on HST imaging of random galaxy fields. They find counts of morphologically normal ellipticals and spirals to be much as expected from no-evolution models, whereas the faint excess arises from steep number counts in the irregular population.

Nevertheless, because we favour a model in which the normalization is high, this does have implications for evolution in the optical. Various workers (e.g. Metcalfe *et al.* 1991, 1995) have argued for a high normalization, based on the good fit of a no-evolution model around $b_{\text{ßJ}} = 19$. This would then imply that the bright counts are incomplete, by a factor of 2 at $b_{\text{ßJ}} = 17$, favouring a prosaic explanation such as that of McGaugh (1994). In this sense, the implication of our result is that the faint blue galaxies may be less dominant than often supposed, and less in need of radical explanations.

References

Aaronson, M. 1978, *Astrophys. J. Lett.* **221**, L103

Bruzual, A.G., Charlot, S. 1993, *Astrophys. J.* **405**, 538

Cowie, L.L., Songaila, A. 1993, Proc 8th IAP meeting, eds B. Rocca-Volmerange *et al.* , Editions Frontieres, Paris, p.147

Cowie, L.L., Gardner, J.P., Hu, E.M., Songaila, A., Hodapp, K.W., Wainscoat R.J., 1994, *Astrophys. J.* **434**, 114

Cowie, L.L., Songaila, A., Hu, E.M. 1995, *Astrophys. J.*, in press

Driver, S.P., Phillipps, S., Davies, J.I., Morgan, I., Disney, M.J. 1994, *Mon. Not. R. Astr. Soc.* **268**, 404

Ellis, R.S. 1990, in *Evolution of the Universe of Galaxies*, Kron R. G., ed., A.S.P. Conference Series, Vol. **10**, p.248

Felten, J.E. 1976, it Astrophys. J. **207**, 700

Gardner, J.P. 1992, *Ph.D thesis*, University of Hawaii,

Gardner, J.P., Cowie, L.L., Wainscoat, R.J. 1993, *Astrophys. J. Lett.* **415**, L9

Glazebrook, K., Peacock, J.A., Collins, C.A., Miller, L. 1994, *Mon. Not. R. Astr. Soc.* **266**, 65

Glazebrook, K., *et al.* 1995, *Mon. Not. R. Astr. Soc.* **275**, L19

Gronwall, C., Koo, D.C. 1995, *Astrophys. J. Lett.* **440**, L1

Gunn, J.E., Oke, J.B. 1975, *Astrophys. J.* **195**, 255

Jenkins C. R., Reid I. N., 1991, *Astron. J.* **101**, 1595

Koo, D.C., Gronwall, C., Bruzual, G.A. 1993, *Astrophys. J. Lett.* **415**, L21

Loveday, J., Peterson, B.A., Efstathiou, G., Maddox, S.J. 1992, *Astrophys. J.* **390**, 338

McGaugh, S.S. 1994, *Nature* **367**, 538

Metcalfe, N., Shanks, T., Fong, R., Jones, L.R. 1991, *Mon. Not. R. Astr. Soc.* **249**, 498

Metcalfe, N., Shanks, T., Fong, R., Roche, N. 1995, *Mon. Not. R. Astr. Soc.* **273**, 257

Mobasher, B., Ellis, R.S., Sharples, R.M. 1993, *Mon. Not. R. Astr. Soc.* **263**, 560.

Peletier, R.F., Valentijn, E.A., Jameson, R.F. 1990, *Astron. Astrophys.* **233**, 62 it Astrophys. J. Suppl. **94**, 461

The Tully-Fisher Relation at Intermediate Redshifts

Matthew A. Bershady[1,2]

[1] Penn State University, State College PA 16802, USA
[2] Hubble Fellow

1 Introduction

Motivated by recent pioneering measurements of galaxy kinematics at intermediate redshifts (Franx 1993, Vogt *et al.* 1993), we have begun a pilot survey to measure rotation curves for blue, star-forming galaxies between $0.05<z<0.35$. The scientific impetus for such measurements is to construct internal velocity - luminosity relations at redshifts substantial enough to make cosmological and evolutionary tests (Kron 1986, van der Kruit & Pickles 1988). Departures from a fiducial relation as a function of redshift are sensitive to both space-time curvature and the evolution of the mass-to-light ratio in galaxies. Because of this ambiguity – curvature vs. evolution – kinematics should be viewed as a new cosmological tool at intermediate redshifts to be combined with additional measurements. In particular, the apparent internal velocities, colors, surface-brightness, and image shape are all unaffected by curvature, so that this ensemble of observables can be used unambiguously to explore galaxy evolution.

Here we discuss a well-defined method of target selection which assures efficient measurement of spatially and spectrally resolved kinematics, namely rotation curves. Spatial resolution is critical since it is difficult to use integrated line-widths to distinguish between, for example, a low-mass star-forming galaxy and a high-mass galaxy with a central star-burst. For 18 of 19 appropriately selected galaxies to $B<20.5$ from Bershady *et al.* (1994), we have successfully measured rotation curves using the KAST spectrograph on the Lick 3m telescope with 1-2 hour integrations for each target. As a sanity check, we present preliminary results for half of our sample: We measure H_0 to be $\sim$ 75 km s^{-1} Mpc^{-1} at a median redshift of 0.15.

2 Target Selection

There are three desiderata for selecting galaxies for rotation-curve surveys at substantial redshift: (i) A similar range of galaxies should be chosen at disparate redshifts on the basis of some objective criteria. (ii) Targets should be optimized for the available instrumental resolution. (iii) Targets should be efficient to observe. There are several well-known correlations for galaxies that when

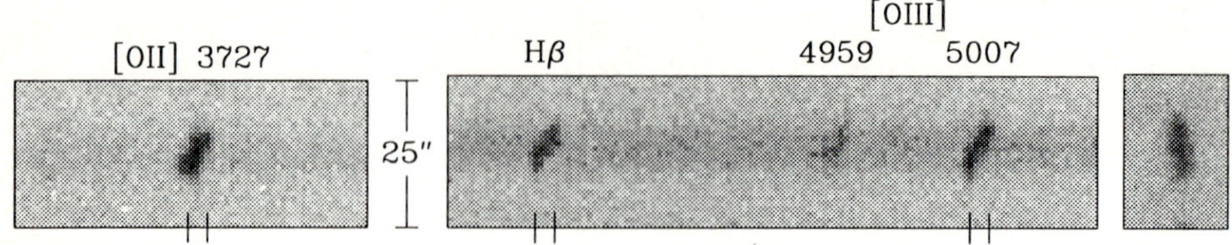

Fig. 1. An example of an observed rotation-curve spectrum for one of our higher-redshift targets at z=0.3 (sa68.5154 from Bershady *et al.* 1994 and Bershady 1995) using the KAST spectrograph and Lick 3m telescope. We derive a maximum rotation velocity of 405±20 km/s, peak-to-peak from [O II], Hβ, and [O III] 5007. [O III] 4959 suffers from being partly on a sky-line. [O II] appears fuzzy because we have not resolved this doublet. A gound-based B band image of the galaxy (to scale), taken in ~ 1" FWHM seeing at the KPNO 4m, is at the far right.

combined, point to a well-defined galaxy type. The correlations are between size and luminosity, color and emission-line strength, color and luminosity, and luminosity and internal velocity. These conspire to make the ideal rotation-curve targets blue, luminous galaxies. How luminous depends on the spatial and spectral resolution of your spectrograph and telescope, and the redshift limit of your survey. How blue depends on your desired efficiency. For our sample, we were able to measure rotation curves one at a time on a 3m telescope at a pace that outstriped the multiplexing advantage of observing red galaxies in clusters at comparable redshifts using a bigger telescope.[1] At the same time, we could spectrally and spatially resolve rotation curves to z=0.3 (see Fig. 1) with 1-2" seeing and R~1500-2500 in the red.

The ideal galaxy for intermediate and high redshift rotation curve studies has spectral type "bm" in the nomenclature of Bershady (1995). This is comparable to luminous "Sc" galaxies. However for the purposes of selecting these galaxies for surveys on 4m-class telescopes, it is much more fruitful to think in terms of spectral types. Depending on how restrictive one makes the above selection, the surface-density of luminous, blue galaxies to B=20.5 is low. Without substantially increasing telescope aperture as well as spatial and spectral resolution, fainter limits will not produce substantially greater surface-densities of viable targets: Fainter galaxies will either be at higher redshift (and apparently smaller), or lower in luminosity (and internal velocity and apparent size) at comparable redshifts. This makes selection via Hubble type rather inefficient since this requires Hubble Space Telescope (HST) images which cover little area. In contrast, ground-based, multi-band imaging can provide photometric redshifts (Connolly *et al.* 1995) and hence spectral classification and luminosities over large areas, ideal for selecting targets for spectroscopic follow-up and pointed HST imaging (when needed). With adaptive optics and/or a 10m-class telescope, target selection strategy can be altered. Here one is optimally exploring a different redshift and/or luminosity regime than should be pursued with a conventional 4m-class telescope.

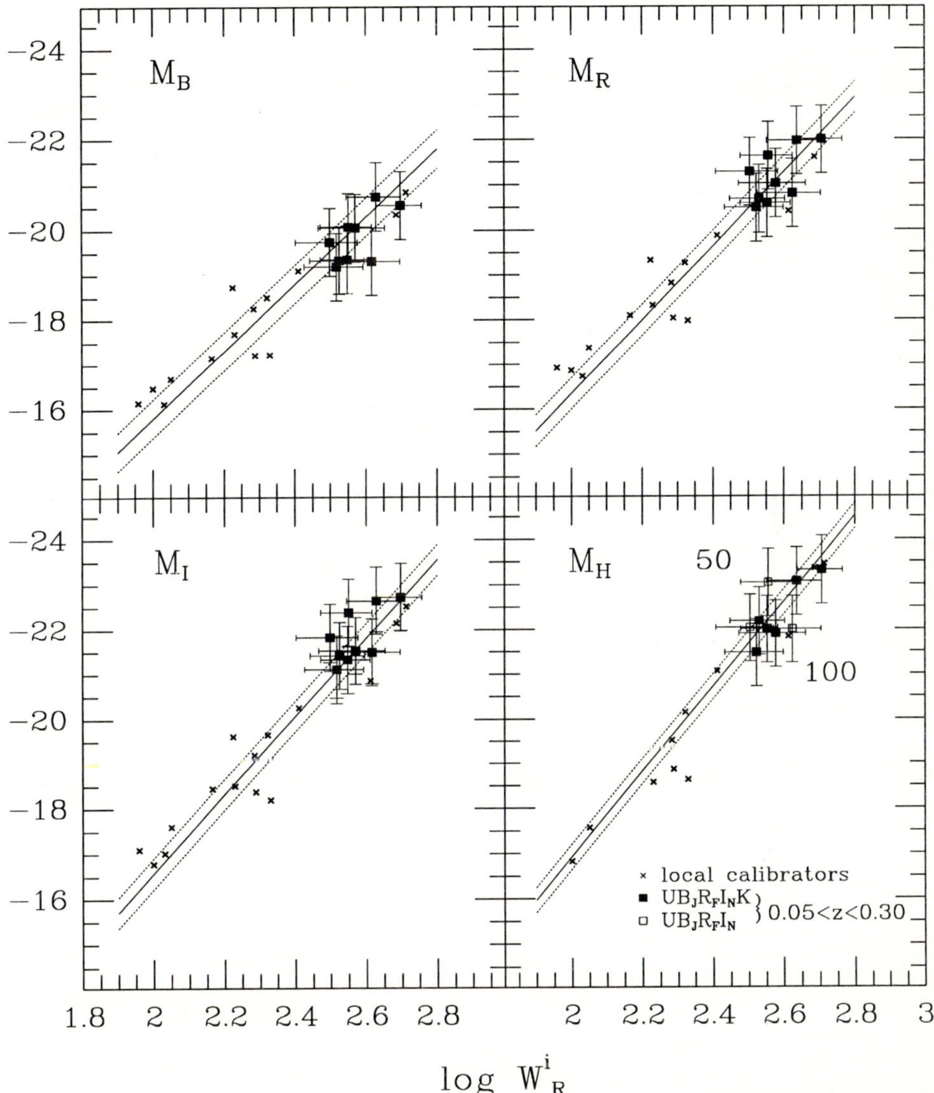

Fig. 2. The Tully-Fisher relation as derived for half our current sample (fully processed). Local calibrating galaxies from [9], with their regression slopes and dispersions are plotted for comparison. Transformation from rotation speed to 21-cm line-widths for our data use formulae in Courteau (1992). "Error-bars" in the y-dimension indicate not photometric uncertainties but different assumed values of H_0, as indicated ($q_0=0.1$). Galaxies in our sample represented by filled boxes have 5-band photometry including the K band; open boxes represent (two) galaxies without K band photometry. Absolute magnitudes in the B, R, I, and H bands are determined empirically by interpolating between observed bands via stellar synthesis models fit directly to the data (Bershady 1995).

3 A Sanity Check

To see whether it is possible to make sensible kinematic measurements at intermediate redshifts, we have transformed emission-line maximum rotation velocities, measured now for half our sample, into 21-cm line-widths using transformations from Courteau (1992). In Fig. 2 we compare our sample to the local calibrators of the Tully-Fisher relation from Pierce & Tully (1992). The errors-bars in the x-axis (velocity) are indeed estimates of measurement errors (rotation velocity, inclination, and transformation uncertainties). However, in the y-axis, the photometric uncertainties are negligible (<10%, since we have deep, multi-band photometry from which we empirically derive κ-corrections). Instead, these error bars represent a range of *assumed* values of H_0 between 50 and 100 km s^{-1} Mpc^{-1}. Relative to the local calibrators, we find average offsets indicating values between 60 and 90 km s^{-1} Mpc^{-1}, depending on band. The median redshift of this sample is 0.15. This result is preliminary since we haven't, for example, corrected for internal extinction (although the correction is negligible in the H band), or corrected for more subtle differences between our photometric system and that used in Pierce & Tully (1992). Furthermore, inclination corrections have been made on the basis of ground-based images and should be checked with higher-resolution data.

Future prospects should be even more promising. The measurement we are actually interested in making is *differential*, and not absolute. That is, we want to study the deviations from, or changes in scatter about *some* fiducial correlation of internal velocity and luminosity as a function of redshift. Instead of using integrated line-widths or transformations to another system imposed by observations of local galaxies, we can instead devise a new standard measurement of internal velocity (and luminosity) that is optimal for intermediate to high redshift studies. One possibility along these lines has already been suggested (Chiba & Yoshii 1995).

The work presented here involves a collaboration with D. Koo and C. Mihos (Lick Observatory). MAB was funded by NASA through grant HF-1028.02-92A from STScI (operated by AURA, Inc. under contract NAS5-26555).

References

Bershady, M.A., Hereld, M., Kron, R.G., Koo, D.C., Munn, J.A., Majewski, S.R. 1994, *Astron. J.*, **108**, 870
Bershady, M A. 1995, *Astron.J.*, **109**, 87
Chiba, M., & Yoshii, Y. 1995, *Astrophys. J.* **442**, 82
Connolly, A., Csabai, I., Szalay, A., Koo, D., Kron, R., Munn, J. 1995, *Astron. J.*, submitted
Courteau, S. 1992, *Ph.D. thesis*, UC Santa Cruz
Franx, M. 1993, *Proc. Astr. Soc. Pac.* **105**, 1058
Kron, R.G. 1986, in *Nearly Normal Galaxies*, ed. Faber, Springer-Verlag, p.300

Pierce, M.J., Tully, R.B. 1992, *Astrophys. J.* **387**, 47
van der Kruit, P.C., Pickles 1988, in *Towards Understanding Galaxies at Large Redshift*, eds. Kron & Renzini, Kluwer Dordrecht, p.339
Vogt, N.P., Herter, T., Haynes, M.P., Courteau, S. 1993, *Astrophys. J. Lett.*, **415**, L95

A Cluster of Galaxies at z = 1.206

Mark Dickinson

Space Telescope Science Institute, 3700 San Martin Dr., Baltimore MD 21218, USA

1 Introduction – Finding Very Distant Clusters

Although only a small fraction of all galaxies are located in rich clusters, substantial effort has gone into finding and studying clusters of galaxies at high redshift over the years. This is in part simply because we can *find* distant clusters – their enhanced surface density on the sky is recognizable in deep images out to quite high redshifts. The traditional approach to finding distant clusters begins with wide–field imaging (historically photographic), identification of candidates as enhancements in the local galaxy density, and then painstaking spectroscopic follow–up to measure redshifts and confirm the reality of the cluster. The most distant clusters from such surveys (e.g. that of Gunn, Hoessel & Oke 1986) extend to $z \approx 0.9$, providing an attractively long redshift baseline for evolutionary investigations. Only recently have deep spectroscopic studies of field galaxies provided significant numbers of non–cluster objects at similar look–back times (cf. Lilly *et al.* 1995). Once a redshift is secured for a few galaxies in a distant cluster, one can assume, at least for statistical purposes, that many others are present at the same distance. A cluster then presents a volume limited sample of galaxies collected together in one convenient place for investigation.

As the most massive collapsed structures in the universe, rich clusters have particular cosmological importance, since they represent a critical scale in the spectrum of mass fluctuations. Their space densities, dynamical properties, and redshift evolution all are highly sensitive to fundamental cosmological parameters, the nature of dark matter, and the details of structure formation in the universe. X–ray selected samples such as the EMSS (cf. Gioia & Luppino 1994) are particularly valuable for such purposes, but again existing x–ray catalogs reach their limit at $z \approx 0.8$. Note that x–ray selection is essentially a variant on the "traditional approach" of cluster hunting described above – the imaging portion is simply carried out at a rather different wavelength. X–ray selection has the tremendous virtue of having an understandable physical basis, however: hot gas marks the presence of a deep gravitational potential.

Clusters have also been avidly studied as laboratories for galaxy evolution. They were arguably the first place where evolution was observed directly, in the pioneering work of Butcher & Oemler (1978, 1984), who noted an increasing fraction of blue galaxies in distant clusters. By contrast, the cores of nearby rich clusters are dominated by red E/S0 galaxies. Their histories too are important, since they probably represent the oldest galactic stellar populations at any epoch. Any measure of spectral evolution in the elliptical population may thus point back to the earliest epoch of galaxy formation.

Only a handful of clusters have spectroscopically confirmed redshifts greater than 0.6, leading most investigators to study the same few systems. It is likely that the optically selected samples, at least, are highly incomplete beyond that redshift. This leads to worries about whether those clusters which *do* enter such samples are representative, or are perhaps unusual in some way that enhances their visibility but which may skew our impressions of galaxy evolution within them. E.g. at $z \approx 1$, optical surveys image galaxies and clusters at blue/UV rest frame wavelengths, and may thus be more likely to select clusters where a substantial fraction of the galaxies are forming stars (i.e. clusters with a high "'blue fraction"). Indeed, clusters dominated by red, early–type galaxies may simply "disappear" from optical images at $z > 1$ because of the very strong k–corrections affecting ellipticals and S0s. Selection in the near infrared would alleviate this concern, but the limitations of current array technology prohibit a general survey for clusters to deep enough limits over a wide enough solid angle.

How then may we go about looking for clusters at $z > 1$? Peter Eisenhardt and I have chosen a targeted approach, using powerful radio galaxies as likely sites to search for distant clusters. It has been known for some time that radio galaxies and radio loud quasars at $z \approx 0.5$ often inhabit rich environments (cf. Yee & Green 1984; Ellingson, Yee and Green 1991; Yates *et al.* 1989; Hill & Lilly 1991; Dickinson 1994). We may hope that this holds true at higher redshifts still, and if so then take advantage of the large number of radio galaxies presently known at $z > 1$ to provide targets for a cluster hunt. The radio galaxy also provides an *a priori* likely redshift for any cluster candidate which is found.

Using the 4m telescope at KPNO, we have been obtaining deep, wide–field infrared and optical images (R, J and K–bands) of the fields around radio galaxies with $0.8 < z < 1.4$. Cluster candidates are selected on the basis of enhanced galaxy surface density in the K–band, and/or by unusual galaxy color distributions. In particular, we look for galaxies with very red optical–IR colors, a likely signature of early–type galaxies at high redshift.

2 The Environment of 3C 324

Figure 1 shows R and K–band images of the field around 3C 324, a radio galaxy at $z = 1.206$, and the site of one of the best cluster candidates found thus far in our survey. The cluster is hardly visible in the R image, whereas the enhancement is quite evident at K. Figure 2 quantifies this impression, showing the radial galaxy overdensity profiles at R and K (normalized to the mean field background level) averaged in annuli around the radio galaxy. While the cluster contrast barely exceeds a factor of 2 in the R–band, even at its center, in the K–band it reaches a peak with $6\times$ overdensity.

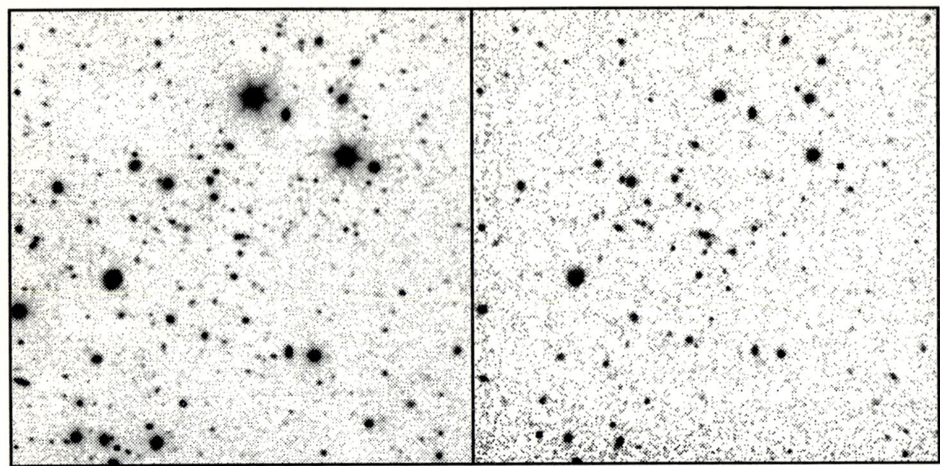

Fig. 1. R (left) and K (right) images of the field around 3C 324, a radio galaxy at z=1.206. The field of view is 2.5 arcmin on a side, corresponding to ~1.5 Mpc (H_0=50 km/s/Mpc, q_0=0.2) at the radio galaxy redshift. The radio galaxy is at field center. The cluster is nearly unrecognizable in the optical image, but becomes visible in the infrared due to a large number of extremely red galaxies grouped around the radio source.

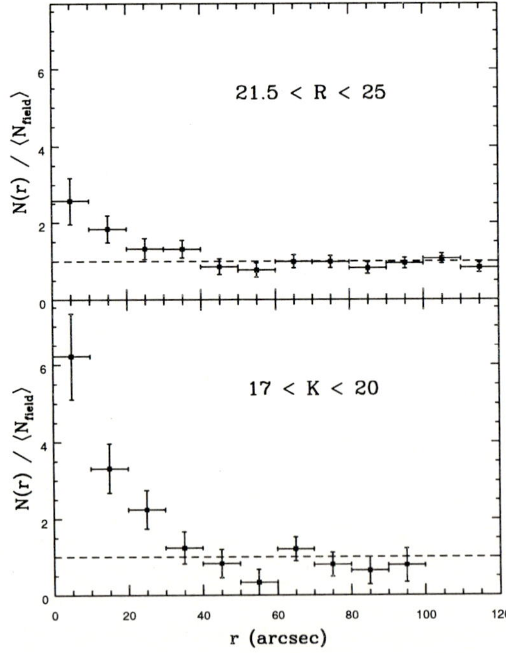

Fig. 2. Normalized galaxy overdensity in the optical and near infrared, averaged in annuli around 3C 324. The cluster contrast is much stronger in the K-band.

This difference in cluster visibility from R to K is explained by figure 3, which shows the distribution of galaxy colors in this field. A prominent spike is evident at $R - K \approx 6$. Objects this red *are* found in field galaxy surveys, but they are uncommon (cf. McLeod *et al.* 1995; Hu & Ridgway 1994). The spike in figure 3 represents a factor of $\gtrsim 10$ enhancement over the number of such red galaxies typically found in a field of this size.

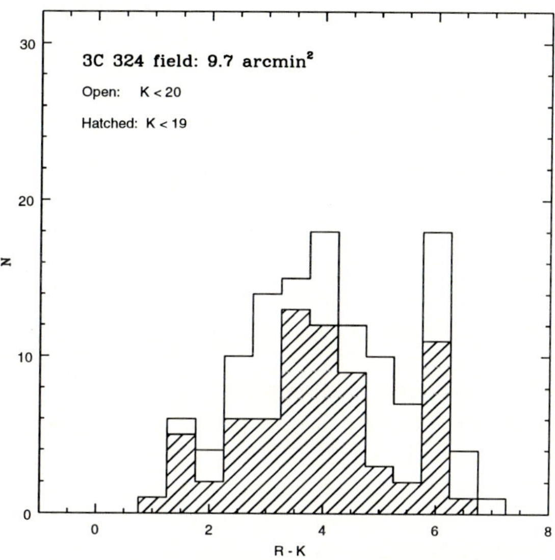

Fig. 3. Histogram of $R-K$ colors for galaxies around 3C 324. The prominent spike at $R-K \approx 6$ consists of faint galaxies with E/S0 morphologies in our WFPC2 images.

3 HST Observations of the 3C 324 Field

In *HST* Cycle 4 we used the WFPC2 to image the field around 3C 324. In order to reach the faint surface brightness limits required to perform quantitative analysis on the faint galaxies of interest, we exposed for 18 hours through the F702W filter ($\lambda_{\rm eff} \approx 6900$Å), The faintest objects detected in these data have $R > 29$, and the surface density of galaxies exceeds 1 million/deg^2. Among this multitude of faint galaxies, most are certainly not part of the cluster (cf. figure 3, recalling that *WFPC2* is an optical camera, not infrared). Without spectroscopy,[1] we can only consider the properties of the galaxies in broad terms, e.g. by examining their morphologies as a function of color (what are the very

[1] Spectroscopy obtained with the Keck telescope after this conference was held has confirmed the presence of galaxies in this field which share the redshift of the radio galaxy, including several of the red ellipticals discussed below.

red galaxies in figure 3 ?), or considering the statistics of their number density and morphological properties.

Figure 4 presents a montage of faint objects drawn from the *HST* image. This is not a statistical sample in any way, but is selected to span the range of morphologies present.

With few exceptions, the very red galaxies (those objects with $(R-K) \approx 6$) have simple spheroidal morphologies. By all indications, they are remarkably ordinary E/S0 galaxies. Their light profiles are generally well represented by $r^{1/4}$–laws, and their sizes and surface brightnesses are consistent with those expected of elliptical galaxies at z=1.2 with only mild passive evolution (Dickinson 1995). These characteristics, combined with the red colors, lead us to believe that these are indeed mature early–type galaxies in a high redshift cluster.

The bluer galaxies exhibit a bewildering variety of morphologies. At faint magnitudes, few galaxies have the morphologies of normal "Hubble Atlas" spirals. Many are irregular, clumpy, or amorphous, and there are a number of "cometary" objects with head–tail morphologies. Many elongated objects with large axial ratios are present.

Several cautions are in order before we leap to the conclusion that cluster galaxies (excepting the ellipticals) have evolved dramatically. The first, already noted, is that many or even *most* of the galaxies in figure 4 are probably not part of the $z = 1.2$ cluster, but are rather interloping field galaxies (foreground and background). At somewhat brighter magnitudes than those considered here, the *HST* Medium Deep Survey has provided statistics for field galaxies subdivided by morphological type (e.g. as presented by Windhorst at this meeting; also Driver *et al.* 1995 and Glazebrook *et al.* 1995). Irregular and peculiar galaxies have very steep $N(m)$ slopes suggesting rapid evolution, while ordinary Hubble sequence galaxies appear to follow the "no–evolution" predictions (albeit requiring a relatively high normalization for their luminosity function). Extrapolating this to fainter magnitudes, we would expect the irregular/peculiar population to dominate very deep images. Moreover, we must be cautious about how even the most ordinary disk galaxies might appear when seen at high redshift, as their rest–frame ultraviolet light shifts into the WFPC2 bandpasses. UV morphologies are probably dominated by the distribution of recent star formation. We know too little about the local universe as it would appear if seen in the near–UV, but can expect that the prominence of spiral arms would be exaggerated relative to bulges and smooth, old disk components. Except for the "cleanest" grand design spirals, the resulting morphology is likely to be more irregular.

4 Red Elliptical Galaxies at z =1.206

The existence of red elliptical galaxies at $z = 1.2$ has immediate cosmological implications. With *HST* morphologies in hand, we may reasonably assume that the red colors of these galaxies primarily reflect their stellar populations, rather than (for example) extinction by dust. Red color is generally associated with old stellar populations. How old might these galaxies be? Age–dating a galaxy from

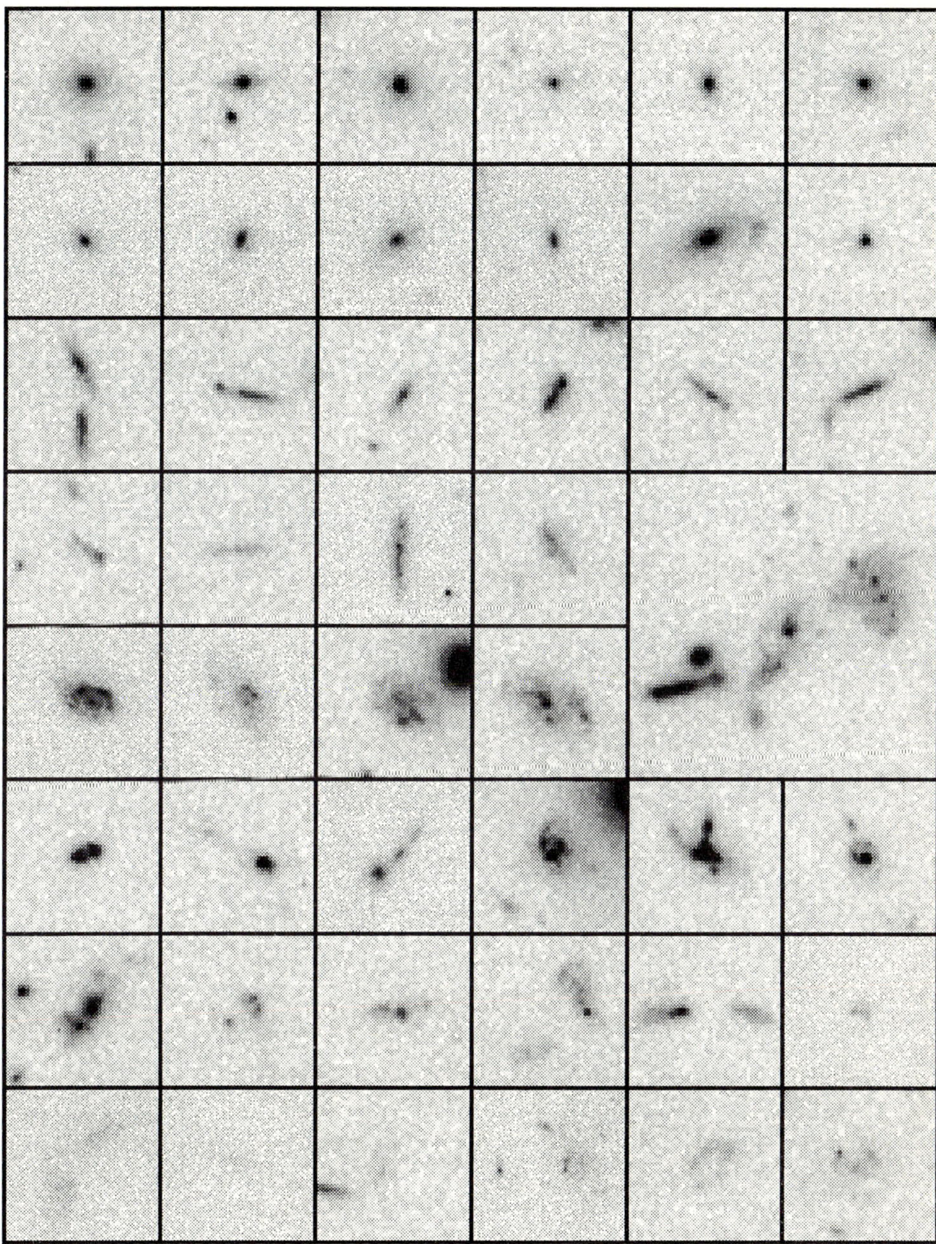

Fig. 4. Montage of galaxies from a deep *HST* image of the 3C 324 field. The first two rows show examples of the very red galaxies from figure 2 which show simple E/S0 morphologies. These are most likely to be members of the cluster around 3C 324. The remaining galaxies demonstrate the range of morphologies visible in the *HST* image. While these are largely selected from the central ~1 arcmin around the radio galaxy, the galaxy overdensity is such that only a fraction of these are probably cluster members.

integrated colors alone is indefinite, since color necessarily reflects a variety of parameters including age, metallicity, the star formation history, and the form of the stellar initial mass function. As an illustration, however, let us assume a very "traditional" model for elliptical galaxies (cf. Bruzual & Charlot 1993), namely a single–burst episode of star formation in a solar metallicity stellar population followed thereafter by simple passive evolution. For a galaxy observed at z=1.206, figure 5 shows the predicted $R - K$ color as a function of galaxy age for various star formation histories. The single–burst model reddens the most quickly, reaching $R - K = 5$ one Gyr after the termination of star formation. After 3.3 Gyr, the model matches the observed mean color of the 3C 324 ellipticals ($\langle R - K \rangle = 5.9$). While it is important to remember that other models, especially with higher metallicities, would yield younger ages, the ~3 Gyr value can be considered as a fiducial. Evidently, the bulk of star formation in the 3C 324 ellipticals ceased several Gyr before $z = 1.2$. Similar conclusions were reached from a sample of $z < 0.9$ cluster galaxies studied by Aragón–Salamanca *et al.* (1993).

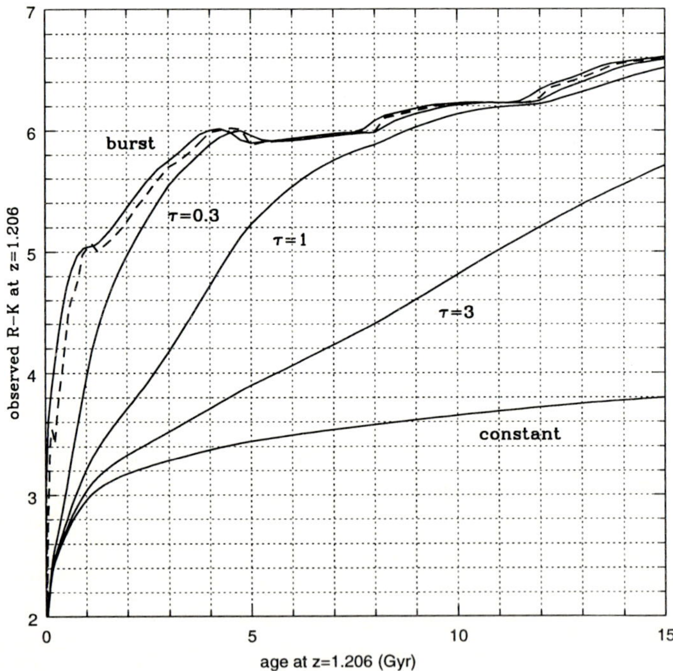

Fig. 5. $R - K$ colors (as observed at $z = 1.206$) for synthetic galaxies with various star formation histories, generated with the population synthesis models of Bruzual & Charlot (1993). Models with exponentially declining declining star formation rates are labeled by their e–folding time τ in Gyr. The single burst models redden most quickly, and reach $R - K = 5.9$ after ~3.3 Gyr.

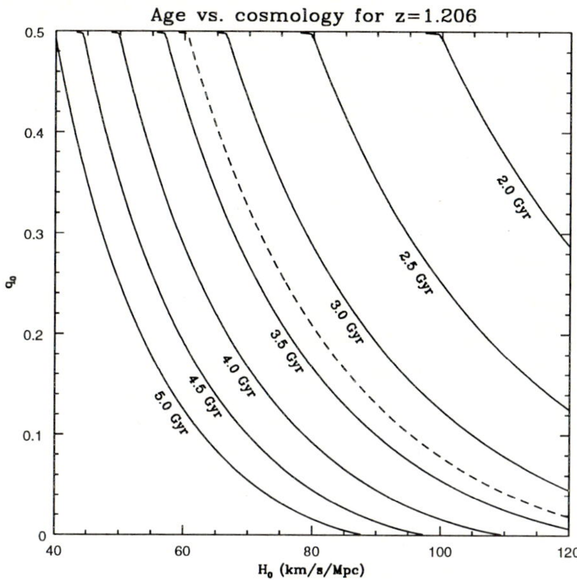

Fig. 6. Loci of constant universal age at $z = 1.206$ vs. the fundamental cosmological parameters in a $\Lambda = 0$ universe. Tracks mark acceptable combinations of q_0 and H_0 which produce the labeled age. The fiducial 3.3 Gyr is marked by a dashed line.

At $z = 1.206$, the Universe must presumably have been old enough to accommodate as its oldest galaxies. Figure 6 shows the age of the universe at this redshift for various combinations of H_0 and q_0. If an age of 3.3 Gyr age is adopted, then closed cosmologies require low values of the Hubble parameter, while $H_0 = 80\,\mathrm{km\,s^{-1}\,Mpc^{-1}}$ universes must be open (or dominated by a cosmological constant). This is a familiar situation: the same conclusions have been reached from considering the oldest stars in our own galaxy and its globular clusters. Globular cluster ages are presumably more reliable (or at least better constrained) than are those of faint galaxies estimated from a single color. What is striking here is not the actual values of H_0 and q_0 which are allowable (since these require considerable faith in the choice of population synthesis models), but rather the persistence of the age problem from nearby stellar clusters out to galaxies at $z = 1.2$. It seems that we are faced either with a genuine cosmological constraint, or with the need to adjust stellar evolutionary models in order to rescale ages at all redshifts.

5 Conclusions

This paper has presented early results from analyses of a single galaxy cluster at high redshift. Further progress depends vitally on larger samples spanning as wide a range of cosmic time as possible, and particularly on spectroscopy to confirm redshifts for individual galaxies in these fields and to study their properties. Out to $z = 1.2$ at least, optical spectra taken with large telescopes can measure indices useful for studying the star formation histories of these galaxies, such as emission line strengths, 4000Å break amplitudes, near–UV spectral energy distributions, and potentially stellar absorption line strengths as well. We may hope to extend such work to still higher redshifts, finding clusters (if they exist) using radio sources and other selection techniques. Gradually, the universe of galaxies at $z > 1$, previously known only via exotica such as quasars and radio galaxies, can be populated with more "normal" objects, providing samples with which to explore the evolution of galaxies in the young Universe.

Acknowledgements: I would like to thank Peter Eisenhardt and Adam Stanford for ongoing collaboration and conversation concerning high redshift clusters, and Stephane Charlot for advice and consent regarding the population synthesis models (and their limitations). I also thank my other collaborators in the *HST* imaging and Keck spectroscopy programs: Hy Spinrad, Arjun Dey and George Djorgovski. Finally, I am especially grateful to Hans Hippelein for his considerable patience in waiting for this manuscript!

References

Aragón–Salamanca, A., Ellis, R.S., Couch, W.J., and Carter, D. 1993, *Mon. Not. R. Astr. Soc.* **254**, 601
Bruzual, G., and Charlot, S. 1993, *Astrophys. J.* **405**, 538
Butcher, H.R., and Oemler, A. 1978, *Astrophys. J.* **219**, 18
Butcher, H.R., and Oemler, A. 1984, *Astrophys. J.* **285**, 426
Dickinson, M. 1994, *Ph.D. thesis*, University of California, Berkeley
Dickinson, M. 1995, in *Fresh Views on Elliptical Galaxies,* ASP Conference Series, eds. A. Buzzoni, A. Renzini & A. Serrano
Driver, S.P., Windhorst, R.A., Windhorst, A., Ostrander, E.J., and Griffiths, R. 1995, *Astrophys. J. Lett.* **449**, L23
Ellingson, E., Yee, H.K.C., and Green, R.F. 1991, *Astrophys. J.* **371**, 49
Gioia, I., and Luppino, G. 1994, *Astrophys. J. Suppl.* **94**, 583
Glazebrook, K., Ellis, R.S., Santiago, B., and Griffith, R. 1995, *Mon. Not. R. Astr. Soc.* **275**, L19
Gunn, J.E., Hoessel, J.G., and Oke, J.B. 1986, *Astrophys. J.* **306**, 30
Hill, G.J., & Lilly, S.J. 1991, *Astrophys. J.* **367**, 1
Hu, E.M., and Ridgway, S.E. 1994, *Astron. J.* **107**, 1303
Lilly, S.J., Tresse, L., Hammer, F., Crampton, D., Le Fevre, O. 1995, preprint
McLeod, B.A, Bernstein, G.M., Rieke, M.J., Tollestrup, E.V. 1995, *Astrophys. J. Suppl.* **96**, 117
Yates, M., Miller, L., & Peacock, J. 1989, *Mon. Not. R. Astr. Soc.* **240**, 129
Yee, H.K.C., and Green, R.F. 1984, *Astrophys. J.* **280**, 79

Galaxy Populations in Intermediate-Redshift Clusters

Paola Belloni and Hermann-Josef Röser

Max-Planck-Institut für Astronomie
Königstuhl 17, 69117 Heidelberg, Germany

1 Introduction

Clusters of galaxies are ideal laboratories to study the evolution of galaxies. High redshift clusters especially provide the opportunity to look back in time and analyse the galaxy populations at early epochs and compare them with those in clusters like Coma and Virgo in our vicinity. Since the seminal work of Butcher & Oemler (1978a) we know that significant evolution has taken place over the past few giga-years, since rich clusters at redshifts of about 0.5 contain significantly more blue (*i.e.* star-forming) galaxies than *e.g.* the Coma cluster (Butcher and Oemler, 1984). Analysing galaxy populations in clusters over a wide range in redshift will thus be a powerful tool in our study of galaxy evolution in general.

Although the Butcher-Oemler effect was discovered by photometry in two filters only, its verification required time consuming slit spectroscopy. A few years ago we began a project at MPIA in Heidelberg to develop tools to analyse the galaxy populations in high redshift clusters more reliably and in a more efficient way than previously possible. The primary goals have been to

— use photometric imaging methods,
— establish a large number of galaxies as cluster members,
— classify each member galaxy according to its type,
— extend the application to redshifts of order unity.

We will briefly describe below our approach to isolate line-emitting galaxies and — in more detail — first results of a study of the early galaxy type populations in two clusters of redshift of about 0.5, including E+A-galaxies. We finally introduce an attempt just recently initiated to survey for clusters at high redshift.

2 Line-Emitting Galaxies Studied with Fabry-Pérot Imaging

As confirmed by spectroscopic follow-up work, the galaxies responsible for the Butcher-Oemler effect come in three different kinds: (1) AGN, which are relatively rare, (2) galaxies with strong narrow emission lines indicative of active star formation and (3) galaxies with an early type continuum onto which strong Balmer absorption lines are superimposed, now called E+A-galaxies (Dressler & Gunn 1992, DG92) of moderately blue colour. In his PhD thesis work G. Thimm (see Thimm *et al.* 1994) used a Fabry-Pérot-interferometer in the prime focus of the 3.5m-telescope on Calar Alto as a tunable filter with a spectral resolution of about 10 Å to survey a field of 4′ diameter centered on the galaxy cluster around the radio source 3C 295 for galaxies of kinds (1) and (2) by means of the [OII]3727 line. As discussed in detail by Thimm *et al.* (1994), all galaxies with spectroscopically detected emission lines were verified and 17 new ones found. The velocity range surveyed was ±5500 km/sec around the mean cluster redshift of $z = 0.458$.

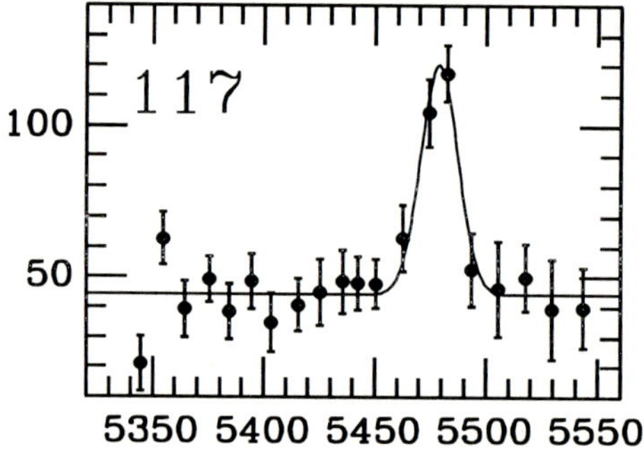

Fig. 1. An example of a newly detected emission line galaxy in the cluster around 3C 295. The spectrum is derived from 20 different settings of the Fabry-Pérot etalon with exposure times of typically 3000 sec each. This line is due to [OIII] at a redshift of 0.470. The galaxy's magnitude is $R = 20.7$ mag and the line has a flux of $(15.8 \pm 3.7) \times 10^{-17}$ erg/cm^2/sec and a FWHM of (1.9 ± 0.4) nm corresponding to an equivalent width of (3.6 ± 0.7) nm (data from Thimm *et al.* , 1994).

From a comparison with published slit spectroscopy of galaxies in the cluster around 3C 295 by Dressler & Gunn (1982, 1983), several advantages of the Fabry-Pérot approach were evident:

1. Fabry-Pérot images detect emission lines much easier than the slit spectra. The detection limit for 3000 sec integration time was 5.6×10^{-17} erg/cm^2/sec for an [OII]3727 line of 1 nm FWHM.
2. No (potentially biased) pre-selection of objects is necessary.
3. A large field is surveyed simultaneously.
4. Combined with narrow-band imaging a reliable classification of the galaxy type is possible (see below).

But for a complete analysis of the population in the cluster also early type galaxies and galaxies of kind (3), to which the Fabry-Pérot method is not sensitive, have to be isolated and their cluster membership verified (the latter was automatically provided by the FPI data with the interferometer tuned to the cluster's redshift). In her PhD thesis work, described in the next section, P. Belloni extended the work by Thimm *et al.* (1994) and developed the tools needed to accomplish this second part of the project.

3 E+A- and Early Type Galaxies from Narrow Band Imaging

For this study the two galaxy clusters Cl 0939+47 ($z = 0.41$) and Cl 0016+161 ($z = 0.54$). have been chosen because of their known high content of active galaxies and in particular of E+A galaxies. The term E+A galaxy (Dressler & Gunn 1983) originates from the fact that their spectra are suggestive of a decayed star-burst in an early-type galaxy: the O – B stars have died out, leaving A and F stars and an underlying old K-giant population. The importance of recognizing the delayed activity shown by E+A galaxies in high redshift galaxy clusters has been discussed *e.g.* by Dressler & Gunn (1990).

Although these two galaxy clusters may be considered among the best studied, only about 30 spectra per cluster are available (Dressler & Gunn 1992), so that any conclusions reached to date about their galaxy populations rely on these few spectra.

3.1 Observations and Data Reduction

All the observations were obtained in November 1991 and 1992 in the prime focus of the 3.5m-telescope on Calar Alto (Spain). The filters used in the observations were selected to examine as accurately as possible the spectral energy distributions of ellipticals and E+A galaxies at the clusters' redshift. Therefore, narrow band filters were selected around the 4000 Å-break, a prominent feature in elliptical galaxies, that allows the redshift to be measured with an accuracy $\sigma_z = 0.010$ (Belloni 1994, Belloni *et al.* 1995). Furthermore, the 4000 Å-break index (hereafter D_{4000}) is a powerful tool for distinguishing a passively evolving normal elliptical from one in which a recent episode of star formation has taken place. Since the analysis of the clusters is performed by means of photometry, the standard application of the D_{4000} index (Spinrad 1986) was not possible. Therefore a similar photometric index $\tilde{D}_{4000}$ has been introduced, which attempts to reproduce Spinrad's one as accurately as possible, using narrow band filters with rest-frame wavelengths around 3850 Å and 4100 Å.

INVENTORY was used to extract a statistically complete sample of galaxies for each cluster field. The two galaxy samples were defined by extracting all objects brighter than 24.5 mag in R above isophotal levels of 25.5 mag/□″ (Cl 0939+472) and 26.0 mag/□″ (Cl 0016+161). To be able to establish the spectral energy distribution for all galaxies in the sample from images taken under varying seeing conditions, the images were analysed following the lines described in Röser & Meisenheimer (1991). This ensures that the counts refer always to the same beam. Relative photometry was established by reference to secondary photometric standard stars in the field (see Belloni 1994) and Belloni *et al.* 1995) for details).

In order to establish the cluster membership and the morphological classification of the galaxies in this sample, the low-resolution spectral energy distributions obtained were fitted with template spectra (Coleman *et al.* 1980). To identify an E+A-galaxy using the same procedure as for the classical Hubble types, a template for its spectral energy distribution is required, which was unavailable in the literature. Moreover, after the burst phase an E+A-galaxy will fade to its passive, original state changing its spectral energy distribution significantly. Therefore such a galaxy cannot reasonably be reproduced by one typical E+A-spectrum but instead requires a sequence of spectra, following its rapid temporal evolution.

3.2 Models of E+A-Galaxies

In modelling E+A-galaxies we made a basic assumption, *i.e.* that they are the result of a strong star formation episode occuring in an early-type galaxy (post-star-burst picture). In building up a post-star-burst galaxy we used the isochrone population synthesis models in Bruzual & Charlot 1993). We assumed that an elliptical formed the bulk of its stars in the first Gyr with a Salpeter initial mass function. At the age of 9.5 Gyr (corresponding to a passively evolved galaxy at the cluster redshift) it experiences a second burst of star formation, which lasts 0.25 Gyr and involves 20% of the initial galaxy's mass. The choice of the free parameters in the models was driven by their ability to reproduce the UV continuum (our bluest filter samples a central rest frame wavelength of 2500 Å) and the Balmer absorption line strengths of the few known post-star-burst galaxies. The grid of post-star-burst spectral energy distributions we used as templates covers a range in age spanning 0.5 to 2.5 Gyr after the second burst. Indeed, no signatures of the burst phase can be detect in galaxies more than 3 Gyr after the burst. Fig. 2 shows the temporal evolution of such models compared to a passively evolving elliptical galaxy (Belloni *et al.* 1995).

3.3 Results

Before entering on the details of the results obtained for the individual clusters, it is worth outlining the efficiency of the narrow-band photometry approach performed here. A systematic analysis of all objects in the cluster field brighter than 22.5 mag resulted in a sample of more than one hundred cluster members per cluster, improving significantly the statistics of any previous work (Tab. 1). Note how E+A-galaxies have significant bluer colours and lower D_{4000} compared to the ellipticals. They are also clearly separated from the spiral population in these properties. Figure 3 shows the quality of the fits for two newly found cluster members, an elliptical and an E+A-galaxy.

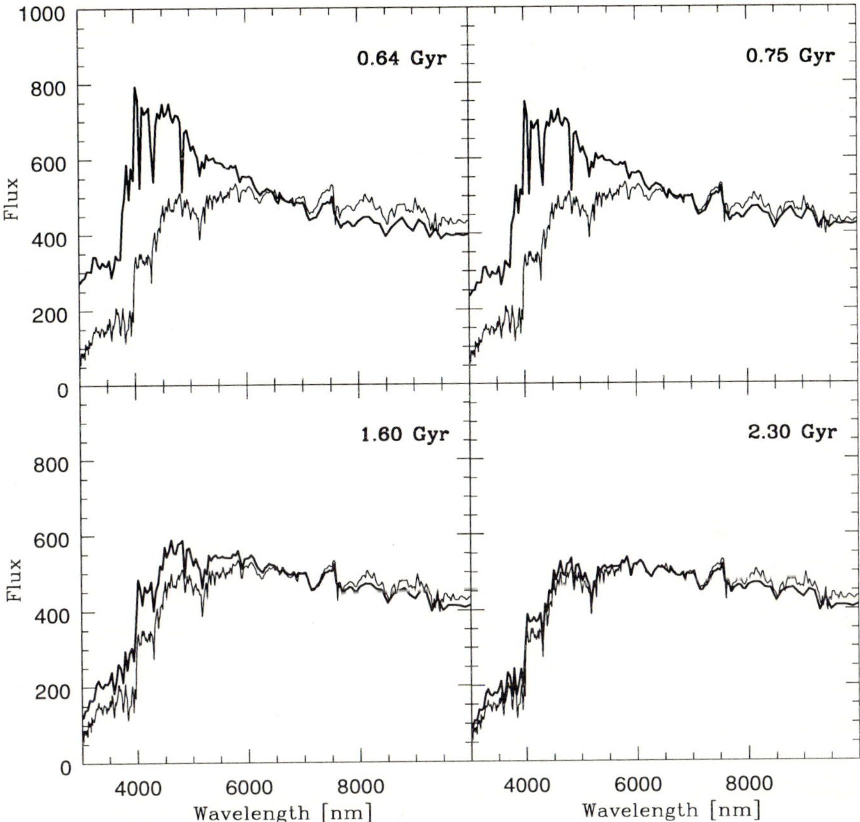

Fig. 2. Spectral evolution of an E+A-model (thick line) compared to an elliptical galaxy (thin line). Note the strong Balmer absorption lines in the young E+As. Three giga-years after the second burst, the spectral energy distribution of an E+A-galaxy can no longer be distinguished from that of a normal elliptical.

Cl 0939+472: The results of the analysis for the Cl 0939+472 field are presented in Tab. 1. The E+A-templates not only successfully identified the 6 previously known E+A, but also allowed the detection of 29 new ones. Thus the post-star-burst galaxies represent $(21 \pm 3)\,\%$ of the whole population. The distinction between morphological galaxy types allows their projected distribution in the cluster to be studied, as shown in Fig. 4a. Whereas the elliptical galaxies show a clear concentration around the cluster center, E+A- and spiral galaxies are more spread out around the field.

The HST morphological information available for two $1' \times 1'$ fields in the cluster (Dressler *et al.* 1994) was combined with our redshift and spectral energy distribution information. Using the Dressler *et al.* (1994) colour-morphology diagram and the photometry given by Dressler & Gunn (1992) we obtained the morphological classification for 85 galaxies in our sample. The results can be summarized as follows (Belloni *et al.* 1995):

Table 1. Galaxy population in Cl 0016+161 and Cl 0939+472. The mean values for colours and $\tilde{D}_{4000}$ are indicated. A colour-magnitude correction (Visvanathan & Sandage 1977, Bower *et al.* 1992) has been applied to the elliptical galaxies.

Cl 0016+161

Spectral type	DG92	This work	$B-R$	$B-I$	$\tilde{D}_{4000}$
Elliptical	19	73	2.80 ± 0.21	4.02 ± 0.24	2.10 ± 0.45
E+A	9	20	2.11 ± 0.34	3.17 ± 0.45	1.74 ± 0.23
Spiral & Im	2	7	1.69 ± 0.34	2.75 ± 0.34	1.18 ± 0.22

Cl 0939+472

Spectral type	DG92	This work	$B-R$	$B-I$	$\tilde{D}_{4000}$
Elliptical	22	110	2.65 ± 0.17	3.75 ± 0.26	2.21 ± 0.35
E+A	6	35	2.03 ± 0.17	3.01 ± 0.32	1.81 ± 0.23
Spiral & Im	2	24	1.37 ± 0.34	2.11 ± 0.34	1.30 ± 0.22

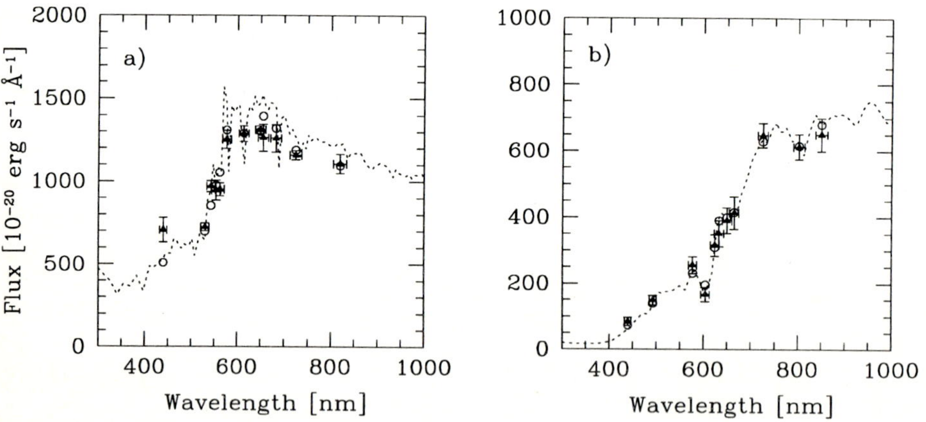

Fig. 3. Reconstructed spectrum using multi-filter photometry. The circles represent the values expected by convolving the theoretical spectral energy distribution with the filters used; the triangles are the observed values. The error bars show the 1σ-error of the photometry.
a) Spectral energy distribution of a newly detected E+A-galaxy in Cl 0939+472, classified as a 0.5 Gyr old post-star-burst. **b)** Same as in a) but for a newly found elliptical member in Cl 0016+161.

- For 85% of the galaxies in Dressler's sample our classification based on spectroscopic properties matches the morphological classification.
- 14 E+A's identified on the Hubble images turned out to be disk-systems (spirals and irregulars) or mergers.

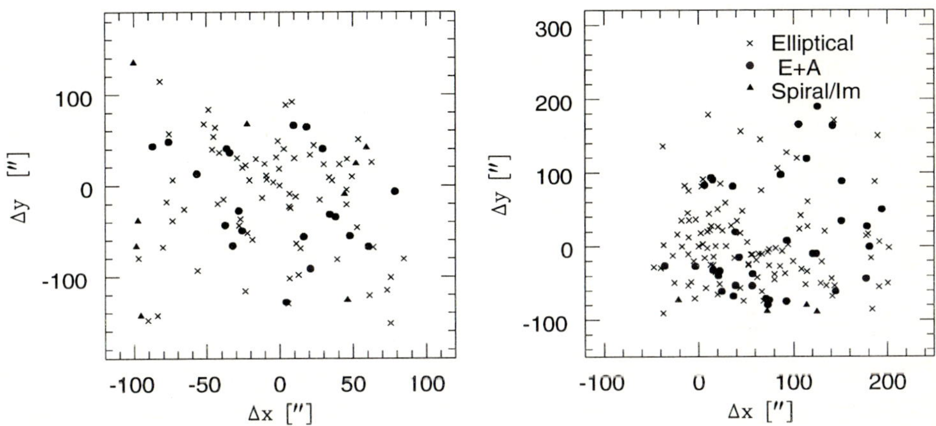

Fig. 4. a) Spatial distribution of the galaxies cluster members (elliptical, E+A and Spiral/Im) in Cl 0016+161. **b)** Same as in a) but for Cl 0939+472.

Cl 0016+161: The analysis of this cluster has been performed with the same approach and tools as the ones of Cl 0939+472. In particular, the same E+A-templates have been used with the assumption that they could describe the appearance of E+A-galaxies in high redshift clusters in general. Out of the 200 objects analyzed, 161 could be classified, representing a success rate of 80 % in the determination of redshifts and corresponding morphological types from the spectral energy distributions.

Note, however, that this analysis of a complete sample resulted in the detection of proportionally more ellipticals than E+A and spiral/Im galaxies, confirming that this cluster is dominated by old early-type galaxies. In Cl 0016+161, spiral/Im galaxies represent only 7% of the whole galaxy population, whereas E+A-galaxies (20 in total, of which 12 were newly found) comprise (21 ± 4) %, thus dominating the percentage of active galaxies. As in Cl 0939+472, E+A and spiral galaxies avoid the center of the cluster (Fig. 4a).

The existence of a foreground cluster projected on the Cl 0016+161 field has been proposed by Ellis *et al.* (1985). Therefore, particular attention has been payed to the redshift and spatial distribution of the non-member galaxies. Their spatial distribution is not clumped, lacking a peak close to the center of the cluster at $z = 0.54$ and fills all of the $5' \times 5'$ area. The foreground component is thus more likely to represent field galaxies than a foreground cluster.

3.4 Discussion

HST morphology, combined with the spectral information allows to shed some light on the trigger mechanism of the star-burst activity revealed in its delayed phases by E+A-galaxies. The fact, that in Cl 0939+472 the majority of these galaxies is not presently interacting but looks like normal late-tye spirals, can be interpreted in two ways:

a) The merger scenario is not the only explanation.
b) Merging is not necessarily ruled out as primary agent of the star-burst activity.
   Indeed, one should not expect E+A-galaxies to still be near their interaction partner, since the cluster velocity dispersion would carry them far away over the 10 Gyr since the burst.

A large disk fraction is expected in the case, in which the star formation bursts are induced by interaction with the inter-cluster medium. In this case, E+A-galaxies are not the result of merging, but are new acquisitions in the cluster being captured from the field prevalently populated by spiral galaxies.

A significant result is also the good agreement between morphological classification derived from multi-filter photometry and that provided by HST images. This points out the effectiveness of this approach in the recognition of morphological types of distant galaxies.

In the light of the results obtained, the question of the nature of the galaxies responsible for the Butcher-Oemler effect in the two clusters studied can now be considered. In Cl 0939+472 the Butcher-Oemler effect is caused more by a population of galaxies having recently undergone bursts of star formation than by normal spiral galaxies. This result contradicts the conclusion of Dressler *et al.* (1994) that the excess blue galaxies are predominantly normal late-type spirals. One reason for this discrepancy is, that the information provided by the galaxies' spectral energy distributions allows us to: 1) recognize, if a disk-like galaxy is a normal spiral- or an E+A-type and 2) estimate the cluster membership of spiral and irregular galaxies in the cluster field. Morphology coupled with colour information only is inadequate for these purposes.

Also Cl 0016+161, despite its red nature, confirms the presence of the Butcher-Oemler effect. Spiral galaxies or bursts of star formation are significantly less common in Cl 0016+161 than in Cl 0939+49. Indeed, the activity is largely confined to the later evolutionary phases of the star-burst phenomenon represented by E+A-galaxies.

4 Future Work

4.1 Clusters at Intermediate Redshifts

The work described in the previous two sections has now paved the way for an application of these methods to other galaxy clusters. In the meantime we have collected data for 5 clusters, all at a redshift around 0.5 (see Tab. 2). Our main goal with this data set will be to compare the galaxy populations in clusters of about the same redshift in order to trace the effects shaping their evolution.

Table 2. Clusters at intermediate redshifts observed with Fabry- Pérot and narrow band imaging.

Name	Redshift	Remarks
Cl 0016+161	0.546	strong X-ray source
Cl 0303+171	0.418	
Cl 0939+472	0.407	very rich
Cl 1447+263	0.374	irregular, many blue galaxies
3C 295 cluster	0.458	strong radio source

4.2 Survey for Intermediate Redshift Galaxy Clusters

The ultimate goal of our project is, however, the study of clusters with redshifts up to about 1. As there are only very few such clusters available in the literature, we started our own survey for high redshift galaxy clusters in order to establish a well defined sample. The survey material will be provided by M.R.S. Hawkins (Edinburgh), who had acquired and digitally stacked three sets of 64 J-, 64 F- and 35 N UK-Schmidt plates of one southern field of about 25 □°. These data have been taken over the last ≈ 15 years, mainly to search for variable QSOs (Hawkins 1993). Following the identification of clusters using standard search algorithms by H. MacGillivray, we plan to develop a search strategy tailored specifically to the needs of this data set. Based on the colour information available from the three data sets we hope to be able to derive also a rough redshift estimate of clusters found from the survey material itself. Following the verification and measurement of the clusters' redshifts we will employ the methods described above in a study of the population in high redshift galaxy clusters.

Acknowledgements: The work described here is a joint effort at the MPI für Astronomie. Hans Hippelein, Ulrich Hopp, Klaus Meisenheimer and Guido Thimm (now at ESO) are part of the team. Further collaborateurs are Gustavo Bruzual (population synthesis calculations), Mike Hawkins and Harvey MacGillivray (cluster search).

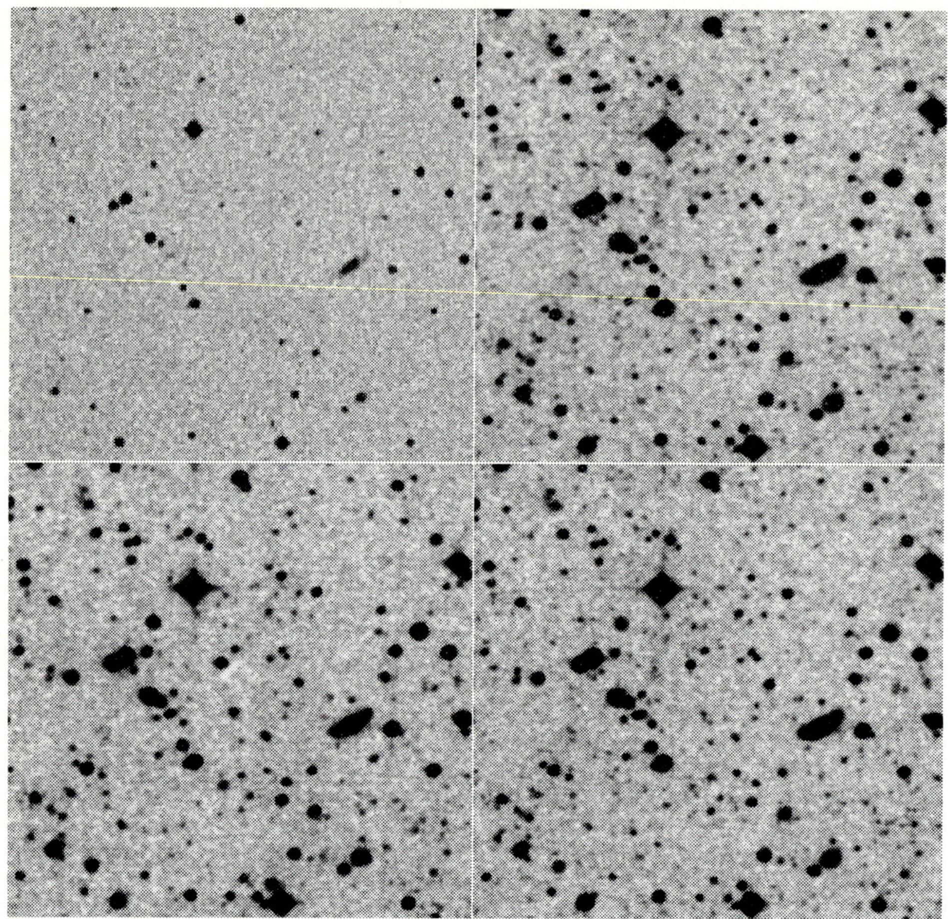

Fig. 5. An area of $5' \times 5'$ from the field covered by the stacked UK-Schmidt plates, that will be used to survey for galaxy clusters at high redshift.
top left: Field from the digitized sky survey
top right: Sum of 64 J-plates of the same field
bottom left: Sum of 64 F-plates
bottom right: Sum of 30 IV-N-plates.

The Sky survey image is based on photographic data obtained using the UK Schmidt Telescope. The UK Schmidt Telescope was operated by the Royal Observatory Edinburgh, with funding from the UK Science and Engineering Research Council, until 1988 June, and thereafter by the Anglo-Australian Observatory. Original plate material is copyright the Royal Observatory Edinburgh and the Anglo-Australian Observatory. The plates were processed into the present compressed digital form with their permission. The *Digitized Sky Survey* was produced at the Space Telescope Science Institute (ST ScI) under U. S. Government grant NAG W-2166.

References

Belloni, P. 1994, *Ph.D. thesis*, Universität Heidelberg

Belloni, P., Bruzual, G., Thimm, G. J., and Röser, H.-J. 1995, *Astron. Astrophys.* **297**, 61

Bower, R. G., Lucey, J. R., Ellis, R. S. 1992, *Mon. Not. R. Astron. Soc.* **254**, 601

Bruzual, G., & Charlot, S. 1993, *Astrophys. J.* **405**, 538

Butcher, H. & Oemler Jr., A. 1978, *Astrophys. J.* **219**, 18

Butcher, H. and Oemler Jr., A. 1984, *Astrophys. J.* **285**, 426

Coleman, G. D., Wu, C.-C., & Weedman, D. W. 1980, *Astrophys. J. Suppl.* **43**, 393

Dressler, A., Augustus Oemler, J., Butcher, H. R., & Gunn, J. E. 1994, *Astrophys. J.* **430**, 107

Dressler, A. & Gunn, J. E. 1982, *Astrophys. J.* **263**, 533

Dressler, A. & Gunn, J. E. 1983, *Astrophys. J.* **270**, 7

Dressler, A. & Gunn, J. E. 1990, in *Evolution of the Universe of Galaxies*, R.G. Kron (ed.), *Astronomical Society of the Pacific Conference Series*, Vol. 10, p.200

Dressler, A. & Gunn, J. E. 1992, *Astrophys. J., Suppl.* **78**, 1

Ellis, R. S., Couch, W. J., MacLaren, I., & Koo, D. C. 1985, *Mon. Not. R. Astron. Soc.* **217**, 239

Hawkins, M. R. S. 1993, *Nature* **366**, 242

Röser, H.-J. & Meisenheimer, K. 1991, *Astron. Astrophys.* **252**, 458

Spinrad, H. 1986, *Publ. Astron. Soc. Pac.* **98**, 269

Thimm, G. J., Röser, H.-J., Hippelein, H., & Meisenheimer, K. 1994, *Astron. Astrophys.* **285**, 785

Visvanathan, N. & Sandage, A. 1977, *Astrophys. J.* **216**, 214

What is 3C 324 ?

Mark Dickinson[1], *Arjun Dey*[2,3] *and Hyron Spinrad*[2]

[1] Space Telescope Science Institute, 3700 San Martin Dr., Baltimore MD 21218, USA
[2] Department of Astronomy, Univ. of California, Berkeley CA 94720, USA
[3] Institute of Geophysics and Planetary Physics, LLNL, Livermore CA 94550, USA

1 Radio Galaxies at High Redshift – Hopes and Fears

At one time, radio sources offered our only access to the universe of galaxies at very large redshifts. Amongst the swarms of faint objects in the sky, powerful radio emission draws our attention and guides our telescopes. Strong emission line spectra allow us to measure redshifts out to $z = 4.25$ (thus far – Lacy *et al.* 1994). The majority of extragalactic, steep spectrum radio sources have faint, spatially extended optical counterparts, and exhibit narrow line emission. Thus we identify them as galaxies, regardless of what AGN activity may have drawn our attention in the first place.

Nearby, powerful radio galaxies are giant ellipticals. This association led to the hope that their high–redshift counterparts could be studied as progenitors – that their evolution could be traced through their colors and magnitudes. The small photometric scatter observed in the infrared Hubble diagram (magnitude vs. redshift relation) out to at least $z \approx 2$ has suggested that radio galaxies are a stable population of intrinsically similar galaxies, perhaps even well–enough behaved for use as cosmological probes (Lilly 1989, McCarthy 1993).

At the same time, some properties of high–z radio galaxies appear to be quite extraordinary. Their optical (i.e. rest–frame UV) magnitudes and colors vary widely (cf. Lilly & Longair 1984), and high–quality images often reveal highly elongated morphologies, sometimes resolved into multiple clumps (cf. Le Fèvre & Hammer 1988).

A breakthrough (or death–blow, for the pessimist) occurred when it was realized that the elongated, complex morphologies closely share the primary axis of the radio source (McCarthy *et al.* 1987, Chambers *et al.* 1987). This was clear evidence that the radio source, a supposedly short–lived phenomenon of the central AGN, exerts a dramatic influence on the host galaxy. Early interpretation (see McCarthy 1993 for references and a review) of this 'alignment effect' centered on massive star–formation induced by the passage of the radio jet. It was suggested that perhaps a large fraction of the stellar mass could be formed in this fashion: perhaps these are true protogalaxies, caught in the act of formation. However, the discovery of strong linear polarization in many objects cast doubt about whether the light we see is starlight at all (di Serego Alighieri *et al.* 1989).

Since the most intriguing and unexplained aspect of high-z radio galaxies is their aligned morphologies, it has long been hoped that imaging with the *Hubble Space Telescope* (HST) would help explain the origins of these peculiar structures by elucidating their kiloparsec-scale details (cf. Miley *et al.* 1992). Longair *et al.* (1995) and Röttgering (this volume, p.97) have presented initial results from an HST imaging survey of high redshift 3CR galaxies. Here we will examine in detail the properties of a single, prototypical example of the alignment effect, taking advantage of a uniquely deep HST image, as well as infrared imaging and deep optical spectroscopy from the Keck 10m telescope.

2 The Case of 3C 324

Figure 1 shows deep optical (R) and infrared (K) images of 3C 324, an aligned radio galaxy at $z = 1.206$. In the rest-frame, the R-band image samples the galaxy light in near-ultraviolet, while the K-band provides an image at $\sim 1\mu$m. In the infrared, 3C 324 appears to be a round, giant elliptical galaxy. In the UV, it is bimodal (at least) and elongated in the general direction of the radio source. Optical polarimetric observations (di Serego Alighieri *et al.* 1993, 1994; Cimatti *et al.* 1995) measure a polarization of 11–18% for 3C 324, with the electric vectors perpendicular to the major axis of the aligned light.

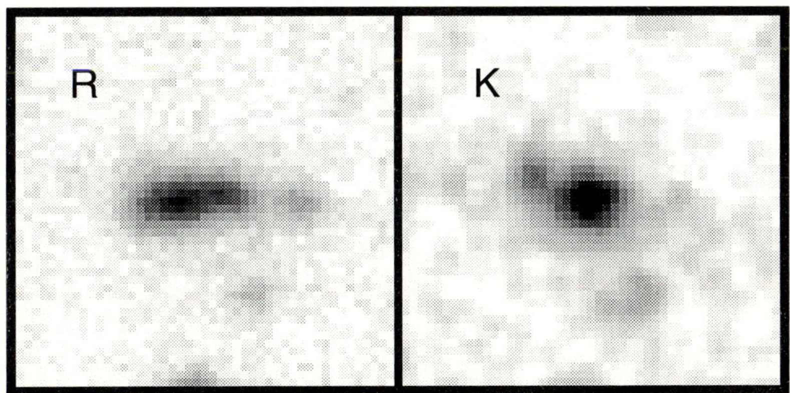

Fig. 1. R and K-band images of 3C 324 ($z = 1.206$), obtained with the Keck and KPNO 4m telescopes, respectively. The field of view is $10'' \times 10''$ for each panel, and the seeing FWHM=$0\farcs85$ for both.

Figure 2*a* presents an HST image of 3C 324 with 5 GHz radio contours superimposed. The galaxy was observed with the Planetary Camera (PC) and F702W filter ($\lambda_{\rm eff} \approx 7000$Å). A very long exposure (32 orbits, $t_{\rm exp} = 18$ hours) was obtained in order to study extremely faint galaxies in a cluster surrounding the radio source. The PC pixel scale, $0\farcs046$/pixel, corresponds to $\sim$500 pc at the redshift of the radio galaxy (for H_0=50 km s^{-1} Mpc^{-1}, q_0=0.1). The F702W filter samples the galaxy light at ~3200Å in its rest-frame. Although several emission lines are included in this bandpass (Mg II, Ne V]), the strongest lines (e.g. [O II] λ3727) are largely excluded. Our spectra (see below) show that line emission contributes only 11% to the integrated flux from the galaxy, although the point-by-point contribution could be smaller or larger depending on the spatial distribution of the emission line gas.

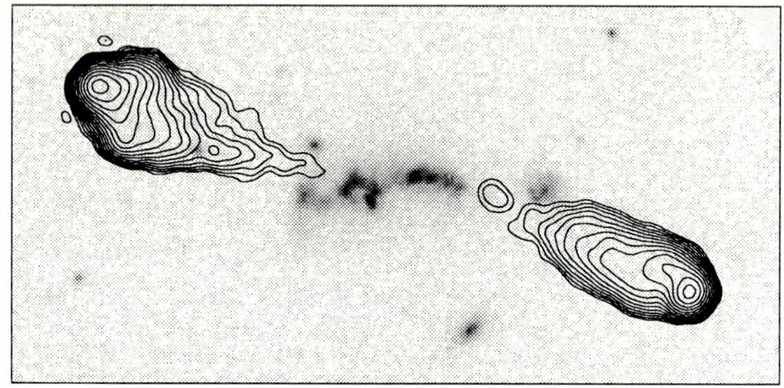

1″ ~ 10 kpc

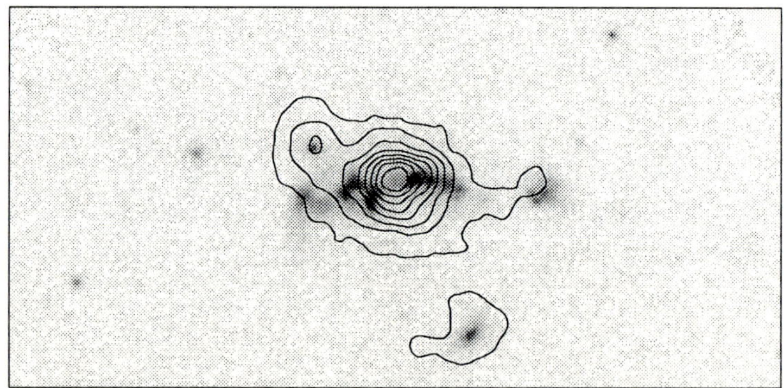

Fig. 2. *(a) Top:* HST Planetary Camera image of 3C 324 through the F702W filter. The contours show the 5 GHz VLA map of Fernini *et al.* 1993. The formal uncertainty in radio-optical registration is ~0.″15 in each coordinate, although systematic errors in the astrometry could be larger. *(b) Bottom:* KPNO 4m *K*-band image superimposed on the HST image. The peak of the *K*-band light lies between the major complexes of rest-frame UV emission.

At ~ 0.″1 resolution, the elongated structure of 3C 324 breaks up into a long, remarkably narrow chain of compact, high-surface brightness clumps, with scale sizes ∼0.″2 to 0.″4 (FWHM), i.e. ∼2 to 4 kpc. There is little point–by–point correspondence between radio and optical features. Considering the axes defined by the radio hotspots, the radio jet (visible on the eastern side), and the UV continuum, there is a sense of counterclockwise rotation as one progresses from large angular scales to small, suggesting a precession of the AGN emission axis.

Figure 2*b* overlays the PC image with contours from the K–band image. While the angular resolutions of the HST and ground–based infrared data differ substantially, it is evident that the K–band light peaks at a local *minimum* in the rest–frame UV emission, between the two major "blob complexes." If this is considered to be the center of the host galaxy, it is apparently heavily obscured at ultraviolet wavelengths.

Figure 3 plots a spectrum of 3C 324 taken with the Keck Telescope and LRIS spectrograph. The slit was oriented along the optical major axis, and the extraction presented here sums nearly all of the light from the galaxy. Several features are notable:

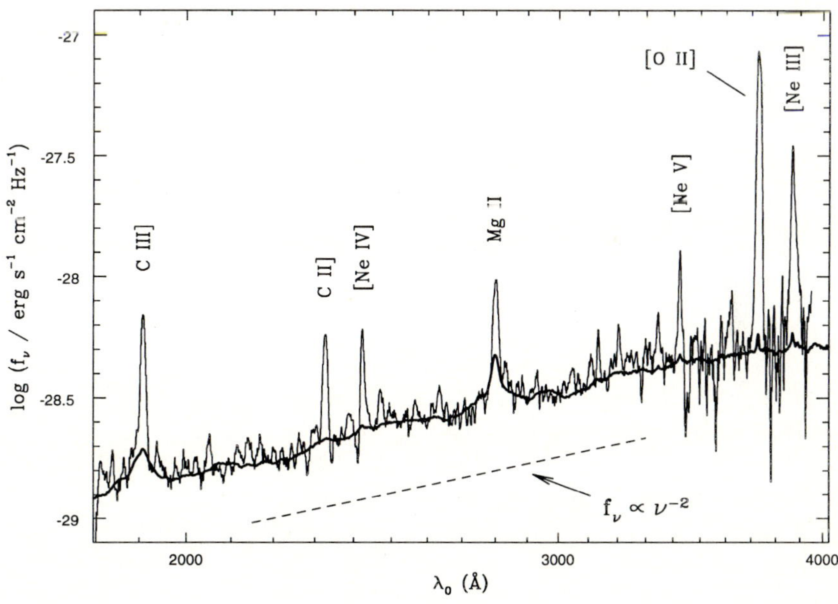

Fig. 3. Log–log plot of a total light spectrogram of 3C 324, obtained with the Keck LRIS. The heavy line superimposes the average LBQS QSO spectrum of Francis *et al.* 1991, artificially reddened to approximately match the slope of the radio galaxy continuum. The Mg II and C III] emission lines in the radio galaxy exhibit broad wings consistent with the presence of underlying, quasar–like broad lines with normal QSO equivalent widths.

- The aligned galaxy continuum is actually quite red. It is reasonably well represented by a power law $f_\nu \propto \nu^\alpha$ with spectral index $\alpha \approx -2$.
- The permitted and semi–forbidden lines (Mg II, C III]) which dominate quasar spectra appear to have broad wings. In Figure 3 we superimpose a composite QSO spectrum over that of the radio galaxy. In order to match the red continuum slope of the radio galaxy, we have artificially "reddened" the quasar by 1.6 powers of ν and scaled the continuum to that of the radio galaxy at $\sim$ 3000Å. The Mg II and C III] lines of the quasar provide a good match to the broad wings seen in the radio galaxy spectrum. The broad features cannot arise from an active nucleus viewed directly, since no such unresolved feature is observed in the HST images. Indeed the presumed location of the nucleus appears to be heavily obscured in Fig. 2*b*. Note that strong, narrow Mg II and C III] emission is superimposed on the broad features, which would partially mask their visibility in lower signal–to–noise data.

3 Interpretation

The strength and position angle of the observed polarization in 3C 324 strongly suggests that some fraction of the UV continuum is scattered. Other mechanisms for producing the polarized light seem untenable. E.g. for synchrotron radiation, one would expect better correspondence between radio and optical components unless the two emission regions produce radically different synchrotron spectra. Moreover, the broad wings seen in Mg II and C III] resemble quasar emission lines with normal equivalent widths, supporting the assumption that most or all of the radio galaxy continuum is scattered AGN light. Previous objections to scattering models based on the lack of broad line emission need not apply to 3C 324 at least. Substantial contribution from additional *in situ* continuum would only dilute the equivalent widths of the scattered broad components, requiring that their *intrinsic* strength be greater than that seen in typical quasars. Similarly strong, broad MgII features have been observed in other radio galaxies, and evidence has been presented demonstrating that Mg II may be polarized (e.g. di Serego Alighieri *et al.* 1994) and spatially extended (Dey & Spinrad 1995) – both clear predictions of the scattering hypothesis.

The surface brightnesses of the clumps in the HST image decline with increasing distance from the galaxy center and presumed AGN location – see Figure 4. This is qualitatively consistent with expectations from a scattering model, although the prediction is not unique. The spatially resolved [O II] emission in our long–slit spectra is strongest at the linear extremes where the continuum surface brightness is lowest, contrary to what might be expected if the continuum morphology and line emission traced recent star formation along the jet axis. In addition, this provides further evidence that line emission is not dominating the morphology of the HST image.

Where is the giant elliptical host galaxy? It apparently dominates the K–band image, but is seemingly invisible at $\lambda_{\rm obs}$ = 7000Å in the HST data. To quantify this, we have simulated a gE host galaxy at $z = 1.2$. An actual brightest

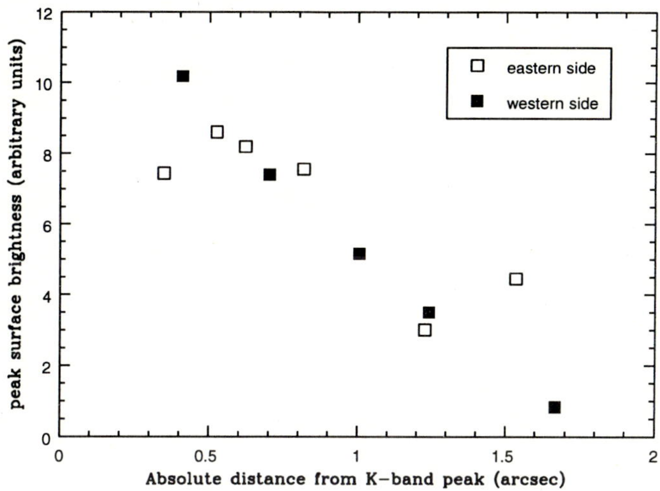

Fig. 4. Peak surface brightnesses (averaged over 0.″3 diameter apertures) of the clumps in 3C 324 vs. their radial distances from the K–band peak. A nearly monotonic decline is observed, particularly in the tight, curving chain of knots to the west of the galaxy center.

cluster elliptical (BCE) at $z = 0.4$ observed with HST by Dressler *et al.* (1994) was artificially "redshifted" out to $z = 1.2$. Angular size was converted assuming $q_0 = 0.5$, and surface brightnesses were set by requiring the host galaxy to have $K = 17.5$ (as we measure from the K images) and $(R - K) = 5.9$, the color of other faint ellipticals in the cluster surrounding 3C 324 (cf. Dickinson 1995 in this volume). Noise was added to match the real PC data. The result is shown in Figure 5. The low surface brightness of the BCE envelope makes it difficult to detect, but the galaxy nucleus should be evident. No such feature is seen in the actual image (cf. Fig. 2*b*), however. We conclude that either the host galaxy is substantially less nucleated than a typical BCE, or that it is obscured by dust. We calculate that extinction with $E(B - V) \approx 0.3$ is sufficient to render the galaxy nucleus invisible in the PC image. This also provides further evidence for the presence of distributed dust in 3C 324, which can also serve as the scattering medium for producing the aligned UV light.

There seems to be no hard evidence against a pure–scattering model for the aligned UV continuum in 3C 324. The present observations cannot confirm or firmly exclude other possible sources, e.g. hot stars, hydrogen recombination continuum, etc. These should be unpolarized, however, and high quality polarimetry (imaging and spectroscopic) should go a long way toward providing a resolution. Keck spectropolarimetry is now in hand (Cimatti *et al.* 1995), and WFPC2 imaging polarimetry with HST is scheduled for Cycle 5.

Several tasks remain before a model for the alignment effect in 3C 324 can be considered convincing:

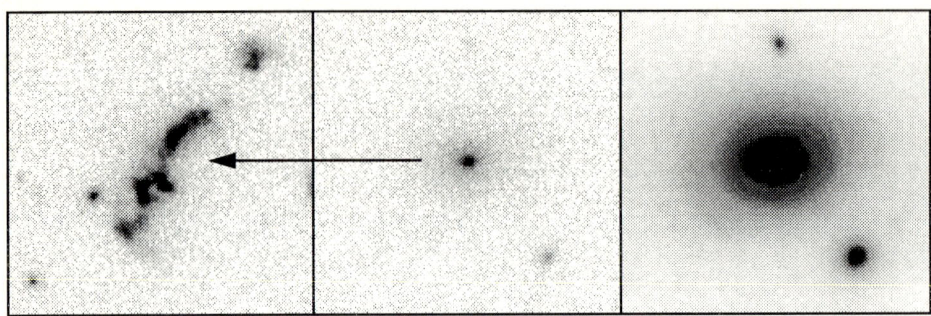

Fig. 5. Simulation of a giant elliptical host galaxy for 3C 324. The left panel shows the HST Planetary Camera image of the radio galaxy. An HST observation of a brightest cluster elliptical (BCE) at $z = 0.4$ is seen at right. The center panel simulates the BCE as it would appear at $z = 1.206$ in the Planetary Camera field.

- It must explain the continuum configuration seen in the HST images. Why are the blobs so compact and colinear? AGN unification models such as that of Barthel (1989) propose a anisotropic "radiation cone" with an opening angle of $\sim 45°$. Let us suppose that the AGN lies at the K–band peak, and shines out in a cone roughly axisymmetric about the radio jet axis. The UV light we observe is certainly confined within such a cone, but does not fill it. Rather, it forms a linear configuration, clumped along the long axis. At first glance, this arrangement may seem to fit more naturally with a cartoon picture of jet–induced star formation than with scattering models. If the bulk of the UV light is scattered, then apparently the scattering medium is also strongly clumped – the "mirrors" are closely confined to an axis which is similar (but not quite identical to) the radio source axis. Does the radio jet play a role in creating this clumpiness? If the UV light is emitted *in situ* by sources along this axis, we are then left to explain the observed polarization and broad line emission.
- Why is the aligned light red? If it is scattered from a hidden but otherwise ordinary central quasar, it must have been reddened. Extinction at the source (e.g. as the AGN shines out through a dusty medium closely surrounding the galaxy center) is possible, but would strongly attenuate ionizing UV photons, posing a difficulty for AGN photoionization models of the extended emission line gas. While simple, optically thin Rayleigh scattering would make the incident spectrum *bluer,* realistic models produce more complex effects, due to multiple scattering, absorption, and varying albedo depending on the grain–size distribution (e.g. Laor 1995) Similar problems face other models for the continuum emission. Spectrophotometry and high angular resolution color measurements over a wide range of wavelengths would be invaluable for constraining the nature of the continuum.
- If the light is scattered, what is the scattering medium? Dust and electrons

are the two popular contenders. The apparent obscuration of host galaxy nucleus and the red color of the aligned UV continuum suggest that dust may be abundant. If dust is responsible for scattering, what is its origin, and why is it distributed throughout a volume tens of kpc across? Is it manufactured in the host galaxy and strewn wide by outflows/winds? Could it precipitate from the surrounding cluster, either from cooling intra-cluster gas or from companion galaxies? Or is the dust somehow manufactured by the radio source itself?

For those who hope to study radio galaxies as *galaxies*, these conclusions may perhaps be disappointing. It appears that the optical light from objects like 3C 324 is entirely dominated by a manifestation (albeit indirect) of an AGN phenomenon. While the alignment mechanism still presents us with fascinating puzzles to solve, its relevance for understanding galaxies, and correspondingly the utility of radio galaxies themselves for studying evolution, is questionable.

The optimist, however, might consider these new results encouraging. If the aligned light is only a reflection of the AGN, then it is in some sense a phantom. Perhaps the underlying galaxy is largely unaffected by the "fireworks," contrary to extreme star-formation scenarios for the alignment effect where the stellar content of the galaxy could be dramatically altered as a consequence of the AGN activity. The infrared properties of radio galaxies, where the aligned light is apparently least important, remain reasonably unaffected and are still a valid subject for evolutionary studies, e.g. via the infrared Hubble diagram.

Moreover, recent work suggests that the optical alignments may be weaker or absent in galaxies with less powerful radio sources (Dunlop & Peacock 1993, but cf. McLeod 1994 for a contrary view). Perhaps this phantom only haunts those few, rare galaxies at the extreme bright end of the radio luminosity function – even the optical properties of weaker radio sources might be safe for study.

Acknowledgements: We would like to thank the conference organizers for providing a wonderful opportunity and locale for this gathering, and the castle staff for keeping the Hexenzimmer well stocked with wine. We also thank the other co-investigators on our HST imaging program: George Djorgovski, Peter Eisenhardt, and Adam Stanford. M.D. acknowledges generous travel support from STScI.

References

Barthel, P.D. 1989, *Astrophys. J.* **336**, 606.

Chambers, K.C., Miley, G.K., and van Breugel, W.J.M. 1987, *Nature* **329**, 604

Cimatti, A., di Serego Alighieri, S., Fosbury, R.A.E., Salvati, M., and Taylor, D. 1993, *Mon. Not. R. Astron. Soc.* **264**, 421

Cimatti, A., Dey, A., van Breugel, W., Antonucci, R., & Spinrad, H. 1995, in preparation

Dey, A., and Spinrad, H., 1995, *Astrophys. J.*, accepted

di Serego Alighieri, S., Fosbury, R.A.E., Quinn, P.J., and Tadhunter, C.N. 1989, *Nature* **341**, 307

di Serego Alighieri, S., Cimatti, A., and Fosbury, R.A.E. 1993, *Astrophys. J.* **404**, 584

di Serego Alighieri, S., Cimatti, A., and Fosbury, R.A.E. 1994, *Astrophys. J.* **431**, 123

Dressler, A., Oemler, A., Sparks, W.B., and Lucas, R.A. 1994, *Astrophys. J. Lett.* **435**, L23

Dunlop, J.S., and Peacock, J.A. 1993, *Mon. Not. R. Astron. Soc.* **263**, 936

Fernini, I, Burns, J.O., Bridle, A.H., and Perley, R.A. 1993, *Astrophys. J.* **105**, 1690

Francis, P.J., Hewett, P.C., Foltz, C.B., Chaffee, F.H., Weymann, R.J., and Morris, S.L. 1991, *Astrophys. J.* **373**, 46

Lacy, M., Miley, G., Rawlings, S., Saunders, R., Dickinson, M., Garrington, S., Maddox, S., Pooley, G., Steidel, C.C., Bremer, M.N., Cotter, G., van Ojik, R., Röttgering, H., and Warner, P. 1994, *Mon. Not. R. Astron. Soc.* **271**, 504

Laor, A. 1995, in preparation

Le Fèvre, O., and Hammer, F. 1988, *Astrophys. J. Lett.* **333**, L37

Lilly, S.J., and Longair, M. 1984, *Mon. Not. R. Astron. Soc.* **211**, 833

Lilly, S.J. 1989, *Astrophys. J.* **340**, 77

Longair, M.S., Best, P.N., and Röttgering, H.J.A. 1995, *Mon. Not. R. Astron. Soc.* **275**, L47

McCarthy, P.J., van Breugel, W.J.M., Spinrad, H., and Djorgovski, S.G. 1987, *Astrophys. J. Lett.* **321**, L29

McCarthy, P.J. 1993, *Ann. Rev. Astron. Astrophys.* **31**, 639

McLeod, B.A. 1994, *Ph.D. Thesis*, University of Arizona

Miley, G.K., Chambers, K.C., van Breugel, W., and Macchetto, D. 1992, *Astrophys. J. Lett.* **401**, L69

Spectrophotometry of Distant Radio Galaxies with the Field Spectrograph TIGER/CFHT

Brigitte Rocca-Volmerange

Institut d'Astrophysique de Paris, 98bis Bd Arago, F-75014 PARIS, France

1 Introduction

It is now well established that one difficulty to interpret the observations of distant galaxies is to estimate the respective contributions of cosmological and evolutionary effects. The recent discovery of a radio-quiet galaxy at redshift $z = 3.4$ (Giavalisco *et al.* 1994) is of high interest but the population of high-z radiogalaxies gives a statistically more significant sample and allow to possibly relate the active nucleus activity to an intense star formation activity, initiating a link between AGN, starburst and galaxies. The radiogalaxies recently discovered at the most remote distances ($z \geq 3.5$) are among the best cosmological targets. However so various features caracterize these galaxies (red stellar energy distribution, huge emission lines, high density of galaxy companions, alignment of ultraviolet and radio axes, large degree of polarisation) that their structures are not simple to understand. Stellar populations will only become the best indicators of evolution of galaxies if these structures are clearly understood from a two-dimension spectroscopy on each image point. The integral field spectrograph TIGER (Bacon *et al.* 1995) is a unique instrument at the CFHT to give details on the nature and velocities of the various components of distant radiogalaxies.We present the observations with TIGER of an intermediate-redshift galaxy 3C 435A ($z = 0.471$) (Rocca-Volmerange *et al.* 1994). The two nebular emission lines [O II], [O III] and blue to red stellar continua of two components are observable, allowing to date the corresponding stellar population with the help of our atlas of synthetic spectra of galaxies (Rocca-Volmerange & Guideroni 1988). The present and past star formation activities and the origin of alignment will be thus analysed in terms of galaxy evolution.

2 Main Properties of Distant Radiogalaxies

2.1 The Stellar Energy Distribution

Spectrophotometry carried out through broad bands shows a surprising feature, preferentially attributed to stars rather than to non-thermal or dust phenomena.

A typical gap of the continuum is observed from the blue to the red (through B and K filters), corresponding in the rest-frame to a plateau in the ultra-violet and a strong increase in the visible (Lilly 1988, Chambers *et al.* 1990). Many interpretations based on populations of evolved stars (giants, supergiants, asymptotic giant branch stars) were successively proposed by various authors. But these interpretations did not take into account the huge nebular emission observed in these galaxies. Recent observations of a sample of distant radiogalaxies with the CGS4 instrument at the UKIRT (Eales *et al.* 1993) show intense [O III] and Hβ lines. These redshifted lines contribute to the K-band emission more than previously thought. After substracting these lines to the observed energy distribution, the stellar continuum fits a model of a very recent age ($\leq$0.5 Gyr), without adding any red population and confirming that at high redshifts, radiogalaxies are dominated by young stars of a starburst (Rocca-Volmerange 1993).

2.2 The Alignment of the UV and Radio Axes

Observed from a large sample of distant radiogalaxies (Mc Carthy *et al.* 1987, Chambers *et al.* 1987), the alignment of the UV axis with radio axis rapidly increases with redshift. For nearby galaxies, this alignment is due to nebular emission van Breugewl *et al.* 1985) while at higher redshifts, the strong increase was attributed to other processes. In many cases, the alignment is established from the central radiogalaxy to a bright galaxy companion near the radio axis and not to a nebular component. That could be an observational bias, surface brightness of galaxies being higher than of nebulosities. Star formation was thus proposed in various models as triggered by the radio-jet (Begelman & Cioffi 1989) or by interactions of the radio plasma with the intergalactic environment (Begelman & Cioffi 1989, Daly 1990, de Young 1989), such models need confirmation from observations.

2.3 Nebular Emission Lines and Polarisation

The Lyα emission line of powerful radiogalaxies reach more than 10^{44} erg s^{-1} with a 100 kpc diameter and a FWHM of 1000 km s^{-1}, as the famous example of the galaxy 3C326.1 (Mc Carthy *et al.* 1987). Other typical lines are enormous as the [O II] line, a possible signature of star formation, and the [O III] and Hβ lines, observed in the near-infrared at high redshifts, better indicators of shocks, AGN or non-thermal effects. The physical origin is not well identified, in particular for the ionized [O II] 3727Å line which frequently is a signature of star formation process. Moreover the degree of polarisation and the orientation of the electric vector are useful indicators of the interaction of the relativistic electrons with the gas, even if recent data clarify the relation with radiogalaxy activity (Cimatti *et al.* 1994).

3 Observations of the Radio Galaxy 3C 435A

We present the results of the first radiogalaxy from a selected sample observed with TIGER at the CFHT from June 1991. The radio galaxy 3C 435A at $z =$

0.471 forms an optical pair with 3C 435B at higher redshift. Radio maps and a low level of polarisation (di Seregho Alighieri *et al.* 1993) bring complementary data. Observations cover the blue (5000-7000Å) and red (6500-8500Å) wavelength ranges with a spatial sampling 0."61 and a spectral sampling 8Å. The average seeing is 0."71. Data reduction has been carried out with the TIGER software package, installed on the MIDAS ESO software by (Rousset 1992).

Fig. 1 shows the [O II] (left) and [O III] (right) emission maps, respectively superimposed with the 1.4 GHz (Mc Carthy *et al.* 1987) and 4.8 GHz (Hutchings 1993) radio isophotes. The alignment is evident as well as the identification of an interfacing zone with the intergalactic medium. The velocity field of the [O II] line is strongly distorted by many components as shown in Rocca-Volmerange *et al.* (1994). The ratio [O II]/[O III] follows a crescent distribution along the radio lobe which peaks at the value 8. A new observational result is from the comparison of the isophote maps respectively for the continua (B and R) and the [O II] and [O III] emission lines. Galaxy companions are well separated from the nebular peaks (Rocca-Volmerange *et al.* 1994). Stellar continua of companions are significantly observed in two of them, others being below the detection limit. Resulting from the sum of dispersed micropupils, these spectra are shown on Figure 2. The continuity of the slope from the V to the R-bands confirms the availability of the data processing. Details of absorption lines show an evolved stellar population, in particular around the active nucleus, the interpretation of which is given below.

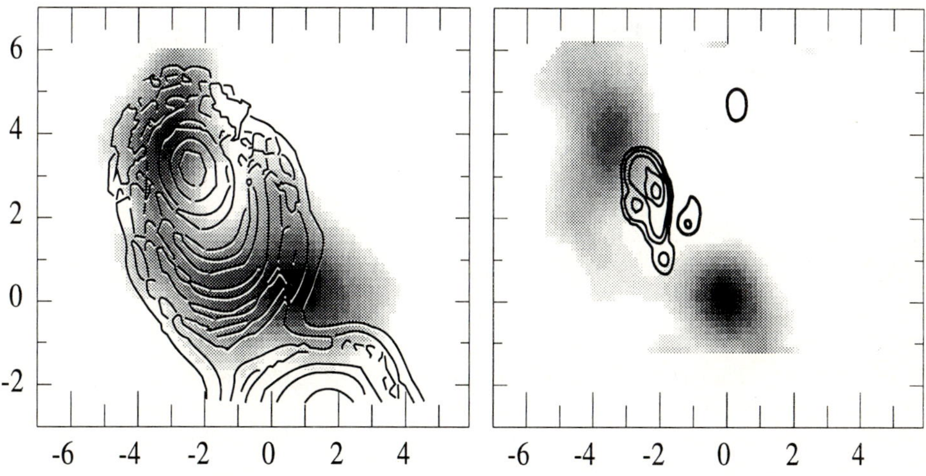

Fig. 1. The gray-scale [O II] and [O III] maps of 3C 435A respectively superimposed to the 1.4GHz (left) and 4.8 GHz (right) radio isophotes.

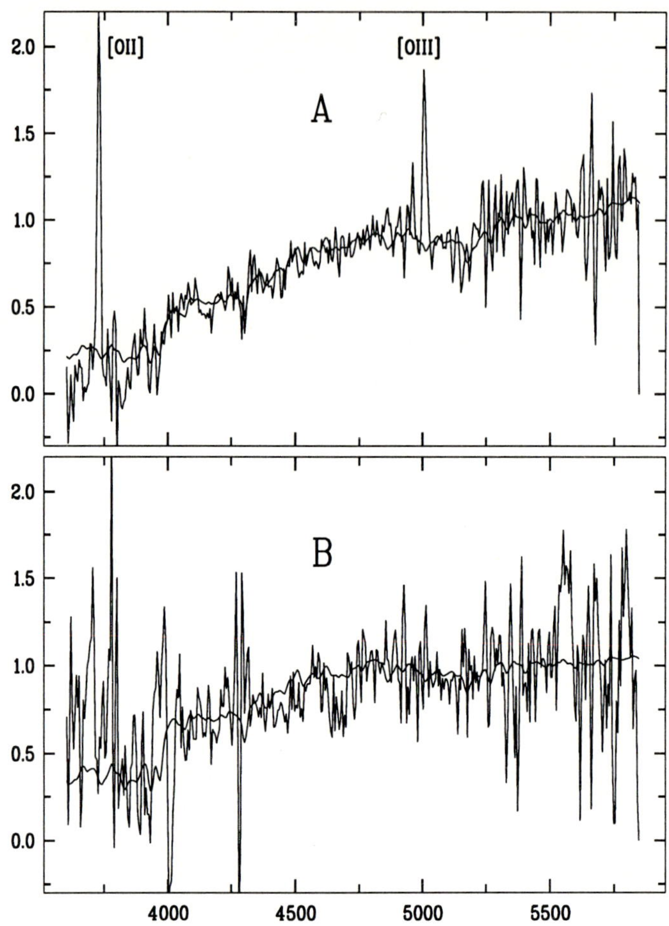

Fig. 2. Spectra of the central A and northern B components of 3C 435A fitted with oursynthetic Atlas of galaxies (Rocca-Volmerange & Guideroni 1988). The stellar energy distributions are in energy/time/area/wavelength

4 Past and Present Star Formation Activities

Several important results on the current and past star formation activities can be derived from these data. The [O II] and radio axis alignment is observed as in nearby radiogalaxies (van Breugel *et al.* 1985). If the [O II] emission line is used as a possible signature of the present star formation, the curvature of the isophotes is not in agreement with a triggering along the central radio jet, as suggested by models. Here the [O II] emission only seems to trace zones of overpressured gas. In that case, the curvature of isophotes following the radioplasma in expansion is in favor of the model of overpressured cocoon (Begelman & Cioffi 1989). The

peripherial shell of ionised gas merely explains the confinment of the radiolobes without any help of magnetic field. This scenario favors a current low level of polarisation as observed by di Seregho Alighieri (1993) without eliminating the possibility of a stronger activity at earlier epochs.
The stellar energy distributions (SED) of the central component A and of the north companion B (Figure 2) are indicators of the past activity. A fit of the SED by a minimisation procedure with our atlas of synthetic spectra of galaxies (Rocca-Volmerange & Guideroni 1988) gives respective ages of 11 Gyrs and 8.5 Gyrs, corresponding to 14 Gyrs and 12 Gyrs at the redshift z = 0 in a low density universe, this datation is based on isochrones of globular clusters and means an old stellar population. A more statistically significant sample of distant radiogalaxies observed at increasing redshifts is in progress with TIGER at the CFHT to relate star formation activity with various powers of radio jets, the environment density and the radiolobe confinment. In particular, the distant radiogalaxy 4C 41.17 at $z = 3.8$ (Chambers *et al.* 1990) has been recently observed with TIGER, data are being processed.

References

Bacon, R., Adam, G., Baranne,A., Courtès, G., Dubet, D., Dubois, J.P., Emsellem, E., Ferruit, P., Georgelin, Y., Monnet, G., Pécontal, E., Rousset, A., Sayède F. 1995, *Astron. Astrophys.*, in press

Begelman, M.C., Cioffi, D.F. 1989, *Astrophys. J. Lett.* **345**, L21

Bithell, M., Rees, M. 1989, *Mon. Not. R. Astr. Soc.* **242**, 570

Chambers, K., Miley, G., van Breugel, W. 1987, *Nature* **329**, 604

Chambers, K., Miley, G., van Breugel, W. 1990, *Astrophys. J.* **363**, 21

Cimatti, A., di Seregho Alighieri, S., Fosbury, R.A., Salvati, M., Taylor, D. 1993, *Mon. Not. R. Astr. Soc.* **264**, 421

Daly, R.A. 1990, *Astrophys. J.* **355**, 416

de Young, D.S. 1989, *Astrophys. J. Lett.* **342**, L59

Eales, S., Rawlings, S., Puxley, P., Rocca-Volmerange, B., Kuntz, K. 1993, *Nature* **33**, 14

Giavalisco, M., Macchetto D., Madau, P., Sparks, W. 1994, StSci preprint n0 885

Hutchings, J., private communication

Lilly, S. 1988, *Astrophys. J.* **333**, 161

McCarthy, P.J., van Breugel, W., Spinrad, H., Djorgovski, S. 1987, *Astrophys. J.* **321**, L29

Rousset, A. 1992, *Thèse de Spécialité de l'Université de St Etienne*, France

Rocca-Volmerange, B. 1993, in *First Light in the Universe*, Proceedings of the 8e IAP meeting, Rocca-Volmerange B. *et al.* , eds, Editions Frontieres, p.283

Rocca-Volmerange, B., Adam, G., Ferruit, P., Bacon, R. 1994, *Astron. Astrophys.* **292**, 20

Rocca-Volmerange, B., Guiderdoni, B. 1988, *Astron. Astrophys. Suppl.* **75**, 93

di Seregho Alighieri, S., Cimatti, A., Fosbury, R.A.E. 1993, *Astrophys. J.* **404**, 584

van Breugel, W., Miley, G., Heckman, T., Butcher, H., Bridle, A. 1985, *Astrophys. J.* **290**, 496

Low-Redshift Constraints on the Formation of Elliptical Galaxies

Ralf Bender[*]

Universitäts-Sternwarte
Scheinerstr. 1, D-81679 München, Germany

1 Introduction

This paper discusses the constraints on the formation processes and ages of elliptical galaxies that can be obtained from (mostly) low redshift observations.

The stellar kinematics of elliptical galaxies provides still the best evidence for significant merging having been involved in their formation (Section 2). The stellar population properties of ellipticals, on the other hand, all point to a rather early formation of the bulk of their stars in a relatively short time scale (Section 3). These two findings together favor a formation picture that is, to first order, consistent with hierarchical structure formation in a cold dark matter Universe.

2 Formation of Elliptical Galaxies via Dissipative Merging

2.1 A Case Study: NGC 4365

The evidence for the formation of luminous elliptical galaxies in dissipative merging is most easily demonstrated by investigating a typical object in some detail.

NGC 4365 is a photometrically normal, luminous elliptical galaxy in the Virgo cluster. However, it is kinematically highly interesting. Fig. 1 shows the rotation velocity and velocity dispersion profiles along the projected major and minor axis (cf. Bender 1990, Surma & Bender 1995). Astonishingly, the main body of this object rotates slowly along the minor axis, while the core (i.e. the region inside of a few hundred parsecs) rotates more rapidly along the major axis. This means that the angular momentum vectors of core and main body are perpendicular to each other. The only plausible explanation for this observation is that the main body of the object is (nearly) prolate and spinning around the long axis while the core is (nearly) oblate and spinning around the short axis (Franx & Illingworth 1990). It is evident that this kind of structure could not have been generated if the galaxy was assembled from mainly gaseous constituents because efficient angular

[*] Visiting Astronomer of the German-Spanish Astronomical Center, Calar Alto, operated by the Max-Planck-Institut für Astronomie, Heidelberg, jointly with the Spanish National Commission for Astronomy.

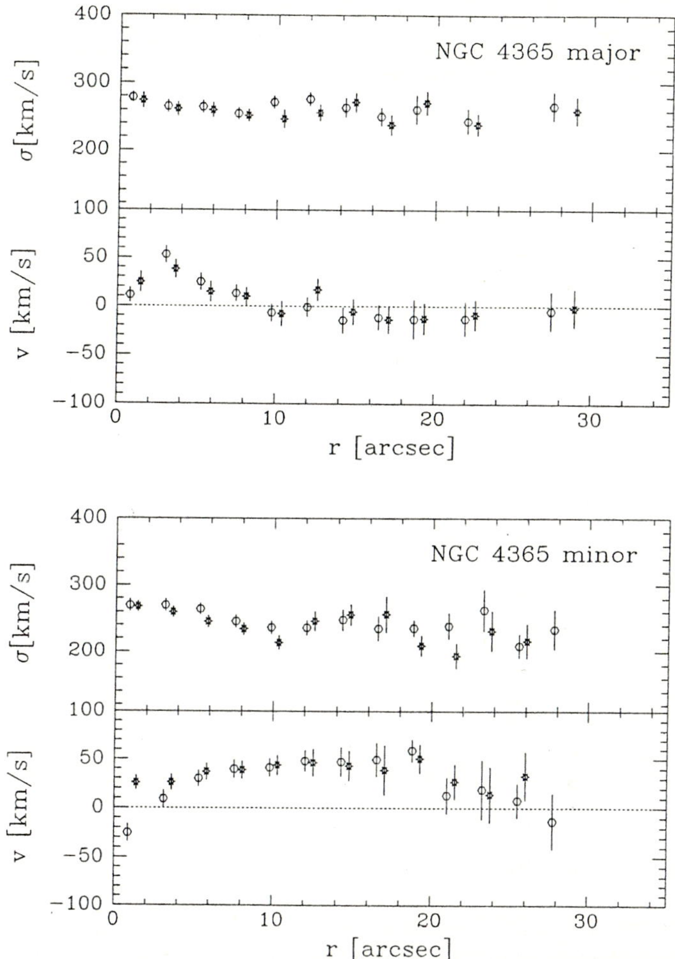

Fig. 1. Rotation velocities v and velocity dispersions σ along the major and minor axis of NGC 4365. Different symbols refer to different sides from the center. The main body and the core of NGC 4365 have angular momentum vectors that are perpendicular to each other.

momentum exchange would have lined up the angular momentum vectors of inner and outer parts. In other words, the merging constituents did already consist, at least partly, of stars.

Detailed analysis of the line-of-sight velocity distribution in the core reveals strong asymmetries. These can very plausibly be interpreted as being due to the superposition of two components: a 'background' component from the main body of the galaxy and a rapidly spinning component with $v/\sigma \approx 1$ which is confined to the core (Surma & Bender 1995). *The rapidly spinning component contributes between 1% and 2% of the total light and up to 50% of the core light.*

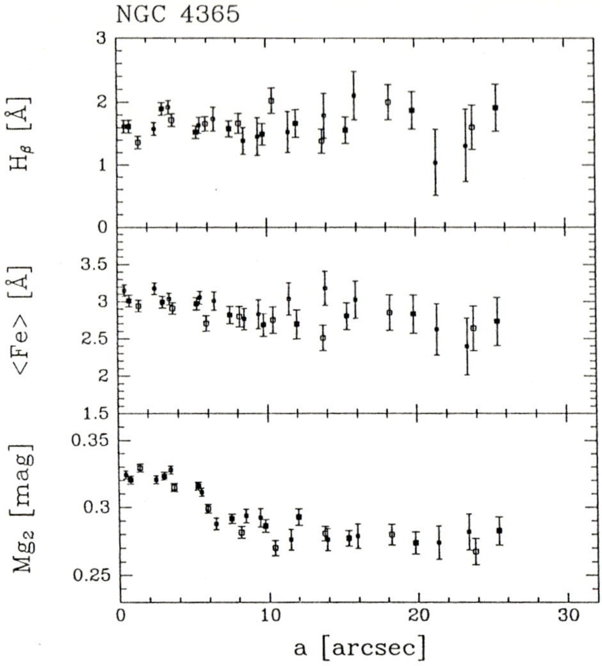

Fig. 2. Mg_2, <Fe>, and H_β line strength profiles of NGC 4365 (for the definition of the line strength indices, see Burstein *et al.* 1984). Bigger symbols refer to the major axis profile; smaller symbols to the minor axis profile which was scaled radially to the major axis using the flattening of NGC 4365. The profiles were folded about the center, open and closed symbols refer to different sides from the center. The kinematically decoupled core is more metal-rich than the main body of NGC 4365.

The radial line-strength profiles of NGC 4365 in Mg (5180Å), Fe (5270Å and 5335Å), and H_β reveal further insight, see Fig. 2. They suggest that the spinning core is very metal-rich. At about the same radius at which the kinematics change, there is a change of slope in the Mg_2 profile as well, indicating that we observe the superposition of a shallow Mg_2 gradient due to the main body of stars and a steeper central gradient due to the presence of the additional (very metal-rich) spinning component (Bender & Surma 1992). In addition, the isochromes seem to be flatter than the isophotes (Surma & Bender 1995) which supports the notion that the metallicity is enhanced only in the rapidly rotating component.

Summarizing, the structure of NGC 4365 suggests that the core formed dissipatively from rapidly rotating and metal-rich gaseous material (possibly via a violent star burst), while the *last* major processes that were involved in the creation of the main body were merging and violent relaxation (this does of course not exclude dissipation in an earlier stage). This formation scenario can be termed *dissipative merging* and is similar to the processes taking place in present-day ultra-luminous IRAS mergers (see Section 1.3).

2.2 Peculiar Cores in Ellipticals in General

NGC 4365 is not very unusual in its properties. Among the nearby *luminous* ellipticals about 1/3 show similarly peculiar kinematics (Bender 1990). This fraction is, e.g., also found in the Virgo cluster: of the nine brightest Virgo ellipticals ($M_B < -20.5$, $H_o = 50\,km/s/Mpc$) three show kinematically decoupled cores (NGC 4365, 4406, 4472). Because of projection statistics and most likely equal frequencies of co- and counter-rotation it then follows that *more than 50% of all luminous ellipticals contain kinematically decoupled cores.* In most of them, the analysis of the central kinematics shows that the peculiar cores are likely to be rapidly spinning thick disks or torus-like components that formed dissipatively and involved substantial star formation (because of their high v/σ and metallicity), for further arguments, see also Section 1.3. These components, like the core of NGC 4365, have masses in the range $10^9 - 10^{10}\,M_\odot$ and radii of a few to several hundred parsecs (Franx & Illingworth 1988, Bender 1988, 1990, Rix and White 1992, Surma & Bender 1995).

Most of the ellipticals with peculiar cores belong to the class of boxy ellipticals or ellipticals with irregular isophotes. There is only one elliptical so far (NGC 1700, Franx *et al.* 1989) which has a counter-rotating core and *significantly disky* isophotes. Boxy ellipticals are in general more luminous than disky ellipticals (e.g., Bender *et al.* 1993). Therefore the natural interpretation of these findings is that mergers that lead to counter-rotating peculiar cores will usually form luminous, mainly boxy, ellipticals. Of course, it is to be expected that pre-existing disks cannot survive major mergers (e.g. Quinn, Hernquist & Fullagar 1993) and, indeed, recent simulations by Steinmetz (1995) show that dissipationless merging produces preferentially boxy isophotes outside the core. Consequently, the observed correlation between peculiar cores and boxy isophotes of the main bodies is very plausible.

A note of caution: The peculiar core kinematics in some ellipticals can possibly be explained in ways different from the above scenario. E.g., the decoupled core could be due to streaming in a triaxial figure, obliquely projected (Binney 1985, Franx *et al.* 1991, Statler 1994a, b), or due to *dissipationless* merging with a small compact elliptical or a bulge dominated S0 (Kormendy 1984, Balcells & Quinn 1990, Balcells & Carter 1993). It seems clear however that these latter scenarios cannot account for the formation of the majority of ellipticals with peculiar cores. The reasons are that rotation amplitudes in the cores are in general too high and that core metallicities are enhanced with respect to the main body (Bender & Surma 1992). Similarly unlikely is the accretion of gas-rich irregulars because they simply do not contain enough gas to form a massive central component. Typically, a much larger amount of gas is needed, like the one found in massive spirals.

Finally, it is also noteworthy that ellipticals with peculiar cores are found in all environments, in rich clusters as well as in small groups; examples for the latter are NGC 5322 (Bender 1988), IC 1459 (Franx & Illingworth 1988), or NGC 3608 (Jedrzejewski & Schechter 1988).

2.3 The Analogy between IRAS Mergers and the Formation of Elliptical Galaxies

A plausible formation scenario for ellipticals with kinematically decoupled cores can be envisaged by inspecting the properties of ultraluminous IRAS galaxies and by N-body simulations of merging spirals. Hernquist & Barnes (1991) showed in their simulations that ellipticals with counter-rotating cores can be generated in spiral-spiral mergers. The model spirals consisted of dark matter, stars and gas in the usual mix. While the stars undergo a process of violent relaxation during merging and form a smooth $r^{1/4}$ main body, the molecular gas is efficiently transported to the center of the merger where it settles into a rapidly rotating thick disk or torus. This result of the merger simulations is consistent with the observations of the molecular gas distributions in IRAS mergers, e.g. NGC 520, Arp 220 or NGC 7252(e.g. Sanders et al. 1988, 1991; Schweizer 1990; Kormendy & Sanders 1992; Wang, Schweizer and Scoville 1992). Both simulations and observations show that the molecular gas tori can be kinematically decoupled from the main bodies of the galaxies, i.e. their angular momentum vectors can be opposite or perpendicular to the ones of the main bodies. The molecular gas masses observed in the centers of luminous IRAS mergers are very similar to those of the counter-rotating cores ($10^9 - 10^{10}\,M_\odot$, Sanders *et al.* 1991); the same is true with respect to the radii which are typically of the order of a few hundred parsecs or smaller.

The high concentration of molecular gas in the center of the merger inevitably leads to violent star formation and forms a rotationally flattened central stellar component. This component is likely to be very metal-rich because the molecular clouds were pre-enriched and also because the IMF may be top-heavy in mergers (e.g., Wright *et al.* 1988, Bernlöhr 1993). In some of the IRAS mergers we can observe this process just now (e.g. NGC 520, Arp 220). Once the IRAS mergers have aged by about 5 Gyrs, the relics of the central starbursts are likely to resemble the decoupled cores observed in ellipticals today, both with respect to kinematics and metallicity (Bender & Surma 1992). In some mergers, newly formed stars may not only be found in the center but also at larger radii. These stars are due to star formation triggered in the early phases of the merging (Fritze-von-Alvensleben & Gerhard 1994) and may contribute to a smooth overall appearance of the line-strengths gradients after several Gigayears.

These considerations show that a *qualitative* understanding of the formation of ellipticals via dissipative merging can be reached in consistency with observations of present day mergers and N-body simulations. However, this plausible analogy between IRAS mergers and the formation of ellipticals does of course neither imply that ellipticals must have formed via merging of spirals nor that they formed late (i.e. at low z) in general. It is equally possible that *both* ellipticals and spheroids formed at higher redshifts by (hierarchical) processes involving *both* merging-induced violent relaxation and dissipation. The relative amount of dissipation varied as a function of luminosity and other protogalactic parameters (like density and environment) and determined the degree of anisotropy of the final object. More luminous ellipticals may on average have assembled from more evolved progenitors (in which most of the baryonic matter had already been transformed into stars) and, thus, not only velocity anisotropy but also kinematic de-coupling between core and main body may have been produced in these objects. The important parameter determining whether the final object would show peculiar kinematics and features in the line-strength gradients is the *ratio between star formation timescale and the timescale over which violent mergings occured.* For a more detailed discussion of these points see Bender & Surma (1992), Bender, Burstein & Faber (1992, 1994).

As the discussion of the next paragraphs will show, it is indeed indicated that most present-day ellipticals formed the bulk of their stars at relatively high redshifts and on rather short time scales.

3 The Mean Ages and Star Formation Histories of Elliptical Galaxies

3.1 Mean Ages from the Stellar Populations of Nearby Ellipticals

Despite the large variety of structural properties, the stellar populations of elliptical galaxies are to first order surprisingly homogenous. Colors and line-strengths are one-to-one correlated to such an extent that it makes sense to discuss their stellar populations in wholistic terms (e.g. Sandage & Visvanathan 1978, Burstein *et al.* 1988; Peletier 1989; Faber *et al.* 1992). It is found empirically that the stellar populations of elliptical and dwarf galaxies are very tightly related to their central velocity dispersions (σ) (e.g., Dressler *et al.* 1987, Bender, Burstein & Faber 1993)

From existing stellar population synthesis models (e.g., O'Connell 1986; Bruzual & Charlot 1993; Worthey 1994) one can estimate the combined scatter in age and metallicity from the observed scatter in Mg_2. Bender, Burstein & Faber 1993 found for *normal ellipticals* that the scatter in age and/or metallicity at a fixed σ must be smaller than 15%. This means that, *for a given σ, normal ellipticals cannot have formed continuosly over the Hubble time* (this is consistent with a recent analysis of Schweizer and Seitzer 1992).[2]

Similar constraints on the range of age and/or metallicity at a fixed σ were reached independently for Coma and Virgo cluster luminous ellipticals by Ellis and collaborators (e.g., Ellis 1992) on the basis of the (V-K)$-\sigma$ correlation, and by Faber *et al.* (1987) and Renzini & Ciotti (1993) on the basis of the small scatter in M/L perpendicular to the fundamental plane. Again, it has to be noted that these arguments apply to luminous ellipticals only.

[2] However, this does not rule out the possibility that ellipticals of different σ are of systematically different ages. Systematic age differences are indeed indicated by the most recent investigation of Gonzalez (1993), low-luminosity Es ($M_T \approx -18$) seem to be systematically younger than giant Es ($M_T \approx -21$). Note that this trend runs opposite to the one expected in a cold-dark-matter model (Kauffman *et al.* 1993).

3.2 Ages of Ellipticals from their Color and Line-Strengths Evolution with Redshift

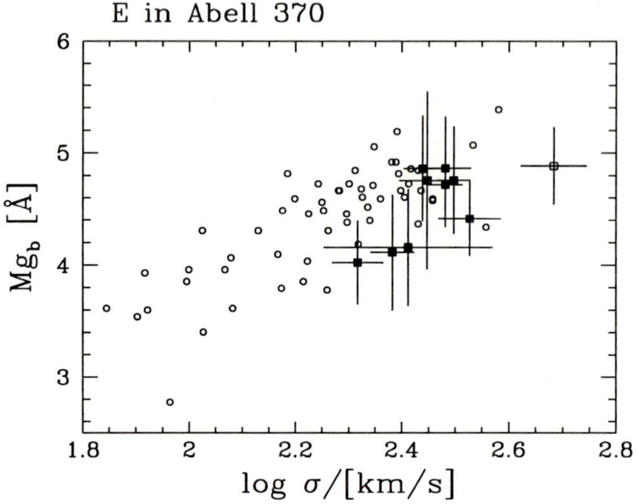

Fig. 3. The $\mathrm{Mg}_b - \sigma$ relation of elliptical galaxies in the Virgo and Coma cluster (open circles) and of cluster ellipticals at $z \approx 0.37$, mostly in Abell 370 (filled squares). Mg_b gives the equivalent width of the Mg lines at 5170Å (rest wavelength) and σ the velocity dispersion of the stars in the ellipticals. Both values refer to mean values within about half the effective radius, the $z = 0.37$ measurements were corrected for aperture effects. The relatively modest weakening of Mg_b with redshift requires the bulk of the stars in cluster ellipticals to have formed at redshifts above 3.

In order to constrain formation ages of ellipticals, Ellis and collaborators (e.g., Ellis 1992, Aragón-Salamanca *et al.* 1993) investigated the evolution of V-K colors of Brightest Cluster Members (BCM) up to $z = 0.9$. From the rather small color evolution they concluded that ellipticals have mostly formed at redshifts of above 5 if $q_o = 0.5$. The reliability of this result relies on a perfect understanding of selection effects. E.g., if systematic metallicity differences between objects at different redshifts existed, then this would have a severe influence on the interpretation of relative ages. Worthey's (1994) evolution models for simple stellar populations illustrate this very clearly. For an old (> 3 Gyr) stellar population, an increase in metallicity by a factor two and a decrease in age by a factor three results in almost identical optical and near IR colors. For typical ages and metallicities one obtains for instance:
An elliptical galaxy of 12 Gyr and [Fe/H]= 0.0 *has the same* $V - K$ *color as an elliptical of 8 Gyr and [Fe/H]=* 0.1.
Yet another striking example: increasing the metallicity of an 8 Gyr elliptical by 0.25 dex makes this object redder by $\Delta(V - K) \approx 0.4$ mag. This rather extreme

sensitivity of colors on metallicity implies that, if, e.g., the colors of high redshift BCM's show only little evolution with respect to 'local' BCM's, it may simply mean that an underlying stronger evolution is compensated for by introducing a redshift-dependent bias to more massive and therefore more metal-rich objects.

Therefore, an unambiguous measurement of the evolution of BCM's with z requires an estimate of their metallicity. This can be done using the Mg$-\sigma$ relation which, as discussed above, constrains the metallicity scatter to be smaller than 15% at a given velocity dispersion. Although a small difference in the Mg$-\sigma$ relation between low and high density environments may exist, this effect does not affect this conclusion significantly. From these considerations it follows that the determination of the Mg$-\sigma$ relation for BCM's at higher redshift would allow a much more reliable estimate of their evolution than colors alone. In addition, the knowledge of the velocity dispersion allows one to test whether an observed sample of high z BCM's contains significantly more massive objects than a local comparison sample.

We are currently measuring the Mg$-\sigma$ relation for a sample of brightest cluster members at redshifts between 0.2 and 0.7 with the 3.5m telescope at Calar Alto. A first result from this project is presented in Figure 3 (Bender, Ziegler and Bruzual 1995). There is clear evidence for evolution, in the sense that, at any given velocity dispersion, the strongest Mg absorption found at $z = 0.4$ is significantly weaker than at $z = 0$ (note that due to sample selection effects (e.g., color), evolution can reliably be traced only with the upper envelope of the object distribution in this diagram). Translating the observed Mg difference with Worthey's (1994) models into a relative change in age indicates that *the bulk of stars in cluster ellipticals formed at redshifts above 3.* This result is roughly consistent with the predicted mean ages of elliptical galaxies in a cold dark matter universe (Kauffmann 1995). However, the number of data points is still too small to allow a discrimation between the standard ($\Omega = 1$) cold dark matter model and a low density CDM model.

In the discussion so far we assumed that the stellar population in ellipticals evolve passively only. If we use the same line of arguments as above but allow for significant low redshift star formation then this implies an even higher age for the bulk of the old stars in these objects.

3.3 Element Ratios and Star Formation Time Scales in Elliptical Galaxies

Recent observations by Peletier (1989), Faber *et al.* (1992) and Davies *et al.* (1993) have shown that in luminous elliptical galaxies Mg is overabundant relative to Fe by roughly a factor three[3]. In a recent PhD thesis, Paquet (1994) could show that, besides Mg, other light elements like Na and CN are overabundant relative to iron as well. These findings imply that the enrichment of ellipticals was in general dominated by Supernovae II (that mainly produce light elements), which in turn means that Supernovae Ia (that produce almost exclusively iron peak elements) did contribute significantly less iron than in the Solar

[3] This relies mostly on measurements inside the half-light radius of the galaxies

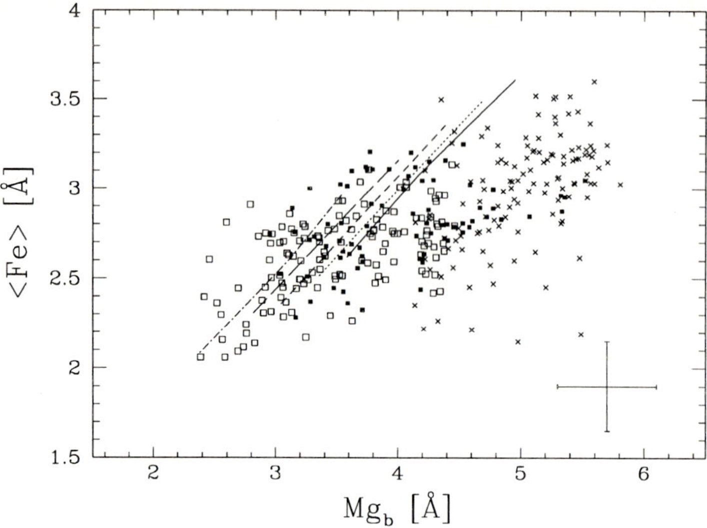

Fig. 4. Local <Fe> vs. Mg_b of S0 disks (open squares), S0 bulges (filled squares) and elliptical galaxies (crosses). The lines represent population synthesis models by Worthey (1994) of ages between 3 Gyrs and 17 Gyrs, metallicity varies along the lines, $-0.3 <$[Fe/H]$< +0.3$. The synthesis models have solar element ratios, as usually do have the disks of S0 galaxies. Elliptical galaxies show Mg overabundance of up to [Mg/Fe]$= +0.5$.

Neighborhood (see, e.g., Truran and Thielemann (1986), Matteucci and Greggio (1986), Faber *et al.* (1992), Matteucci (1993), Truran and Burkert (1994). It is important to note that ellipticals with peculiar cores do indeed show the same overabundance in light elements as luminous ellipticals on average (Surma & Bender 1995, Paquet 1994). So far, we found only one in a sample of five peculiar core ellipticals which showed solar element ratios (NGC 5322). Therefore, the following arguments apply equally to peculiar core ellipticals and to ellipticals in general.

The prevalence of Supernovae II and in turn the light element overabundance in ellipticals can be due to the following effects: (a) a star formation time scale smaller than 10^9 years (SNI explode in significant numbers only after about 1 Gyr after star formation started), (b) a top heavy initial mass function, (c) a reduced frequency of binary stars (leading to fewer SNI events), (d) selective mass loss mechanisms that resulted in a more efficient loss of SNI elements. Options (c) and (d) are rather unlikely because one expects the binary frequency to be determined by the local process of star formation rather than by global galaxy properties and because selective mass loss processes are likely to work, if at all, only in low mass galaxies (see Gilmore and Wyse 1991). However, also option (b) does not work well, if the overabundance in ellipticals approaches

[Mg/Fe]$\approx$ 0.5 dex. Both from the abundance analysis of recent SNII events as from the abundance pattern of Galactic halo stars, it follows that this [Mg/Fe] overabundance is produced by SNII alone without any significant input of Fe from SNI. Turning the argument around, this means that in most luminous ellipticals ([Mg/Fe]$\approx$ 0.3) one can allow for a only rather modest enrichment by SNIa events. The consequence is that *star formation time scales for the bulk of the stars in ellipticals most likely were shorter than roughly 2Gyr.* A moderately top-heavy IMF and *significant* star formation extending over more than 2Gyr is unlikely to solve the overabundance problem because after 2Gyr the Fe enrichment via SNI would start to reduce [Mg/Fe] below the observed value.

4 Conclusions

Most of the arguments given in the previous sections are still rather qualitative. Nevertheless, the following conclusions can be reached with reasonable confidence:

- Elliptical Galaxies formed from merging of massive progenitor objects that consisted partly of stars and partly of gas.
- The bulk of the stars in a very large fraction of massive ellipticals formed at redshifts above three.
- The star formation time scale for the bulk of the stars in elliptical galaxies was most likely shorter than about two Gigayears.
- Conclusions (1) to (3) do not contradict the hypothesis that late *major* mergers do happen and do form ellipticals. The structure of present-day ellipticals is actually similar to the expected final evolutionary stage of present-day mergers. However, late mergers are a minority among the global population of the ellipticals.
- Conclusions (1) to (3) do neither contradict the existence of *minor* accretion or merging events that may lead to the E+A phenomenon (see, *e.g.*, Belloni & Röser 1995). However, these minor events are unlikely to add a large fraction of mass to the existing underlying old population.

Acknowledgements: I thank David Burstein, Sandy Faber, Guinevere Kauffmann, John Kormendy, Roberto Saglia and Peter Surma for many interesting discussions. I am grateful to Alexander Paquet and Bodo Ziegler for allowing to use data obtained for our common projects prior to publication. Last but not least, many thanks go to Hans Hippelein and his colleagues at MPIA for the organization of a stimulating workshop and to the Max-Planck Society for financial support.

References

Aragón-Salamanca, A., Ellis, R.S., Couch, W.J., Carter, D. 1993, *Mon. Not. R. Astr. Soc.* **262**, 764

Balcells, M., Quinn, P. 1990, *Astrophys. J.* **361**, 381
Balcells, M., Carter, D. 1993, *Astron. Astrophys.* **279**, 376
Belloni, P., Röser, H.J. 1995, this conference
Bender, R. 1988, *Astron. Astrophys.* **202**, L5
Bender, R. 1990, in *Dynamics and Interactions of Galaxies*, ed. R. Wielen, Springer Verlag Heidelberg, p.232
Bender, R., Burstein, D., Faber, S.M. 1992, *Astrophys. J.* **399**, 462, BBF1
Bender, R., Burstein, D., Faber, S.M. 1993, *Astrophys. J.* **411**, 153, BBF2
Bender, R., Surma, P. 1992, *Astron. Astrophys.* **258**, 250
Bender, R., Burstein, D., Faber, S.M. 1994, in it Panchromatic View of Galaxies, eds. G Hensler *et al.* , Editions Frontieres, Gif-sur-Yvette, p.99
Bender, R., *et al.* 1993, in *Structure, Dynamics and Chemical Evolution of Elliptical Galaxies*, ESO/EIPC workshop, eds. J. Danziger *et al.* , European Southern Observatory, München
Bender, R., Ziegler, B., Bruzual, G. 1995, *Astrophys. J.*, submitted
Bernlöhr, K. 1993, *Astron. Astrophys.* **270**, 20
Binney, J. 1985, *Mon. Not. R. Astr. Soc.* **212**, 767
Bruzual, G, Charlot, S. 1993, *Astrophys. J.* **405**, 538
Burstein, D., Faber, S.M., Gaskell, C.M., Krumm, N. 1984, *Astrophys. J.* **399**, 462
Burstein,D., Davies, R.L., Dressler, A., Faber, S.M., Stone, R.P.S., Lynden-Bell, D., Terlevich, R., Wegner, G. 1988, in *Towards Understanding Galaxies at Large Redshift*, eds. R.G. Kron & A. Renzini, Kluwer Dordrecht, p.17
Davies, R.L., Sadler, E., Peletier, R. 1993, *Mon. Not. R. Astr. Soc.* **262**, 650
Dressler, A., Lynden-Bell, D., Burstein, D., Davies, R.L., Faber, S.M., D., Terlevich, R., Wegner, G. 1987, *Astrophys. J.* **313**, 42
Ellis, R. 1992, in *The Stellar Populations of Galaxies*, IAU Symp. 149, eds. B. Barbuy & A. Renzini, Kluwer Dordrecht
Faber, S.M., Worthey, G., Gonzalez, J. 1992, in *The Stellar Populations in Galaxies*, IAU Symp. 149, eds. B. Barbuy & A. Renzini, Kluwer Dordrecht
Faber, S.M., Dressler, A., Davies, R.L., Burstein, D., Lynden-Bell, D., Terlevich, R., Wegner, G. 1987, in *Nearly Normal Galaxies, From the Planck Time to the Present*, ed. S.M. Faber, Springer Verlag, New York, p.175
Franx, M., Illingworth, G. 1988, *Astrophys. J.* **327**, L55
Franx, M., Illingworth, G., Heckman, T. 1989, *Astrophys. J.* **344**, 613
Franx, M., Illingworth, G. 1990, in *Dynamics and Interactions of Galaxies*, ed. R. Wielen, Springer Verlag Heidelberg
Franx, M., Illingworth, G., de Zeeuw, T. 1991, *Astrophys. J.* **383**, 112
Fritze-von Alvensleben, U., Gerhard, O. 1994, *Astron. Astrophys.* **285**, 775
Gilmore, G., Wyse, R.F.G. 1991, *Astrophys. J.* **367**, L55
Gonzalez, J. 1993, *Ph.D thesis*, University of California at Santa Cruz
Hernquist, L., Barnes, J.E. 1991, *Nature* **354**, 210
Jedrzejewski, R., Schechter, P. 1988, *Astrophys. J.* **330**, L87
Kauffmann, G., White, S.D.M, Guiderdoni, B. 1993, *Mon. Not. R. Astr. Soc.* **264**, 201
Kauffmann, G. 1995, *Mon. Not. R. Astr. Soc.*, in press
Kormendy, J. 1984, *Astrophys. J.* **287**, 577
Kormendy, J., Sanders, D.B. 1992, *Astrophys. J.* **390**, L53
Matteucci, F., Greggio, L. 1986, *Astron. Astrophys.* **154**, 279
O'Connell, R. 1986 in *Stellar Populations*, eds. C. Norman *et al.* , Cambridge University Press
Paquet, A. 1994, *Ph.D thesis*, University of Heidelberg

Peletier, R. 1989, *Ph.D thesis*, University of Groningen
Quinn, P.J., Hernquist, L., Fullagar, D.P. 1993, *Astrophys. J.* **403**, 74
Renzini, A., Ciotti, L. 1993, *Astrophys. J. Lett.* **416**, L49
Rix, H.-W., White, S.D.M. 1992, *Mon. Not. R. Astr. Soc.* **254**, 389
Sandage, A., Visvanathan, N. 1978, *Astrophys. J.* **223**, 707
Sanders, D.B., Scoville, N.Z., Sargent, A.I., Soifer, B.T. 1988, *Astrophys. J.* **324**, L55
Sanders, D.B., Scoville, N.Z., Soifer, B.T. 1991, *Astrophys. J.* **370**, 158
Schweizer, F. 1990, in *Dynamics and Interactions of Galaxies*, ed. R. Wielen, Springer Verlag Heidelberg, p.232
Schweizer, F., Seitzer, P. 1992, *Astron. J.* **104**, 1039
Statler, T.S. 1994a, *Astrophys. J.* **425**, 500
Statler, T.S. 1994b, *Astrophys. J.* **108**, 111
Steinmetz, M. 1995, this conference
Surma, P., Bender, R. 1995, *Astron. Astrophys.* **298**, 405
Truran, J., Thielemann, F. 1986 in *Stellar Populations*, eds. C. Norman *et al.* , Cambridge University Press
Truran, J., Burkert, A. 1994, in *Panchromatic View of Galaxies*, eds. G Hensler *et al.*, Editions Frontieres, Gif-sur-Yvette, p.389
Worthey, G. 1994, *Astrophys. J. Suppl.* **95**, 107
Wright, G.S., Joseph, R.D., Robertson, N.A., James, P.A., Meikle, W.P.S. 1988, *Mon. Not. R. Astr. Soc.* **233**, 1

Constraints from Element Abundances in the Galaxy

Thomas Gehren

Institut für Astronomie und Astrophysik der Universität München
Scheinerstr. 1, 81679 München, Germany

1 From Primordial Nucleosynthesis to Galaxy Formation

Among the oldest witnesses of the young universe we identify the low-mass stars in the spheroidal components of the Milky Way. Since the pioneering work of Eggen *et al.* (1962) there is no doubt that these objects are nearly as old as the Galaxy, and despite the intricate problems with the post-main sequence evolution of low-mass stars there can be no doubt that the Milky Way halo cannot be much younger than extragalactic matter seen at large redshifts. A number of exciting but often contradictory results have been obtained recently from photometry and spectroscopy of metal-poor stars in the Galaxy, and they begin to set constraints on the early evolution of baryonic matter in the universe.

Standard calculations of primordial nucleosynthesis in the early universe agree upon that elements beyond helium are not produced in substantial amounts. Chemical enrichment of most of the elements heavier than helium by nucleosynthesis in supernovae therefore is the key to the present interpretation of the first stages of galaxy formation. Suppose the formation of the Galactic halo starts with a *metal mass fraction* $Z = 0$. The gravitational contraction of the Galaxy produces one or more stellar generations, and thus leads to a *prompt initial enrichment* of the remaining processed gaseous matter that feeds the disk. The most metal-poor disk stars are known to have $Z \approx 0.001 - 0.003$. Imagine that all heavy elements are produced in supernovae events of type II, and that the stellar generations follow a standard *initial mass function* such as *e.g.* observed in the solar neighbourhood (Salpeter 1955, Miller & Scalo 1979). Then, for each $M_\odot$ of gas reprocessed in a SN II (for which we may assume a lower mass limit of 8 $M_\odot$), $300 M_\odot$ vanish in form of *unevolved subdwarfs* with masses $M_{sd} \leq 1 M_\odot$. These numbers follow straight from integrating the initial mass function (IMF) over its high- and low-mass tails.

Let the initial mass of the gas producing the Milky Way *disk* be $10^{11} M_\odot$. To contaminate this amount of *primordial* gas with heavy elements from SNe II requires $3\,10^7$ SNe II, where for simplicity a mean mass of $25 M_\odot$ has been adopted, which returns $\approx 4 M_\odot$ of heavy elements (Maeder 1992). This amounts to a total of $7.5\,10^8 M_\odot$ going into SNe II *plus* $2\,10^{11} M_\odot$ being lost to subdwarf

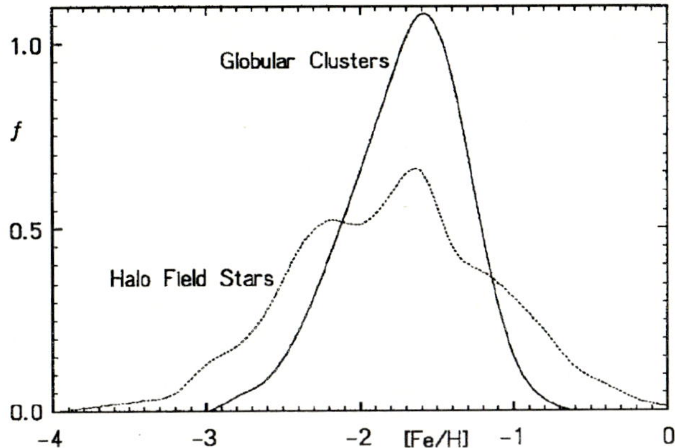

Fig. 1. Distribution of heavy element abundances in stars of the Galactic halo. The data are from Carney (1993, halo field stars), and from the compilation of Webbink (1985, globular clusters)

remnants. Since only $\approx 10^{10} M_\odot$ of subdwarfs are observed the above assumptions are contradictory, and we have to conclude that *the standard IMF is not valid for low Z gas.*

This is already a very strong constraint on the IMF since – irrespective of the history of star formation – it tells us that the first stellar generations must have consisted essentially of high-mass stars. We can extend these estimates to the scenario of Brown *et al.* (1991) who describe the self-enrichment of heavy elements in globular clusters. Here, the initial generation again forms SNe II that explode and return most of their mass to expanding remnants. Thus colliding shells build up in which the globular clusters are finally formed, with heavy element abundances of $Z \geq 10^{-4}$ (see Fig. 1). Using the same assumptions as above, the enrichment of a typical cluster mass of $10^6 M_\odot$ would require 30 SNe of $25 M_\odot$ each. A standard IMF then would imply a low-mass tail of $2\,10^5 M_\odot$ of which 10% must be spectral types between sdG and sdK. Note that the heavy element abundances of these *precursor stars* would have to be between $Z = 0$ and $Z = 10^{-4}$. This pre-globular cluster generation is often referred to as Population III stars, and both estimates given above require a special IMF for such a population. At variance with Brown *et al.* the high number of SNe II necessary to provide the observed enrichment of the following populations definitely requires an effective low-mass cutoff of the Pop III IMF above $1 M_\odot$.

2 The Most Metal-Poor Stars in the Galaxy

From his recent proper motion survey Carney (1993) has derived a new abundance distribution of metal-poor stars. Comparison with globular cluster data of Webbink (1985) makes it obvious that the most metal-poor stars are found among the *field stars* and not in globular clusters (cf. Fig. 1). During the last years a considerable number of such extreme halo stars have been found, and a few of them have been analyzed spectroscopically. Whereas stars with metal abundances of [Fe/H] < −2 such as the standard HD 140283 have been known for some time, the number of stars with [Fe/H] < −3 is really small. Typical abundances for metal-poor stars are given in Table 1.

Table 1. Abundances of the most metal-poor field stars

Object	Type	[Fe/H]	[O/Fe]	[Mg/Fe]	Reference
G77-61	dwarf	−5.5	< 1.6	1.2	Gass *et al.* 1988
CD −38°245	giant	−4.5		0.2	Bessell and Norris 1984
22885-96	giant	−4.2		0.7	Norris *et al.* 1993
22876-32	turnoff	−4.1		0.6	Norris *et al.* 1993
G64-12	turnoff	−3.3		0.1	Norris *et al.* 1993
			< 1.7		Abia & Rebolo 1989
HD 140283	subgiant	−2.7		0.3	Nissen *et al.* 1994
			0.6		Bessell *et al.* 1991
mean halo	subdwarf		0.5	0.4	

While the abundances in Table 1 are uncertain in particular for the extreme stars they give a coarse orientation about the abundance ratios found in the oldest halo stars. It suggests that an approximate *lower limit* of the halo field star heavy element abundance is found around $Z = 10^{-4} Z_{\odot} \approx 1.7\,10^{-6}$. Assuming that this abundance corresponds to the *minimum metal enrichment*, a single SN II of 25 $M_{\odot}$ would be enough to contaminate a gas mass of $M \approx 2\,10^6 M_{\odot}$. The exact values depend on the details of stellar evolution; here $Z = 0.001$ models have been used instead of $Z = 0$. Thus, the minimum metal abundance found in the halo corresponds nicely to the typical Jeans mass of $M_J \approx 10^6 M_{\odot}$ derived by Fall & Rees (1985) for the hierarchical clustering model. Again, the IMF of the first Galactic stars seems to be peaked near 25 $M_{\odot}$ with only small contributions from masses below $15 M_{\odot}$. Further restrictions can be derived from abundance *ratios* found in extremely metal-poor stars.

2.1 Abundance Ratios in Halo Stars

Reliable information about abundances in the Galactic halo is obtained mostly from spectroscopic analyses of nearby stars. This is due to fact that cool stars (redder than the *turnoff*) require high-resolution spectroscopy ($R > 20000$) to yield accurate results. Except for very rough estimates such as obtained from broad-band photometry our present knowledge about the halo comes from stars within ≈ 1 kpc distance from the Sun; yet most of these field halo stars are relatively faint. Apart from corresponding observational errors, however, there exists a number of methodical constraints and uncertainties that contribute strongly to the reliability of the interpretation,

(a) *Stellar parameters* cannot be measured directly, thus modelling with stellar atmospheres is unavoidable. Not any two groups of researchers seem to agree on assumptions about
  - what *physical principles* to include in the models, or
  - how to approximate or determine the necessary stellar parameters T_{eff}, $\log g$, v_{turb}

(b) Atomic and molecular data are often uncertain; for many elements a consistent treatment of the non-thermal excitation and ionization in cool stellar atmospheres is missing

(c) Abundance analyses are affected by systematic errors that generally do not produce normal error distributions

(d) Often results found in the literature are strongly biased by the use of different *types* of atmospheric models to represent the Sun (with an empirical solar model as reference) and the metal-poor stars (with flux-conserving models such as those published by Kurucz 1979)

Consequently, there is no agreement about the basic solar data with which the metal-poor stars are compared. Even the solar Fe/H abundance is heavily debated with values found between the *meteoritic* log(Fe/H) = 7.51 and log(Fe/H) = 7.67 as obtained by Blackwell's group using the empirical solar atmosphere of Holweger & Müller (1974). Although the same atomic data would lead to [Fe/H] ≈ 7.54 when computed from a flux-conserving ODF-blanketed model, the higher value was used as the solar [Fe/H] abundance in Kurucz' (1992) new models. However, if the abundance data are treated (and corrected) individually, a few significant patterns emerge among which those describing the abundances of oxygen and magnesium are most interesting.

The oxygen abundance and, in particular, the oxygen/iron abundance ratio is expected to trace the contributions of different sites and processes of heavy element nucleosynthesis. Usually referred to as SN II and SN I two sites are identified that differ substantially in their respective iron and oxygen production. Whereas SNe II (and particularly those of masses $> 20 M_\odot$ produce mostly oxygen and only a small amount of iron, SNe of type I are thought to leave behind a significant fraction in heavy elements such as iron. Thus, in principle the O/Fe ratio observed in a halo star should represent the fractional contribution of both types of SNe at any epoch of Galactic chemical evolution. However, there are differences between abundances obtained from different spectral regions as is evident from Fig. 2. Although a final conclusion has not yet been arrived at, it is likely that Abia & Rebolo's (1989) results systematically overestimate the oxygen abundances. This is supported by the analyses of Spite & Spite (1991) for a subset of Abia and Rebolo's stars using instead the forbidden [O sc i] line. Consequently, the mean [O/Fe] ratio in extremely metal-poor stars should be

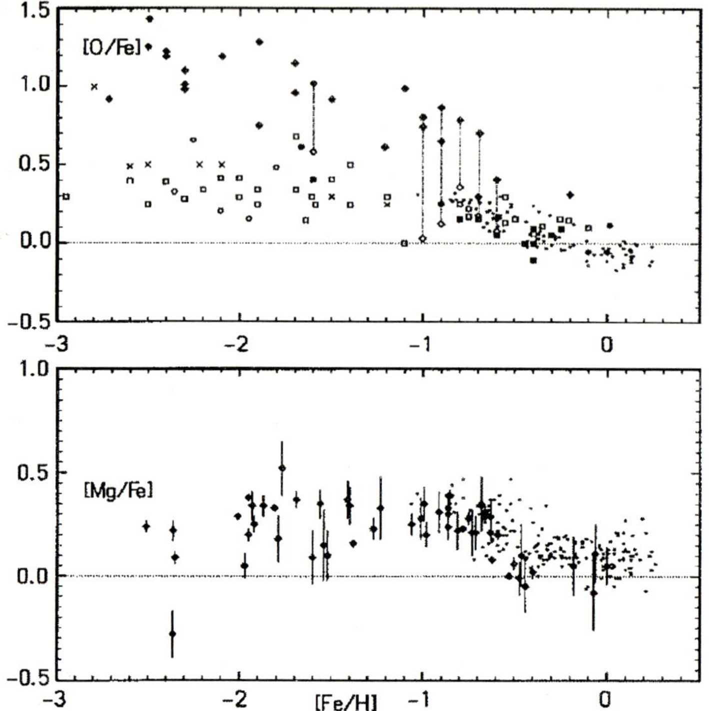

Fig. 2. Logarithmic abundance ratios as a function of overall metal abundance. Dots refer to disk stars of Edvardsson *et al.* (1993). *Top*: Oxygen/iron abundance ratios. Dotted lines combine oxygen abundances as obtained from the red triplet at 777 nm (Abia & Rebolo 89, filled diamonds) and from the forbidden line at 630 nm (Spite & Spite 1991, open diamonds) for the same star. Other data are from Gratton & Ortolani (1986, circles), Barbuy (1988, squares; see also Barbuy & Erdelyi-Mendes 1989), and Bessell *et al.* (1991, crosses). *Bottom*: Mg/Fe ratio as determined by Fuhrmann *et al.* (1994)

more like 0.5. Similar results are obtained for magnesium. Here the data from recent analyses of Fuhrmann *et al.* (1994) lead to a maximum abundance ratio of [Mg/Fe] ≈ 0.4 for stars more metal-poor than [Fe/H] ≈ -0.6, in general agreement with the results of Edvardsson *et al.* (1993).

Both results support the notion that Galactic chemical evolution depends on two characteristic time scales corresponding to the stellar evolution of stars with masses above 8 $M_\odot$ leading to SNe II with their pronounced enrichment of both O and the α-rich element Mg, and of stars with masses of a few $M_\odot$ only that end as SNe I and produce more Fe. Typical evolution times for SNe II thus would be around 10^7 yr and lead to immediate enrichment with O and Mg, whereas SNe I would require at least a few 10^8 or even more than 10^9 yr to contaminate the interstellar medium with Fe. It is not straightforward to explain the O/Fe ratio for the most metal-poor stars in the Galaxy because stellar nucleosynthesis has been calculated mostly for Pop I matter ($Z \geq 0.001$). Candidates for $Z = 0$ nucleosynthesis are the *pair instability* supernovae. El Eid *et al.* (1981) and Woosley & Weaver (1982) both come to the conclusion that roughly 25% of a very massive 200 $M_\odot$ hypernova are returned as ^{16}O. Since even the most extreme subdwarfs show some Fe, however, the number of such hypernovae during the initial stage of Galaxy formation must have been very small.

Theoretical models for type II supernovae of a slightly more metal-rich initial composition $Z = 10^{-2} Z_\odot$ with $M_{\mathrm{SN\,II}} = 25 M_\odot$ have been computed by Woosley & Weaver (1982). Comparison with more recent models of Woosley (1987, $Z = Z_\odot$), Thielemann *et al.* (1993, $Z = Z_\odot$) and Maeder (1992, $Z = Z_\odot, Z = 0.001$) confirms that mass fractions of 0.1, 0.01, and 0.002 are returned from a 25 $M_\odot$ SN II as oxygen, magnesium, and iron, respectively. A single type of a 25 $M_\odot$ supernova thus would lead to abundance ratios of [O/Fe] = 0.77 and [Mg/Fe] = 0.84. One result is that both the 200 $M_\odot$ hypernova with $Z = 0$ and the 25 $M_\odot$ SN II with $Z = 10^{-2} Z_\odot$ lead to [O/Mg] ≈ 0 as is observed in the halo stars. Closer inspection of course requires a proper adjustment of the IMF in order to fit the oberved [O/Fe] and [Mg/Fe]. Wyse & Gilmore (1993) claim that this can be achieved by an appropriate modification of the IMF taking into account the whole spectrum of SN II masses from 8 to 80 $M_\odot$. The chemical evolution of the proto-galactic matter is thus directly constrained by the O/Fe and Mg/Fe ratios observed in extremely metal-poor stars with [Fe/H] < -3.

The above considerations allow some very speculative estimates about the mass distribution among Pop III stars. In particular the notion of a hypernova population with stars of 200 $M_\odot$ can be shown to lead to high oxygen abundances. Therefore the observed O/Fe ratio in *extremely* metal-poor stars is an important number that constrains the Pop III IMF. Two limiting cases are suggested by the observations (cf. Fig. 2).

- All stars with [Fe/H] < -1.8 have [O/Fe] ≤ 0.5. This is the most likely case at present as derived from the observation of [O I] 630 nm and the ultraviolet OH lines. It requires a $Z = 0$ population with masses mainly around 15 $M_\odot$ in order to retain the low O/Fe ratio. No massive stars are allowed. Both SNe II above 25 $M_\odot$ and hypernovae of 200 $M_\odot$ would produce a significantly higher O/Fe ratio.
- All stars with [Fe/H] < -1.8 show [O/Fe] ≥ 1.0, which corresponds to the results of Abia & Rebolo (1989). Approximately half of the halo stars ($5\ 10^9 M_\odot$) would get their heavy elements from a population of massive stars. With a mean $Z = 5\ 10^{-5}$ and oxygen contributing at least half to that number we require a production of more than $10^5 M_\odot$ of oxygen in a Pop III generation. The latter could consist of the type of hypernovae described above or of the high-mass tail of the SNe II, where the *number* of events necessary would be a few thousand.

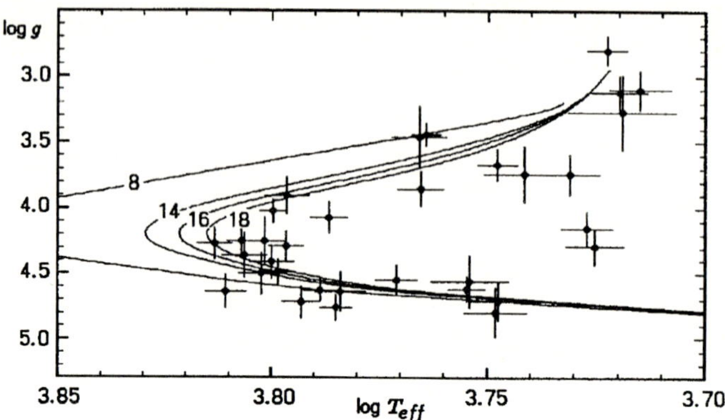

Fig. 3. HR diagram of metal-poor stars with [Fe/H] ≤ -1.6. The individual objects are corrected for the difference in abundance with respect to the mean value. Isochrones from VandenBerg & Bell (1985) are shown for a mixing-length parameter $\alpha = 1.6$. Error bars correspond to a full account of model uncertainty and stellar parameters. Note the large fraction of evolved stars

2.2 Early Evolution of the Galaxy

The strong correlation between metal abundances in stars and their *kinematic* properties first interpreted by Eggen *et al.* (1962) has been rediscussed in the light of the new data obtained in recent surveys. Whereas emphasis was put on examining the bias introduced by proper motion selection (Ryan & Norris 1993), only little effort was devoted to another source of systematic errors, the stellar distances. The kinematic results from the surveys of Sandage & Fouts (1987) or Carney & Latham (1987) have been based on the assumption that the stars involved are *subdwarfs*. At variance with the arguments usually found in favour of the subdwarf hypothesis, high-resolution spectroscopic analyses of more than 100 subdwarfs (a subsample of the above surveys) show that a non-negligible fraction of the sample (probably 50% or more) consists in fact of *subgiants*. The corresponding results of Axer *et al.* (1995) are seen in Fig. 3.

Many of these stars are evolved off the main sequence with surface gravities up to a factor 10 less than on the main sequence, which translates to a change in distance by a similar factor. Thus stars such as HD 140283 with $T_{eff} \approx 5800K$ are not found at a subdwarf location near the main sequence (with $\log g = 4.7$ but on the subgiant branch with $\log g \approx 3.5$. In fact only a very few standard subdwarfs such as HD 19445 or G64-12 turn out to be unevolved. This result affects both the stellar abundance analyses (through the surface gravity) and their kinematic properties (through the influence of the distance on the transverse velocity). It is the latter that systematically changes the space velocities of the

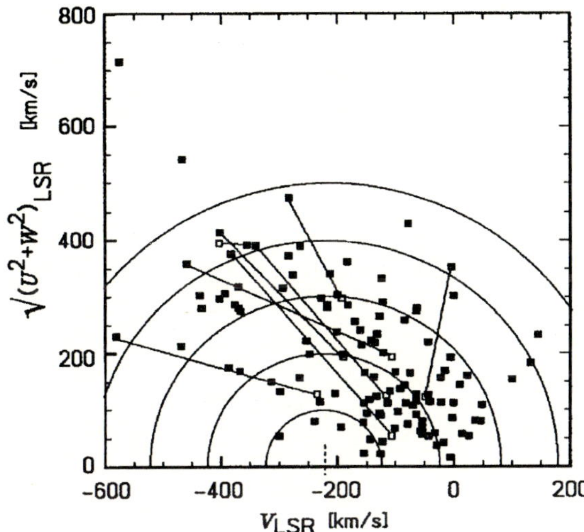

Fig. 4. Toomre diagram of typical proper motion survey stars. Open squares refer to main sequence fitting of distances, filled squares represent the HRD positions found from analyses using spectroscopic methods. Straight lines connect entries for the same star. The two stars with space velocities above 500 km/s are spectroscopic binaries

halo stars since the distances are always *underestimated.* This is demonstrated for a subset of typical survey stars in Fig. 4.

The mean angular momentum of metal-poor stars, often used as a function of metal abundance to indicate the presence or absence of another *discrete* population such as the *thick disk* (cf. Carney 1993) may therefore need considerable corrections before that debate can be finished. In particular, the kinematic changes affect the value of the Galactic escape velocity as much as the question, at what metal abundance the Galactic disk population starts.

3 Abundance Gradients in the Galactic Disk

It is not only the Galactic halo for which abundance information is hard to obtain; similar uncertainties affect our understanding of the evolution and the chemical enrichment of the Galactic disk. Two observations are particularly intricate,

- the average metal abundance found in young stars is *lower* than that of the Sun, and
- the galactocentric variation of the metal abundance found from the spectroscopy of young stars is at variance with the abundance gradient observed in H II regions and in other types of stars

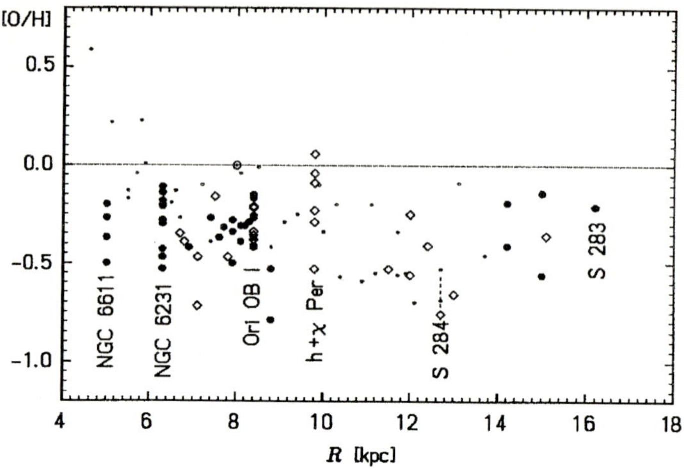

Fig. 5. Oxygen abundances determined from emission lines in H II regions (Shaver *et al.* 1983, dots), and from B stars in young open clusters and associations (Kilian *et al.* 1994, 1995: filled circles; Gehren *et al.* 1985, Lennon *et al.* 1990, Kaufer *et al.* 1995: diamonds)

In fact the hot stars in young open clusters and OB associations in Fig. 5 do *not* show an abundance gradient. The corresponding data for other elements such as C, N, Mg, Al, Si, S, and Fe all have negligible abundance gradients. The spectroscopic results show, however, that oxygen in B stars is underabundant with respect to the solar value, and a similar deficiency is found for C, Mg, and Si. At present, a solution of the discrepancy is not at hand. After careful refinement of spectroscopic methods including non-LTE line formation the B star results are now very reliable, even more so as they refer to similar stars on a baseline of 10 kpc. On the other side, the evidence in favour of an abundance gradient comes from a number of independent methods for such different objects as H II regions (Shaver *et al.* 1983), Cepheids (Harris 1981), and open cluster giants of various ages (Friel & Janes 1993).

It is both reassuring and confusing that open cluster spectroscopy of cool giants obtained by Friel and Janes leads to a general *underabundance* of the heavy elements with respect to the Sun, very similar to that found in B stars, but also shows a clear heavy element abundance gradient with ≈ -0.01 dex/kpc. The Cepheid photometry of Harris yields another factor of 2 higher heavy element abundances, with an abundance gradient similar to that of the cluster giants. Both analyses cannot precisely define the contributions of single elements to the heavy element abundance. Figure 5 seems to suggest that the oxygen abundance gradient established by H II regions is present only inside $R \leq 6$ kpc whereas at larger distances from the Galactic center the mean oxygen abundance is constant and similar to that found in B stars. However, this does not hold for nitrogen abundances which differ by ≈ 0.4 dex at $R \geq 10$ kpc. The very few associations for which both stars and H II regions have been analyzed lead to moderate but systematic abundance differences when compared with the intrinsic abundance scatter found among cluster B stars. Since present spectroscopic analyses are *differential* with respect to each other this scatter is now definitely outside any methodical and observational errors with C, N, O, and Si abundances differing by a factor 3 within a single cluster.

Notwithstanding such evidence it is hard to explain why the abundance gradient in H II regions should be different from that in the corresponding young stars by which they are excited. In particular, there is no reason why the nitrogen abundance should be systematically lower in H II regions with respect to stars of the same association (which is the case for S 284 and S 285). It is as enigmatic that oxygen behaves oppositely inside 6 kpc. Neither oxygen nor nitrogen is believed to condense into an interstellar dust phase, and even if some other elements form dust, it is not evident why that should depend on galactocentric distance. The abundance gradient – if it is real – may well be an event that is local in space and time. It must be connected with the comparatively low heavy element abundances found among nearly all of the young objects (including H II regions). The factor of 2 underabundance of young stars with respect to the older disk and the Sun shows that chemical evolution must be a complicated highly non-local process on scales of kiloparsecs. Missing uniformity of stellar abundances in clusters seems to indicate that even on parsec scales mixing is incomplete. Thus the gradient may well be the result of a preceding *non-uniform enrichment*. This impression is emphasized when the abundances are compared for objects outside the solar circle ($R \geq 8$ kpc).

Part of the confusion connected with the chemical evolution of the Milky Way both in the halo and in the disk seems to emerge from a diversity of methods and data. Thus, after a decade of exciting survey work and statistical evaluations that have isolated some important problems in understanding the very early evolution of the Galaxy it is now the time to step back and continue with detailed high-resolution spectroscopy that makes use of all the recent improvements in modelling stellar structure, stellar atmospheres and line formation.

References

Abia, C., Rebolo, R. 1989, *Astrophys. J.* **347**, 186
Axer, M., Fuhrmann, K., Gehren, T. 1995, *Astron. Astrophys.*, in press
Barbuy, B. 1988, *Astron. Astrophys.* **191**, 121
Barbuy, B., Erdelyi-Mendes, M. 1989, *Astron. Astrophys.* **214**, 239
Bessell, M., Norris, J. 1984, *Astrophys. J.* **285**, 622
Bessell, M.S., Sutherland, R.S., Ruan, K. 1991, *Astrophys. J. Lett.* **383**, L71
Brown, J.H., Burkert, A., Truran, J.W. 1991, *Astrophys. J.* **376**, 115
Carney, B.W., Latham, D.W. 1987, *Astron. J.* **93**, 116
Carney, B. 1993, in *Galaxy Evolution: The Milky Way Perspective*, ASP Conf.Ser. 49, 83

Edvardsson, B., Andersen, J., Gustafsson, B., Lambert, D.L., Nissen, P.E., Tomkin, J. 1993, *Astron. Astrophys.* **275**, 101

Eggen, O.J., Lynden-Bell, D., Sandage, A. 1962, *Astrophys. J.* **136**, 748

El Eid, M., Fricke, K.J., Ober, W.W. 1983, *Astron. Astrophys.* **119**, 54

Fall, S.M., Rees, M.J. 1985, *Astrophys. J.* **298**, 18

Friel, E.D., Janes, K.A 1993, *Astron. Astrophys.* **267**, 75

Fuhrmann, K., Axer, M., Gehren, T. 1994, *Astron. Astrophys.*, submitted

Gass, H., Liebert, J., Wehrse, R. 1988, *Astron. Astrophys.* **189**, 194

Gehren, T., Nissen, P.E., Kudritzki, R.P., Butler, K. 1985, in *Production and Distribution of C, N, O Elements*, eds. I.J. Danziger, F. Matteucci, and K. Kjär, ESO, Garching, p.171

Gratton, R.G., Ortolani, S. 1986, *Astron. Astrophys.* **169**, 201

Harris, H.C. 1981, *Astron. J.* **86**, 707

Holweger, H., Müller, E.A. 1974, *Sol. Phys.* **39**, 19

Kaufer, A., Szeifert, T., Krenzin, R., Baschek, B., Wolf, B. 1995, *Astron. Astrophys.*, in press

Kilian, J., Montenbruck, O., Nissen, P.E. 1994, *Astron. Astrophys.* **284**, 437

Kilian-Montenbruck, J., Gehren, T., Nissen, P.E. 1995, *Astron. Astrophys.*, in press

Kurucz, R.L. 1979, *Astrophys. J. Suppl.* **40**, 1

Kurucz, R.L. 1992, Opacity Distribution Functions, private communication

Lennon, D.J., Dufton, P.L., Fitzsimmons, A., Gehren, T., Nissen, P.E. 1990, *Astron. Astrophys.* **240**, 349

Maeder, A. 1992, *Astron. Astrophys.* **264**, 105

Miller, G.E., Scalo, J.M. 1979, *Astrophys. J. Suppl.* **41**, 513

Nissen, P.E., Gustafsson, B., Edvardsson, B., Gilmore, G. 1994, *Astron. Astrophys.* **285**, 440

Norris, J.E., Peterson, R.C., Beers, T. 1993, *Astrophys. J.* **415**, 797

Ryan, S.G., Norris, J.E. 1993, in *Galaxy Evolution: The Milky Way Perspective*, ASP Conf.Ser. 49, 103

Salpeter, E.E. 1955, *Astrophys. J.* **121**, 161

Sandage, A., Fouts, G. 1987, AJ 93, 74

Shaver, P.A., McGee, R.X., Newton, L.M., Danks, A.C., Pottasch, S.R. 1983, *Mon. Not. R. Astr. Soc.* **204**, 53

Spite, M., Spite, F. 1991, *Astron. Astrophys.* **252**, 689

Thielemann, F.-K., Nomoto, K., Hashimoto, M. 1993, in *Origin and Evolution of the Elements*, eds. N. Prantzos *et al.* , Cambridge, Univ. Press, p.297

VandenBerg, D.A., Bell, R.Acite. 1985, *Astrophys. J. Suppl.* **58**, 561

Webbink, R.F. 1985, in *Dynamics of Star Clusters*, eds. J. Goodman and P. Hut, Dordrecht, Reidel Publ., p.541

Woosley, S.E. 1987, in *Nucleosynthesis and Chemical Evolution*, eds. J. Audouze *et al.* Geneva Observatory

Woosley, S.E., Weaver, T.A. 1982, in *Supernovae, a Survey of Current Research*, eds. M. Rees and R.J. Stoneham, Dordrecht, Reidel Publ., p.79

Wyse, R.F.G., Gilmore, G. 1993, in *Galaxy Evolution: The Milky Way Perspective*, ASP Conf. Ser. **49**, 209

Dissipative Collapse of a Non-Rotating System

Christian Theis and Gerhard Hensler

Institute of Astronomy and Astrophysics, University of Kiel, Olshausenstr. 40, 24098 Kiel, Germany

1 Summary

We investigate the collapse of a dynamically hot, homogeneous sphere of 10^{11} $M_\odot$ consisting of gaseous clouds. Energy dissipation is applied by means of inelastic cloud collisions. Contrary to initially hot dissipationless models the mass distribution follows after 3-4 free-fall timescales a de Vaucouleurs law. The orbit-dependent collision rate of the clouds results in a dominance of the circular orbits and, therefore, in a low and in some region negative anisotropy of the velocity dispersion which is in contrast to dissipationless models.

2 Introduction

Cosmological simulations show that the temperature of a smoothly distributed gaseous phase in protogalaxies never exceeds a few 10^5 K during the early galaxy formation period (Katz & Gunn 1991; Steinmetz & Müller 1993). Comparing the according cooling timescale of 10^7 yrs for a metal-free gas with a density $n = 10^{-3}\mathrm{cm}^{-3}$ with a typical free-fall timescale τ_{ff} of 10^8 yrs one can conclude that the gas should fragment quickly and form gaseous clouds of a typical mass of 10^6 $M_\odot$ (Fall & Rees 1985; Ikeuchi & Norman 1991). After the fragmentation the dynamics of the protogalaxy is determined by the energy dissipation due to inelastic collisions of these clumps.

From a computational point of view it was and is impossible to simulate the formation and evolution of a system of 10^5 to 10^6 clouds directly. Therefore, two different approximations have been used: The first approach used a hydrodynamical model for the dynamical evolution and added a statistical description for the cloud collisions (Larson, 1969; Burkert & Hensler 1988). The cloud-cloud collision rate was calculated by the local dissipation timescale $\tau_{\mathrm{diss}} = 1/(n\sigma A_{\mathrm{C}})$ with the cloud number density n, the relative velocity dispersion σ and the geometrical cross-section $A_{\mathrm{C}} = \pi R_{\mathrm{C}}^2$. However, these models were restricted due to several simplifying assumptions like 1) a high symmetry of the system, 2) a fixed shape of the velocity distribution, 3) a fixed cloud mass (and hence no mass dependent cross-section), 4) a local estimate of the collision rate despite of a large mean free path of 5-10 kpc and 5) a fully inelastic treatment of even only grazing collisions.

All these shortcomings can in principle be solved by the second approach, the N-body simulation (Carlberg *et al.* 1989; Abadi *et al.* 1990). The first models, however, were restricted to $10^3 - 10^4$ particles prohibiting an identification of a single cloud with one particle. Therefore, the collisional cross-section was treated like a free parameter. Additionally, only the distance between two clouds determined whether an inelastic collision takes place, by this neglecting the influence of the relative velocity. Extending Hernquist's (1987) TREE-code for inelastic collisions it was possible to increase the number of particles up to the order of 10^5 and to introduce a more detailed model for cloud collisions (Theis & Hensler 1993). This allows to follow the orbits of each single cloud and the build-up of a cloud mass spectrum by coalescing collisions.

In this paper we want to focus on the early collapse of elliptical galaxies. Because in ellipticals almost no cool gas has been observed, most studies on their dynamics use a dissipationless, purely stellar collapse. E.g., van Albada (1982) showed that in dissipationless models it is necessary to start with a dynamically cold configuration (virial coefficient $\eta_{\rm vir} \equiv 2T/|U| \lesssim 0.1$ with kinetic energy T and potential energy U) and a clumpy distribution of matter in order to end up with a de Vaucouleurs-like surface density profile. The strong collapse leads finally to a large positive anisotropy $A \equiv 1-(\beta/\alpha)^2$ of the velocity dispersion in radial (α) and tangential (β) direction. However, the high phase-space density in ellipticals requires at least some amount of dissipation (Carlberg 1986). Additionally, the de Vaucouleurs law is observed down to 1/10 of the effective radius $r_{\rm eff}$ (Burkert 1993) while dissipationless models fail the $R^{1/4}$-law inside $0.6-0.8 r_{\rm eff}$. Moreover, the origin of the required low initial virial coefficient is unclear: Why do we not observe ellipticals resembling the models of the dissipationless collapse for a virial coefficient larger than 0.1 or a smoother initial mass distribution?

Here we want to investigate the influence of dissipation on the density and anisotropy profile. In the next section we briefly describe the dissipational scheme and in the final section we show and discuss the results.

3 The Dissipation Scheme

Energy dissipation is introduced into the TREE-code by sticky collisions between two particles, if they fulfill several conditions: First, the orbits of the clouds must result in a sufficient overlap for the coalescence of two particles. This geometrical cross-section is given by

$$A_{\rm g} = \eta_{\rm ov}^2 \cdot \pi R_{\rm c}^2 \cdot \left(1 + \frac{2\,G(m_1+m_2)}{\eta_{\rm ov} R_{\rm c} v_{1,2}^2}\right) . \tag{1}$$

with the relative velocity $v_{1,2}$ of the two colliding clouds. The overlap factor $\eta_{\rm ov} \approx 0.2$ prohibits a fully inelastic treatment in the case of grazing collisions, whereas the second term in brackets gives a correction for gravitational focusing. However, in contrast to relaxed galactic disks this correction is almost negligible for typical cloud masses and velocities in protogalaxies.

If an interaction between two particles can geometrically take place, it is, secondly, checked whether the maximum angular momentum of a single final cloud exceeds the relative orbital angular momentum of the two particles. In that case, the two clouds are replaced by one cloud with the same center of mass data of the former two particles. The mass-radius relation of the clouds is chosen according to observations in the Galaxy (Rivolo & Solomon 1987)

$$R_C \approx 95 \cdot (\frac{M_C}{10^6\,\mathrm{M}_\odot})^{0.5}\,\mathrm{pc}. \tag{2}$$

In order two prevent a single cloud to grow infinitely, an upper mass limit of $10^8\,\mathrm{M}_\odot$ for sticky collisions is applied. If the cloud mass exceeds this value, no further inelastic collisions are allowed and the dynamical evolution is dissipationless like for stars. Therefore, this criterion can be interpreted as a crude implementation of star formation. For more details see Theis & Hensler (1993).

4 Results

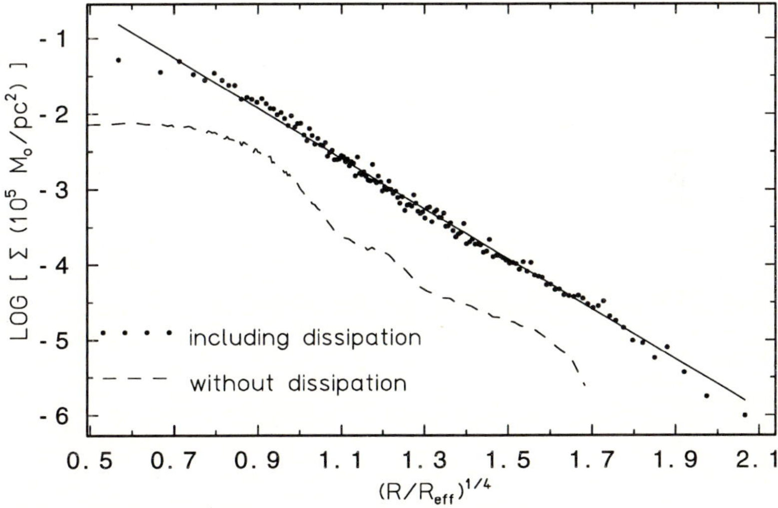

Fig. 1. Radial distribution of the surface density of models starting with a virial coefficient of $\eta_{\mathrm{vir}} = 0.5$ after 5 τ_{ff}: Model with dissipation (filled circles) and without dissipation (dashed line) as well as a de Vaucouleurs-law (solid line). The effective radius of the dissipative model is $r_{\mathrm{eff}} \approx 2.9$kpc.

We start with a homogeneous mass distribution of $10^{11}\,\mathrm{M}_\odot$ inside a 20 kpc sphere. The initial isotropic and homogeneous velocity dispersion of 80 km/s corresponds to a virial coefficient of $\eta_{\rm vir} = 0.5$. The numerical timestep amounts to 1% of the free-fall time of $\tau_{\rm ff} = 1.5 \times 10^8$ yrs. Initially we use $N = 65536$ equal mass particles. This gives an initial cloud mass of $M_{\rm C} \approx 1.5 \times 10^6\,{\rm M}_\odot$, a volume filling factor of 1.3% and a mean cloud-to-cloud distance of 0.5 kpc. The smoothing length of the potential is chosen to 0.5 kpc.

After a strong collapse within the first free-fall time the system relaxes to quasi-equilibrium within the next 2 $\tau_{\rm ff}$. The total number of particles is reduced to 50% of the initial value, whereas the dissipated energy is twice the amount of the initial total energy. In this late phase the radial distribution of the surface density follows a de Vaucouleurs law with an effective radius $r_{\rm eff} \approx 2.9\,{\rm kpc}$ (Fig. 1). It should be noted that even far inside $r_{\rm eff}$ the $R^{1/4}$ law can be found. There is only a small deviation inside 0.2 $r_{\rm eff}$. However, the cloud radius of $R_{\rm C} = 420\,{\rm pc}$ of single clouds at the upper mass limit is not small compared to this spatial scale. Therefore, a more refined cloud model and collisional scheme is required to investigate the innermost 500 pc. If one neglects any dissipation during the evolution no de Vaucouleurs law is found in agreement with van Albada's results (1982). Contrary to dissipationsless simulations dissipative models evolve to a $R^{1/4}$ profile not only for small virial coefficients but also for systems initially close to virial equilibrium.

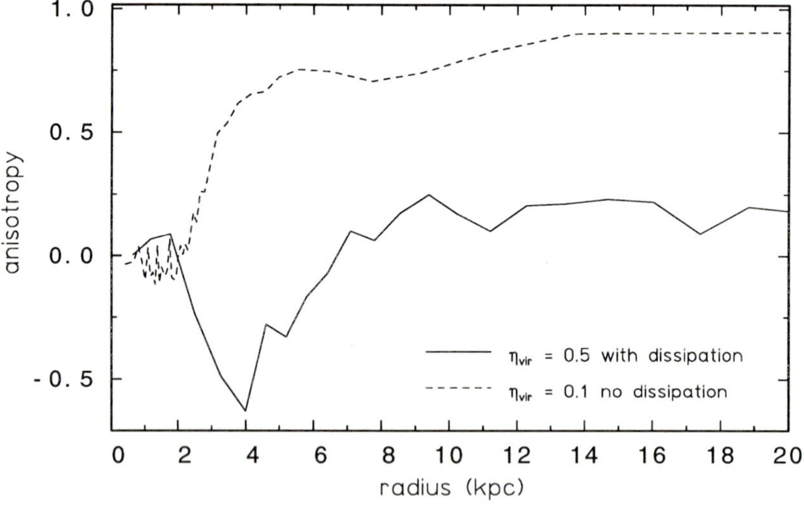

Fig. 2. Radial distribution of the anisotropy A of the velocity dispersions after 5 $\tau_{\rm ff}$: Model with dissipation and a virial coefficient of $\eta_{\rm vir} = 0.5$ (solid line) and a violently collapsing model ($\eta_{\rm vir} = 0.1$) without dissipation (dashed line).

Considering the anisotropy A there is an additional principal difference between dissipationless and dissipative models (Fig. 2). Whereas the former builds an isotropic core and a large positive anisotropy in the outer region (van Albada 1982), the latter has generally a smaller anisotropy and even a *negative* minimum near the half-mass radius. The dominance of the tangential orbits can be understood by the fact that radial orbits pass the dense center where the chance for an inelastic collision is strongly increased. By this, particles on circular orbits will survive and dominate the velocity distribution.

Acknowledgements: This work was partly supported by the DFG under grant He 1487/4-1 (C.T.). The calculations were performed at the Leibniz-Rechenzentrum of the Bayerische Akademie der Wissenschaften, Munich.

References

Abadi, M. G., Lambas, D. G. & Mosconi, M. B. 1990, *Astrophys. J.* **360**, 343
Burkert, A., Hensler, G. 1988, *Astron. Astrophys.* **199**, 131
Burkert, A. 1993, *Astron. Astrophys.* **278**, 23
Carlberg, R.G., Lake, G., Norman, C.A. 1986, *Astrophys. J. Lett.* **300**, L1
Carlberg, R.G. 1986, *Astrophys. J.* **310**, 593
Fall, S.M., Rees, M.J. 1985, *Astrophys. J.* **298**, 18
Hernquist, L 1987, *Astrophys. J. Suppl.* **64**, 715
Ikeuchi, S., Norman, C. 1991, *Astrophys. J.* **375**, 479
Katz, N., Gunn, J.E. 1991, *Astrophys. J.* **377**, 365
Larson, R.B. 1969, *Mon. Not. R. Astr. Soc.* **145**, 405
Rivolo, A.V., Solomon, P.M. 1987, in *Molecular Clouds in External Galaxies*, Dickman, R.L., Snell, R.L., Young, J.S. (eds.), 42
Steinmetz, M., Müller, E. 1993, *Astron. Astrophys.* **268**, 391
Theis, Ch., Hensler, G. 1993, *Astron. Astrophys.* **280**, 85
van Albada, T.S. 1982, *Mon. Not. R. Astr. Soc.* **201**, 939

Mergers and Galaxy Evolution

Ray G. Carlberg

Department of Astronomy, University of Toronto, Toronto M5S 1A7, Canada

1 Summary

Galaxy merging is the late time manifestation of the galaxy formation process and likely significantly effects $z < 1$ galaxies. A "maximum reasonable rate" model for merging finds a ~ 2 mag K band increase in the luminosities of dwarf galaxies so that they contribute significantly to the faint counts, with spirals and ellipticals being far less affected. The median K and I redshifts stabilize (and even decrease slightly) at $z \simeq 0.6$ beyond I=21 or K=19. The B redshifts continue to rise (although strongly dependent on the UV spectral evolution). Such rapid merging predicts that at $z = 1$ the characteristic galaxy mass is reduced to $\sim 30\%$ of the $z = 0$ value. To rule out this model requires good sampling beyond $z = 1$.

A theoretical complication for even a minimal merger rate, which reduces $z = 1$ masses to 2/3 of current epoch values, is that infall of a single satellite having 10% of a disk's mass may destroy thin disks. Using completely self-consistent n-body simulations, we show that the primary response of a disk to "cosmological" satellites up to 20% of the disk mass is to *tilt* the disk with a temporary warping.

2 Merger Rates

The inevitability and significance of merging was highlighted in an influential article (Toomre 1977) which emphasized merging as a source of transformation within the Hubble sequence from disk to spheroidal galaxies. Most aspects of Toomre's argument remain valid today. First, long tidal tails drawn from a *stellar* disk are reliable indicators that the galaxies are on relative orbits sufficiently energetically bound that dynamical theory (and a huge amount of simulation data) indicates the two galaxies will merge within 1-2 rotation periods, typically 0.5 Gyr. Second, the merger rate is expected to rise into the past, Toomre giving an energy distribution argument for a rate increase of $(1+z)^{5/2}$. Third, a lower limit on the present day merger rate is 11 completely "doomed" pairs in approximately 4000 galaxies, which translates to a rate of approximately 0.005 Gyr^{-1}. Integrated to $z = 1$ this reduces the mass of an average galaxy about 10% (for $t_0 \simeq 13$ Gyr). Hence, basic issues are: 1) how to recognize a merger, 2) the current epoch merger rate, and 3) the redshift dependence of the merger rate.

A somewhat separate line of investigation is to predict the outcome of realistic merger events. A "major merger" (nearly equal mass galaxies) inevitably creates a system with essentially no disk and more stars on the orbits characteristic of a spheroidal galaxy (Barnes & Hernquist 1992). The result of a "minor merger", involving a disk system such as the upcoming LMC-Galaxy merger, is considerably less certain, and is discussed below. Star formation and ISM alterations are less certain in general, but a general fate for disk gas is to be reverse torqued by a merger induced bar, and very quickly moved into the central region (Barnes & Hernquist 1991).

2.1 Empirical Merger Rates

At low redshift, estimates of the merger rate fall into two categories. Morphological methods use images of galaxies to determine which systems soon will soon merge and which ones are likely merger remnants (which gives the time integral of the merger rate). These methods are strongly dependent on a "training set" of merging galaxy images, recently considered for HST images (Mihos 1995). Estimates using this technique use the Arp Atlas (Toomre 1977) to give 0.005 Gyr^{-1} and the Arp & Madore Atlas (Carlberg & Couchman 1989) with a looser requirement for "tidal features" gives 0.02 Gyr^{-1}. The difficulty with morphological classification for mergers is that the best indicators depend on tails and shells, both of which are generally very low surface brightness features which require imaging data of very high quality. Furthermore, there will almost certainly be phases of the merger which are not morphologically recognized as a merger-in-progress (Mihos 1995).

An alternate empirical estimate is to use the basic dynamical result that all pairs of galaxies separated by a galaxy diameter or less and having a relative velocity less than the escape velocity at the edge of the disk will merge in 1-2 orbital times, under the important assumption that external tides can be neglected. A useful, but somewhat arbitrary, definition of the maximum separation to constitute a close pair is $20h^{-1}$ kpc. At this distance visible disks will interact significantly, and, external tides should be insignificant. Using this definition, the UGC catalogue of nearby galaxies has 2.3% of its galaxies in low velocity, close pairs, implying a merger rate of 0.046 Gyr^{-1} (Carlberg *et al.* 1994). At $z \simeq 0.4$ the fraction of close pairs (with little information about relative velocities) is 8.5 to 10% (Carlberg *et al.* 1994, Yee & Ellingson 1995), implying a strong rise in the merger rate, $\sim (1+z)^4$. These studies find no evidence for the expected blueing of the colours of close pairs (Larson & Tinsley 1978), although this could be a consequence of dilution with projected galaxies, and the absence of multi-band data.

2.2 Theoretical Merging Rates

A semi-empirical approach to estimate the merger is to calculate the rate of dynamical friction inspiral. The density of neighbours comes from the galaxy-galaxy correlation function. Only those pairs which are moving at relative velocities such that they are on nearly parabolic orbits merge. That is, rich galaxy clusters have few internal mergers, in spite of the high galaxy density. The basic approach is to compute the rate of inspiral of companion galaxies taking the circular velocity from a Fisher-Tully relation, the mass of companions from the galaxy luminosity function, the density of companions from the galaxy correlation function, and an assumption about the initial orbital ellipticity. For reasonable choices of these parameters, these calculations find that to a time of 5 Gyr a typical galaxy will have merged with approximately $20 \pm 10\%$ of its mass (Bahcall & Tremaine 1988, Tóth & Ostriker 1992, Huang 1995), implying a merger rate of 0.04 Gyr^{-1}. This rate is about twice what one infers from morphological indicators, but, not all merging galaxies are expected to have clearly recognizable merger characteristics. Extrapolating (at a uniform rate in time) to redshift 1, this implies that an average galaxy will have a mass that is 67% of the present day values.

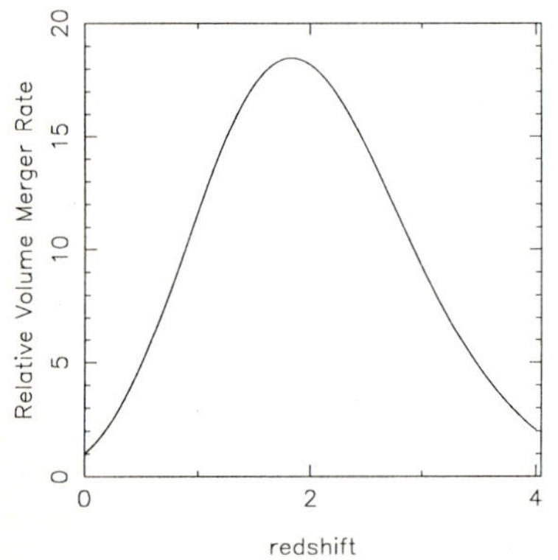

Fig. 1. The volume merger rate as a function of redshift, relativethe redshift zero rate. The location of the peak is set by thefluctuation amplitude on galaxy scales. The peak can be interpreted as the time of assembly of most galaxies. The speed of the rise at low redshift depends on Ω and the fraction of pairs moving slowly enough to be gravitationally captured.

A theoretical merger rate was derived in Carlberg (1990), which a superior later calculation (Lacey & Cole 1993) finds to be an overestimate for dark matter halos of galaxies. The Lacey & Cole (1993) calculation finds that toredshift 1 the average galaxy halo mass will be reduced to about 35% of its present day value (Navarro *et al.* 1994), which serves as a useful upper limit (and possibly the correct value) to the merger rate of galaxies, since galaxies can merge no faster than their dark halos. The situation for the visible components of galaxies, which are likely not well described by the Press-Schechter theory (Press & Schechter 1974), is far less clear. For a practical calculation of the effect of merging on galaxies, the rate will be normalized using the empirical data, extrapolated using my functional form, which to a redshift of 0.5 is approximately equal to $(1+z)^4$, flattening off at higher redshifts as illustrated in Figure 1.

3 Merging and Faint Galaxies

The impact of merging on galaxy properties at earlier epochs is potentially large, however, at the moment there is relatively little evidence for any change in the underlying bulk stellar populations of galaxies (Lilly *et al.* 1995). The morphological dependence of the merger models of Carlberg & Charlot (1992) and Carlberg (1992) are displayed below as an example of a "maximum merger model". As others (Cowie *et al.* 1991, Broadhurst *et al.* 1992) have found, a significant change in the galaxy population, as would be expected from rapid merging is consistent with quite a lot of the data. The idea here is to push the model far enough that a conflict with observational data may emerge. The model has a luminosity dependent density evolution, arising from the increase in gas content to later Hubble types. Mergers are assumed to induce a starburst in much of this gas, resulting in evolution of the luminosity function along the lines suggested on other grounds (Broadhurst *et al.* 1988, Cowie *et al.* 1991, Broadhurst *et al.* 1992). The main parameter of interest here is the redshift zero merger rate, which is set at 0.04 Gyr^{-1}, as indicated by the pair counts and dynamical friction infall calculations discussed above. The counts as a function of morphological type are shown in Figure 2. The rapid rise in the dwarf population is similar to the HST data (Glazebrook *et al.* 1995). Of particular note is the plateau in the median redshift for red selected galaxies. The $n(z)$ does remain broad, with extending to $z = 2$ or so with reasonable numbers.

Table 1. Maximal Merging Median redshifts

B	$z_{1/2}$	V	$z_{1/2}$	I	$z_{1/2}$	K	$z_{1/2}$
20	0.18	20	0.21	17	0.26	15	0.20
21	0.25	21	0.26	18	0.36	16	0.31
22	0.34	22	0.31	19	0.47	17	0.45
23	0.45	23	0.39	20	0.57	18	0.59
24	0.55	24	0.50	21	0.62	19	0.65
25	0.64	25	0.61	22	0.61	20	0.60
26	0.71	26	0.71	23	0.59	21	0.52
27	0.77	27	0.72	24	0.58	22	0.50
28	0.81	28	0.71	25	0.57	23	0.51
29	0.83	29	0.75	26	0.57	24	0.54
30	0.84	30	0.78	27	0.58	25	0.57

3.1 Recapitulation

This "maximal merging" calculation implies such drastic evolution beyond redshift one (galaxies have a characteristic masses only 30% of the current epoch values, and the median redshifts are very low) that compared to a no-merger model the differences are easily detectable in surveys beyond $K \sim 20$ (bands of

I and bluer having reduced sensitivity since the 4000Å break crosses through at $z \simeq 1$). Lower redshift data is less clearly interpreted, since mass decreases are partially offset with luminosity increases. If (or when) the predictions are found to be in conflict with observations, the main implication is that the extrapolation of the merging rate with redshift is too large. The simplest and most natural way to make a substantial change in the merger rate is to put $\Omega \simeq 0.2$ (Carlberg 1990), although this requires a wholesale re-examination of many aspects of observational cosmology for consistency. It should be noted that even a constant merger rate predicts that galaxy masses at $z = 1$ will be reduced to 67% of current epoch values.

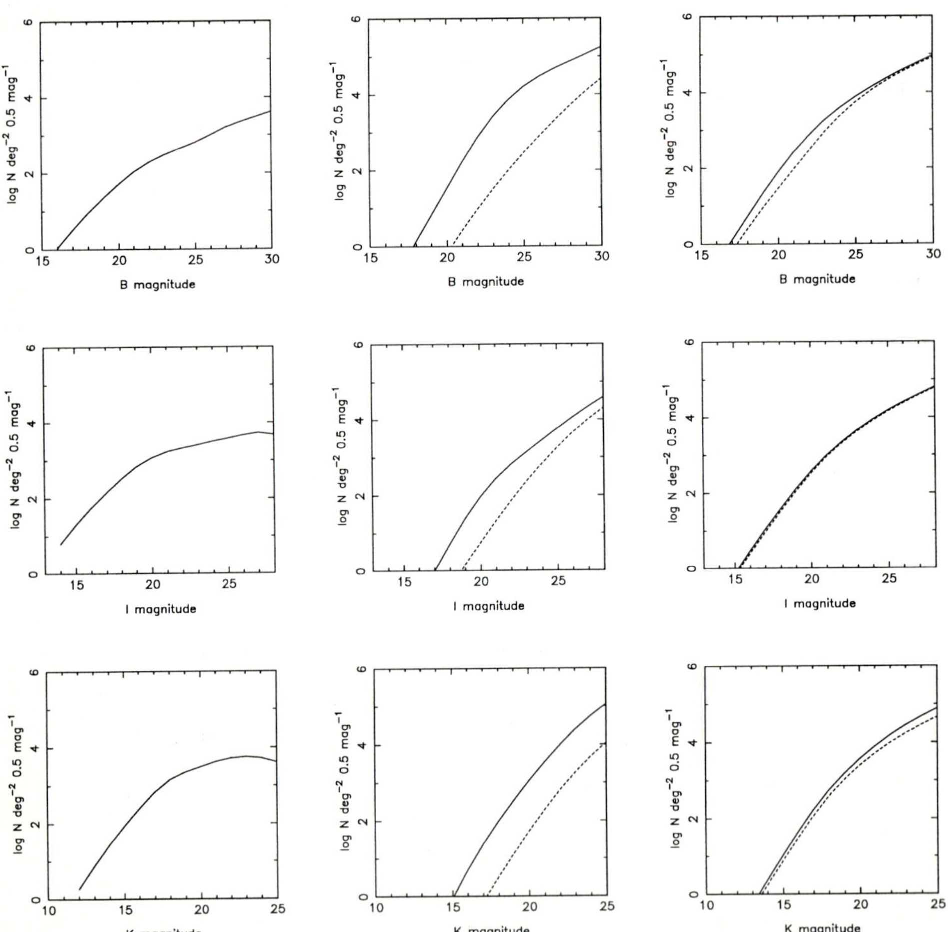

Fig. 2. The dependence of faint galaxy counts on morphology. The first column is E/S0, the second Irr, the third S. The dashed line is evolution with merging but no luminosity changes, the solid line shows the luminosity enhancements due to starbursts. E/S0 types are assumed to have no gas. The numbers Irr types become comparable to S types around $I = 22$.

4 Merging and Thin Disks (with Siqin Huang)

Galactic disks are remarkably thin, typically having a ratio of horizontal to vertical scale lengths of about 10 to 1, which implies that the ratio of kinetic energy in horizontal to vertical motions is about 100 to 1. Hence globally small energy additions could give dramatic vertical structure changes. Several calculations have found that a satellite having 10% of the disk mass spiraling into the disk (Quinn & Goodman 1986, Quinn *et al.* 1993, Tóth & Ostriker 1992) will unacceptably thicken the disk. As noted above, straightforward dynamical calculations find that a typical disk must absorb about 2 ± 1 such satellites. There is no doubt that if a 10% satellite (possibly even a 1% satellite) enters a disk, the disk will be unacceptably thickened.

There are several reasons to believe that thin disks may be quite robust, as long as the satellite substantially dissolves prior to entry into the disk. The strong self-gravity of a disk causes all of its dynamical frequencies to be much higher than the perturbing frequencies of the satellite. For instance, in the solar neighbourhood of our galaxy the ratio of the vertical oscillation frequency to the orbital frequency is a factor of 5, so an exterior satellite would be seen as an adiabatic potential variation. Nevertheless the satellite does exert a torque on the disk stars. The issues are whether a satellite is greatly tidally stripped before it reaches the disk, and, whether the disks stars respond incoherently (leading to disk thickening), or, coherently (leading to some combination of warp/tilt/precession). In particular, it has been suggested (Binney 1990) that warps indicate the recent addition of misaligned angular momentum.

Our approach to the problem is in the spirit of a reasonable counterexample, the calculation being done with a completely self-consistent n-body simulation. This is a challenging problem for n-bodies, mainly because large numbers of particles are needed in order to keep unwanted two-body heating at a level below the dynamics of interest, and, the simulation must be run for many disk rotations (because satellite infall is slow compared to disk dynamical times). The simulations below have been done with particle numbers ranging from 60,000 to 200,000 to check for two-body effects, shorter time steps, and an alternate time stepping scheme. No substantial differences in the conclusions result. The system consists of a disk, a halo, and a satellite, all composed of active particles. The "live" halo is particularly important, since in its absence the transfer of angular momentum from the disk to the satellite drives the satellite away (unless the satellite is started very close to the disk, Quinn & Goodman 1986). The heating of the disk by the halo is significant, but comparable (or less than) the observed rate of increase in the velocities of disk stars. The satellite structure is derived from "cosmological" initial conditions. That is, it is assumed that the density profile of the galaxy's halo and the satellite are self-similar, with the velocity dispersions scaling as $\sigma \propto M^{1/3}$. This is a crucial assumption, which others have generally not made. It implies, whatever the mass profile, that once the satellite has spiraled through 90% of the galaxy's dark halo, say from 100kpc down to 10kpc, then the satellite will have had about 90% of its mass tidally removed, greatly reducing its impact on the disk.

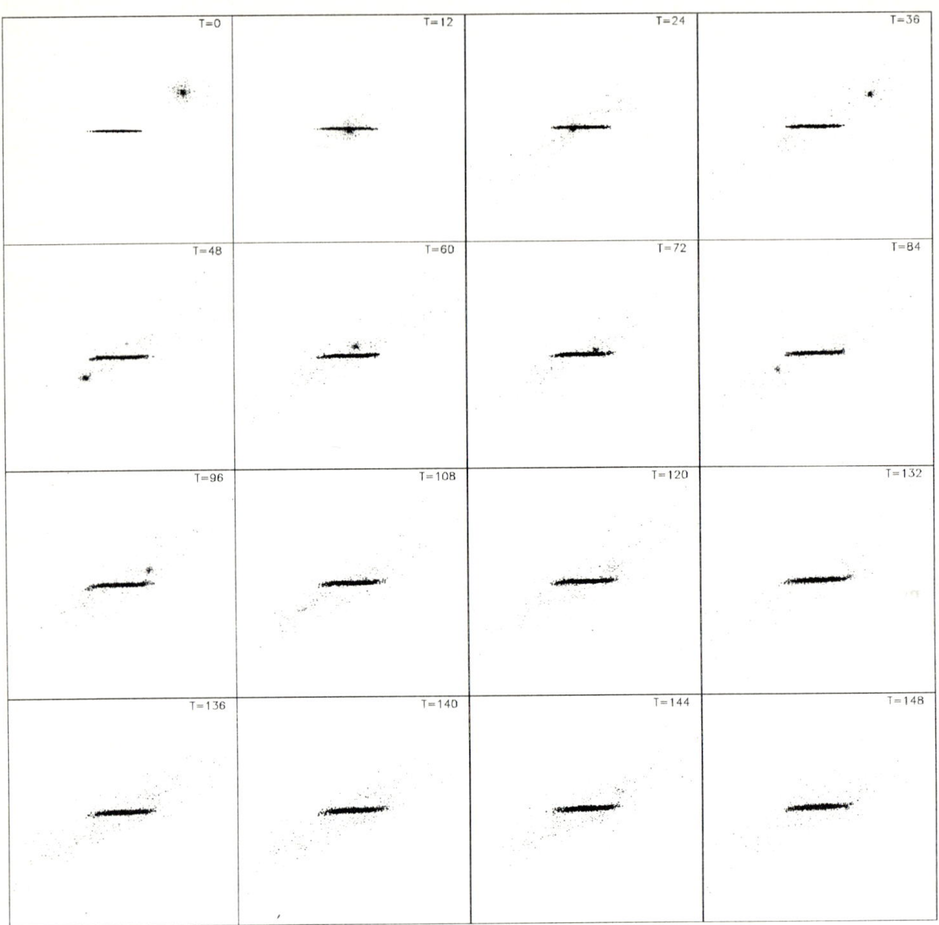

Figure 3: The infall of a 10% satellite onto a disk. The halo particles are not plotted. The primary response of the disk is a tilt of about 4° with essentially no precession. A transient warp is visible. A counter-rotating satellite tilts the disk about half as much in the other direction.

We find that the primary response of a thin disk to "cosmological" low mass satellite infall (up to 20% of the disk mass) is to develop a temporary warp, tilt over (with little precession Nelson & Tremaine 1995) all with negligible thickening. The outcome for a particular satellite will depend a great deal on the mass profile of the satellite itself. Only the rare, dense dwarfs, such as M33, will be able to cause a lot of disk disruption. Noting these caveats, our conclusion is that the predicted infall rate of typical low density dwarfs can be accommodated without unreasonable disk thickening.

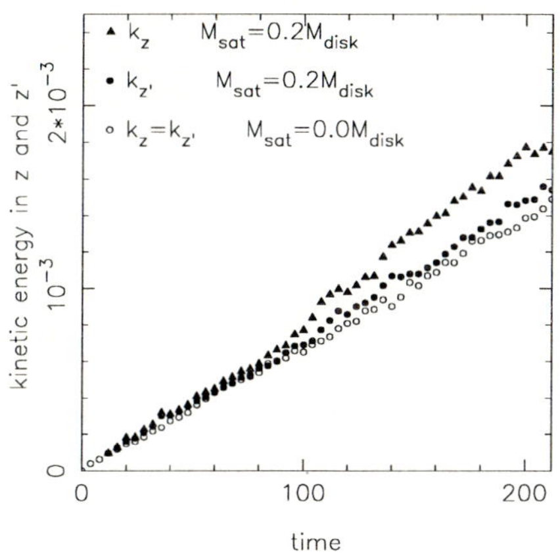

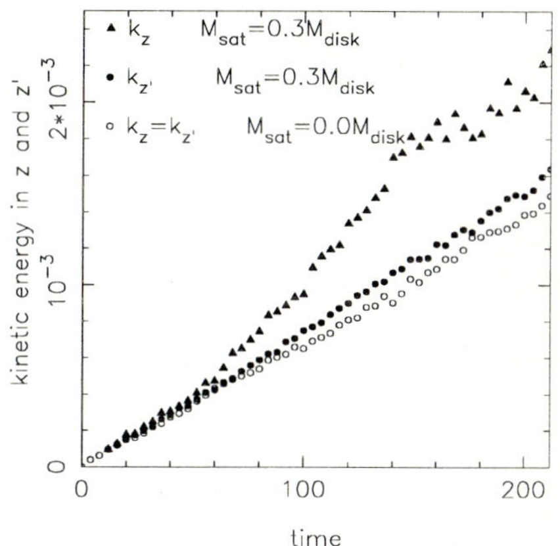

Figure 4: Disk kinetic energy in vertical motions versus time for 20% and 30% satellite inspiral with a no-satellite control model. The vertical motions are calculated in the initial frame and the current principle axis frame (z'). At the initial disk half mass radius the circular velocity is nearly 1 unit and the rotation period about 4 time units. The 20% satellite barely heats the disk. The 30% satellite causes about a 20% thickening inside the half mass radius, and a doubling to tripling beyond that radius. The energy is mainly absorbed as a coherent tilting of the disk.

5 Conclusions

The role of merging in the evolution of galaxies must be significant, but the details remain unclear. This situation will change dramatically over the next few years. First, deeper redshift surveys will soon constrain any dramatic evolution of the galactic mass function. Second, detailed studies of individual galaxies at moderate redshift will measure their scale lengths and kinematic properties to directly examine masses to check whether they are dramatically lower. Third, simulations now have the dynamical resolution to answer questions involving the merging of galaxies in a cosmologically realistic setting. In particular, simulations can usefully guide morphological studies of faint galaxies and constrain the redshift-merger rate relation. It should be emphasized that if the time to $z = 1$ is $\simeq 8.5$ Gyr, and if the merger rate remains constant at $0.04\,\mathrm{Gyr}^{-1}$, then galaxy masses will at most be 67% of current epoch values. Studies of the evolution of clustering of galaxies will be key components of constraining the merger-redshift relation. Unless the understanding of dark halo densities and dynamics is severely flawed mergers *must* be significant at higher redshifts.

Acknowledgements: The disk galaxy simulations are are a substantial part of Siqin Huang's Ph. D. thesis work. This research is supported by NSERC of Canada.

References

Bahcall, S.R. & Tremaine, S. 1988, *Astrophys. J. Lett.* **326**, L1

Barnes, J.E. & Hernquist, L. 1991, *Astrophys. J. Lett.* **370**, L65

Barnes, J.E. & Hernquist, L. 1992, *Ann. Rev. Astr. Astrophys.*, **30**, 705

Binney, J.J. 1990, *Ann. Rev. Astr. Astrophys.* **30**, 51

Broadhurst, T.J., Ellis, R.S. & Shanks, T. 1988, *Mon. Not. R. Astr. Soc.* **235**, 827

Broadhurst, T.J., Ellis, R.S. & Glazebrook, K. 1992, *Nature* **355**, 55

Carlberg, R.G. 1990, *Astrophys. J.Lett.* **359**, L1

Carlberg, R.G. 1992, *Astrophys. J. Lett.* **399**, L3

Carlberg, R.G. & Charlot, S. 1992, *Astrophys. J.* **397**, 5

Carlberg, R.G. & Couchman, H.M.P. 1989, *Astrophys. J.* **340**, 47

Carlberg, R.G., Pritchet, C.J., & Infante, L. 1994, *Astrophys. J.* **435**, 540

Cowie, L.L., Songaila, A., & Hu, M. 1991, *Nature*, 354, 460

Glazebrook, K., Ellis, R.S., Santiago, B., & Griffiths, R. 1995, *Mon. Not. R. Astr. Soc.* **275**, L19

Huang, S. 1995, *Ph. D. thesis*, in preparation

Lacey, C. & Cole, S. 1993, *Mon. Not. R. Astr. Soc.* **262**, 627

Larson, R B. & Tinsley, B.M. 1978, *Astrophys. J.* **219**, 46

Lilly, S., Crampton, D., Hammer, F., Olivier, F, & Schade, D. 1995, preprint

Mihos, J.C. 1995, *Astrophys. J. Lett.* **438**, L75

Navarro, J.F., Frenk, C.S., & White, S.D.M. 1994, preprint

Nelson, R. & Tremaine, S. 1995, *Mon. Not. R. Astr. Soc.*, in press

Press, W. H. & Schechter, P. 1974, *Astrophys. J.* **187**, 425

Quinn, P.J. & Goodman, J. 1986, *Astrophys. J.* **309**, 472

Quinn, P.J., Hernquist, L, & Fullagar, D.P. 1993, *Astrophys. J.* **403**, 74

Toomre, A. 1977, in *Evolution of Galaxies and Stellar Populations,* ed. B.M. Tinsley & R.B. Larson, (Yale Observatory: New Haven) p.401

Tóth, G. & Ostriker, J.P. 1992, *Astrophys. J.* **389**, 5

Yee, H.K.C. & Ellingson, E. 1995, *Astrophys. J.* **445**, 37

Are Ellipticals Formed by Merging Spirals?

Matthias Steinmetz and Stefan Buchner

MPI für Astrophysik, Postfach 1523, 85740 Garching, Germany

1 Introduction

Based on an idea of Toomre (1977), the merging of two spiral galaxies is still one of the most popular models for the formation of elliptical galaxies. However, it is unlikely that *all* ellipticals are formed by mergers, and some of the arguments imposed by Ostriker (1980) against the merging hypothesis are still a matter of debate. Within the last decade, two observations shed much light on the formation process of elliptical galaxies:

1. Davies *et al.* (1983) showed, that there exist two classes of ellipticals: The less luminous ones (and also the bulges of spiral galaxies) seem to rotate moderately, i.e. their flattening can easily be explained by their rotation. The more luminous ones, however, seem to be slow rotators, which are flattened by an anisotropic velocity dispersion. The rotational signature of a elliptical can be rather well characterized by the quantity

$$v/\sigma^* := \frac{(v/\sigma)}{(v/\sigma)_0} \approx \frac{v}{\sigma} \cdot \sqrt{\frac{1-\varepsilon}{\varepsilon}} \tag{1}$$

   v and σ being the rotational speed and the velocity dispersion, respectively. According to the tensor viral theorem, $(v/\sigma)_0 \approx \sqrt{(1-\varepsilon)/\varepsilon}$ is that ratio of rotation velocity and velocity dispersion one would expect for a rotationally flattened system. Less massive ellipticals tend to have $v/\sigma^* \approx 1$, whereas more massive ones have $v/\sigma^* \ll 1$.
2. However, the mass of a galaxy seems to be only a poor discriminator. Over nearly two orders of magnitude in mass, one can find both types of ellipticals in the Davies *et al.* sample. In 1988, Bender found, that the isophote shape of ellipticals provides a much better defined discriminator. Decomposing the deviations of the isophotes from an ideal ellipse into a Fourier series, the so–called a_4 coefficient, which is the coefficient of the fourth cosine, characterizes, whether the isophotes are deformed towards a more disk like object ($a_4 > 0$, *disky*) or a more box like shape ($a_4 < 0$, *boxy*). Surprisingly, all slow rotators seem to have boxy isophotes, whereas disky ellipticals seem to

have $v/\sigma^* \approx 1$. Furthermore, all symptoms which are usually identified with a history involving a major merging event can be found in the class of boxy, but not of disky ellipticals. For example, the boxy ellipticals often show a kinematically decoupled core, are radio loud and X-ray luminous (Bender *et al.* 1989). For the later discussion we also mention, that a_4 coefficients between -2% and +5% are observed, most of the ellipticals, however, are nearly perfect ellipses with $|a_4| < 1\%$.

Different attemps have been made to explain the difference between disky and boxy ellipticals. Rix & White (1990) assumed an exponential disk in the central regions of a $R^{1/4}$ ellipsoid. They showed that for some viewing angles, the isophote deviate towards disky, whereas for most viewing angles, the deviations from ideal elliptic shape would be undetectable. Similarily, Ryden (1992) assumed an additional non–ellipsoidal secondary light distribution with the shape of a box or a cylinder. Assuming a light contribution of 16% in the non–ellipsoidal component, diski- and boxiness at the 1% level can be observed for some viewing angles.

The isophote shapes of elliptical galaxies formed in N–body simulations were investigated by Stiavelli *et al.* (1991, dissipationless collapse), Governato *et al.* (1993, merging spheroids) and Heyl *et al.* (1994, merging spirals). Though starting from different initital conditions, they agree in the conclusion, that dependent on the viewing angle, they can observe disky and boxy isophotes, i.e. diski- and boxiness could be interpreted as a pure projection effect. However, the simulations of Heyl *et al.* exhibit, that the isophote shape can be strongly affected by an insufficient particle number of the N–body simulations: One of their simulations based on 64K particles exhibits a large number of projections with a $a_4 = 2\% - 5\%$. These very disky objects, however, nearly vanish if a simulation with 256K particles is analyzed.

In none of these works (analytical as well as numerical) one has looked for correlations with kinematic properties of the galaxy, i.e. it was never investigated, whether the disky objects are rotationally flattened. More recently, Scorza & Bender (1995) showed, that the assumption of a central disk component might explain the disky isophote shape, but the residual ellipsoidal object is still moderatly rotating and its flattening can almost be explained by its rotation. The residual ellipsoidal object is, therefore, kinematically different to most of the boxy elliptical which are slow rotators.

In this work we want to address the question, whether the isophote shape of merger remnants correlates with their kinematical properties. For that purpose we performed a series of highly resolved N–body simulations of merging spiral galaxies in order to look for signatures in their isophote shape.

2 Simulations and Analysis

Our simulation technique is similar to that reviewed in Barnes & Hernquist (1992) and Heyl *et al.* (1994). In the first place, a stationary spiral galaxy with a dark halo is constructed according to the moment expansion method of Hernquist

(1993b). We took masses similar to those used in other related publications (i.e. Milky Way sizes objects), an exponential disk of 5.6 10^{10} $M_\odot$ and a scale length of 3.5 kpc, and a six times more massive isothermal halo with a core radius of one disk scale length and an exponential cutoff at about 30 kpc. Two of these spiral galaxies were set at a distance of 30 disk scale lengths on a parabolic orbit in the x–y plane with a perigalacticon of about 8 kpc. Three different orbital geometries were investigated, a merging of two galaxies edge on, face on and a third one where the galaxies are tilted towards the orbital plane by 30°. In all cases, the galaxies are corrotating the z–component of the internal spin of the galaxies being parallel to the orbital angular momentum. The N–body simulations themselves were performed with a direct summation N–body code with a resolution of 65.536, 131.072 and 262.144 particles on the special purpose hardware GRAPE (Sugimoto *et al.* 1989). The gravitational softening was chosen to be 100 pc for the disk and 500 pc for the halo particles, respectively.

After a simulation is finished, we follow a strategy similar to that outlined in Governato *et al.* (1992) and Heyl *et al.* (1994). An artificial CCD image is created. For that purpose, every simulation seen from 48 different, randomly selected viewing angles was projected on a 80 × 80 grid with a side length of 40 kpc (which corresponds to 6-7 effective radii). This pixel map is analyzed by the image processing system MIDAS. To define the isophotes and the deviations from an ideal elliptical shape we benefit from the subroutines kindly made available to us by Ralf Bender. We took an isophote spacing of one magnitude, the isophote parameters being averaged over the three inner isophotes. According to Bender *et al.* (in preparation), a peculiarity test were performed comparing the a_4 coefficient with the measure ξ , given by

$$\xi = \sqrt{a_3^2 + b_3^2 + a_5^2 + b_5^2} \tag{2}$$

which means that ξ is dominated by the largest non-axisymmetric deviation. An isophote is assumed to be non regular and, therefore, thrown away in the further analysis, if $a_4 < \xi$. Concerning the kinematic properties of the merger remnant, the rotational velocity is determined at r_e, the velocity dispersion within the inner $0.5\,r_e$, r_e being the half mass or effective radius. In order to estimate the errors due to discreteness noise we have performed a bootstrap analysis as recently proposed by Heyl *et al.* (1994)

The evolution of the two spiral galaxies is quite similar to that observed by Barnes and Hernquist (1992). After a very dynamical epoch of rapid merging, the systems relax into an ellipsoidal merger remnant with an effective radius of about 5-10 kpc and a $r^{1/4}$ surface brightness law. The surface brightness, however, exhibits a cutoff near the center at about $0.25 - 0.5\,r_e$, i.e. well above the gravitational softening. Such a cutoff is not observed in elliptical galaxies (Burkert 1993). As pointed out by Hernquist (1993a), this cutoff can be avoided if the progenitors have a bulge component. Another possibility might be the inclusion of dissipative processes (Mihos & Hernquist 1994, Steinmetz & Müller in preparation).

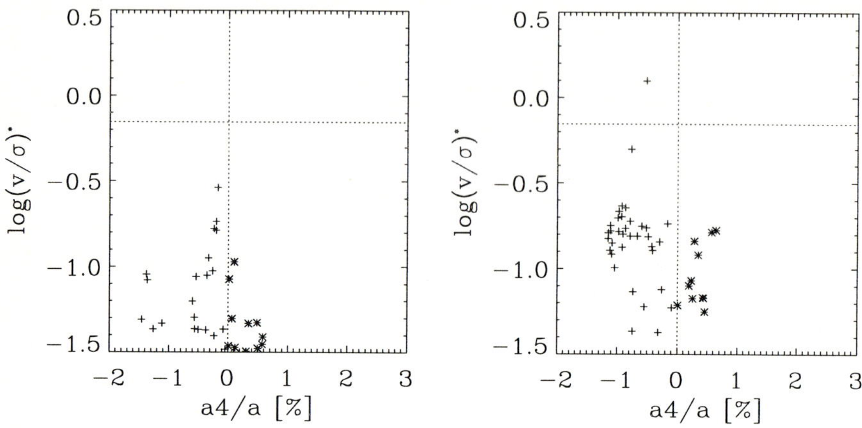

Fig. 1. $\log(v/\sigma)^*$ as a function of the a_4/a coefficient for two different models, a face on merger (left), the merger between two spirals tilted to the orbital plane by 30° (right).

3 Results

Our main results can be summarized as follows:

- Consistent with other simulations (Barnes & Hernquist 1992, Hernquist 1993a), merger remnants are slow rotators with a $v/\sigma^* \lesssim 0.3$ (Fig. 1). Their flattening cannot be explained by rotation but must stem from an anisotropic velocity dispersion. This also holds for the orbital geometry with the highest angular momentum, where the orbital angular momentum and the intrinsic spins of the spirals are aligned. Angular momentum is efficiently transported to the dark haloes. Because of the absence of rotationally flattened merger remnants, *projection effects cannot account for the difference between disky and boxy ellipticals*, at least not within the merging scenario.
- The merging of two edge on spirals yields peculiar results: the results strongly depend on the particle number and on the time difference between the merging event and the "observation". Furthermore, more than 50% of the isophotes seem to exhibit an irregular shape, i.e. higher moments are of the same order of magnitude as the a_4 coefficient. For the other two, more likely configurations, less than 10% of the isophotes show such irregularities. The results also seem to be stable against changes in the particle number and the selected time after the merging event.
- Though most of the remnants seem to have isophote deviations of less than 0.5%, a general tendency towards boxy isophotes is found (Fig. 1). In agreement with observations, boxiness larger than 2% was never found. Only a

few projections have diskiness higher than 0.5%, none of them higher than 1%. However, all these "disky" ellipticals rotate much too slow to be flattened by their rotation ($v/\sigma^* \lesssim 0.3$). This must be compared with observed disky ellipticals, which exhibit a_4 coefficients up to 5% and which are rotationally flattened ($v/\sigma^* \gtrsim 0.7$). We conclude, that rotationally flattened, disky ellipticals cannot be formed in a collisionless merging event.

In summary, merger remnants show a remarkable similarity to observed boxy ellipticals, but neither their isophote shape nor their kinematic properties fit within what is observed for disky ellipticals. The simulations, therefore, favour those models, in which the difference between disky and boxy ellipticals are interpreted as a dichotomy in their origin.

References

Barnes, J.E., Hernquist, L. 1992, *Ann. Rev. Astr. Astrophys.* **30**, 705

Bender, R. 1988, *Astron. Astrophys.* **193**, L7

Bender, R., Surma, P., Döbereiner, S., Möllenhoff, C., Madejsky, R. 1989, *Astron. Astrophys.* **217**, L35

Burkert, A. 1993, *Astron. Astrophys.* **278**, 23

Davies, R.L., Efstathiou, G., Fall, S.M., Illingworth, G., Schechter, P.L. 1983, *Astrophys. J.* **266**, 41

Governato, F., Reduzzi, L., Rampazzo, R. 1993, *Mon. Not. R. Astr. Soc.* **261**, 379

Heyl, J., Hernquist, L., Spergel, D.N. 1994, *Astrophys. J.* **427**, 165

Hernquist, L. 1993a, *Astrophys. J.* **409**, 548

Hernquist, L. 1993b, *Astrophys. J. Suppl.* **86**, 389

Mihos, J.C., Hernquist, L. 1994, *Astrophys. J. Lett.* **437**, L47

Rix, H.-W., White, S.D.M. 1990, *Astrophys. J.* **362**, 52

Ryden, B. 1992, *Astrophys. J.* **386**, 42

Scorza, C., Bender, R. 1995, *Astron. Astrophys.* **293**, 20

Stiavelli, M., Londorillo, P., Messina, A. 1991, *Mon. Not. R. Astr. Soc.* **251**, 57p

Sugimoto, D., Chikada Y., Makino J., Ito T., Ebisuzaki T., Umemura M. 1990, *Nature* **345**, 33

Toomre, A. 1977, in *The Evolution of Galaxies and Stellar Populations*, eds B.M. Tinsley & R.B. Larson, p. 401, New Heaven, Conn.Yale University.

The Structure of Dark Matter Haloes in Dwarf Galaxies

Andreas Burkert

Max-Planck-Institut für Astronomie,
Königstuhl 17, D-69117 Heidelberg, Germany

1 Summary

It is shown, that the dark matter halos of a sample of seven dwarf spiral galaxies, which have been selected from the literature because of their dark matter dominated rotation curves, represent a one-parameter family with self-similar density profiles. The global halo parameters like the finite central dark matter densities or the total mass inside a scale radius r_0 depend on r_0 through simple scaling relations. This result can be explained by the cold dark matter model if one assumes that all halos formed from density fluctuations with the same primordial amplitude.

2 Introduction

It is generally assumed that there exists an invisible matter component which contains 90% to 99% of the total mass of the Universe and which interacts with the visible matter through gravity. The nature and origin of this dark matter (DM) component is one of the most interesting and up to now unsolved problems in astrophysics. Dark matter has been detected through its gravitational force on the baryonic component not only in massive galaxy clusters but also in dwarf galaxies with several orders of magnitude smaller masses. These low-mass DM halos indicate the existence of a non-relativistic cold dark matter component (CDM) which is generally assumed to consist of some kind of non-baryonic and dissipationless DM particle (Dubinski & Carlberg 1991, Navarro *et al.* 1995). CDM simulations indeed predict the existence of extended DM halos around galaxies (Frenk *et al.* 1985). They also lead to halo density distributions which produce constant outer rotation curves, in agreement with the observations.

The observations of constant circular velocities in the outer regions of many spiral galaxies (Casertano & van Gorkom 1991) have lead to the conclusion that the DM halos are isothermal with an r^{-2} density profile in the relevant radius range. A frequently adopted DM density distribution is the so called modified isothermal profile (Begemann *et al.* 1991)

$$\rho(r) = \frac{\rho_0}{1 + (r/r_c)^2} \tag{2}$$

where r_c and ρ_0 is the core radius and the central dark matter density, respectively.

Equation (1) predicts a finite central DM density. CDM simulations with high spatial resolution (Dubinski & Carlberg 1991, Navarro *et al.* 1995) lead however to DM halos with an inner density cusp where the density diverges as $\rho \sim r^{-1}$, implying an infinite density in the center. Unfortunately it is difficult to verify such central DM cusps observationally as galaxies are in general gravitationally dominated by their visible baryonic components in the innermost regions. The inferred DM profiles then depend strongly on the assumed baryonic mass-to-light ratio. They are also affected by secular dynamical processes in the dissipative baryonic component which makes it difficult to compare their present structure with the dissipationless CDM calculations.

Fortunately there exists a class of dwarf spiral galaxies which are completely dominated by dark matter, even in their innermost regions. These objects represent ideal candidates for an investigation of the inner structure of low-mass DM halos. Flores & Primack (1994) and Moore (1994) have shown that these galaxies have rotation curves which seem to rule out central DM cusps and which are in good agreement with equation (1). This contradiction with the predictions of cosmological models is an interesting and up to now unsolved cosmological puzzle.

Here we investigate the dark matter rotation curves of a sample of seven dwarf spiral galaxies which have been selected from the literature according to the requirement that their rotation curves are clearly DM dominated at their core radius (Burkert, 1995).

3 Are Dark Matter Halos Similar?

Cosmological models with non-baryonic, dissipationless DM particles predict halos with similar structure. Navarro et al. (1995) find profiles which can be fitted nicely by the function

$$\rho(r) = 300\bar{\rho}r_{200}^3 \times r^{-1}(r + 0.25r_{200})^{-2} \tag{3}$$

where $\bar{\rho}$ is the mean density of the universe at the epoch when the DM halo formed and r_{200} denotes the radius of the DM sphere of mean overdensity of 200.

This prediction is tested in Figure 1 which shows the DM mass distribution $M(r)$ of the four best studied and completely DM dominated dwarf spiral galaxies DDO154 (open triangle, Carignan & Beaulieu 1989), DDO105 (open square, Schramm 1992), NGC3109 (open circle, Broeils 1990) and DDO170 (starred, Lake *et al.* 1990). Note that all four profiles follow very nicely the same universal mass relation. The cosmological calculations (dotted and dot-dashed curves) predict however too much mass at small radii as a result of their central density

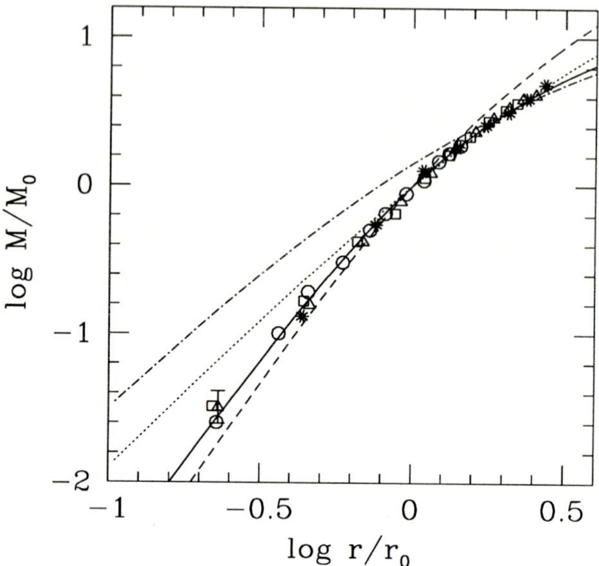

Fig. 1. Dark matter mass profiles, derived from the observed rotation curves are shown for the following dwarf spiral galaxies: DDO154 (open triangle, Carignan & Beaulieu 1989), DDO105 (open square, Schramm 1992), NGC3109 (open circle, Broeils 1990) and DDO170 (starred, Lake *et al.* 1990). The errorbar at the innermost triangle represents the observational uncertainty in determining the rotational velocity and the mass at these radii. The isothermal fit (equation 1) is shown as dashed curve, assuming a core radius $r_c = r_0$. The solid line shows the revised profile (equation 3). The dotted and dot-dashed curves show DM profiles as predicted from CDM calculations (equation 2). The best fit to the data (dotted line) is achieved if one assumes $r_{200} = 17.5 \times r_0$ which for an $\Omega = 1, h = 0.5$ CDM universe corresponds to a formation redshift of $z = 0.6$. The dot-dashed curve shows the mass profile as predicted from the CDM calculations if $r_{200} = 5 \times r_c$ with a corresponding formation redshift of $z = 1.5$.

cusp. Note also that the modified isothermal profile fails to fit the observations in the outer regions where it rises somewhat too fast due to the linear divergence of mass M with radius r.

The solid line shows $M(r)$ for the density distribution

$$\rho_{DM}(r) = \frac{\rho_0 r_0^3}{(r + r_0)(r^2 + r_0^2)} \tag{4}$$

which fits the observations very nicely over the whole observed radius range. ρ_0 and r_0 are free parameters which represent the central DM density and a scale radius, respectively. This revised density profile resembles an isothermal profile in the innermost regions ($r < r_0$). Its mass distribution diverges however only

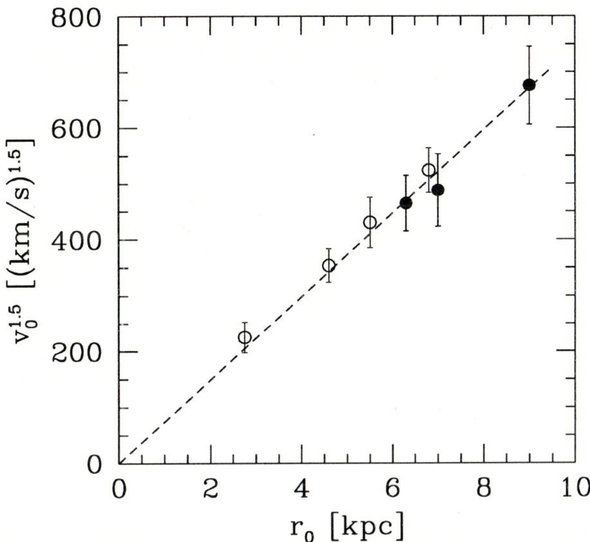

Fig. 2. The scaling relation between the rotational velocity v_0 of the DM haloes at r_0 is shown as function of the scale radius r_0 (equation 3). The open circles show the four best known DDO galaxies which have been used in Figure 1. The filled circles show three additional galaxies for which the contribution of the baryonic component to the rotation curve at r_0 had to be subtracted. The error bars show the observational uncertainty in determining v_0 as quoted in the literature. The dashed line shows a fit (used in equation 4) through the data points.

logarithmically for large r, in better agreement with CDM calculations (equation 2).

4 The Scaling Relations for Dark Matter Halos

Equation 3 contains two free parameters: ρ_0 and r_0. It is interesting to investigate the question whether these parameters are indeed free or whether they are correlated with each other. Figure 2 shows the observed rotational velocity v_0 at the scale radius r_0 of the DM rotation curves for the present sample of 7 galaxies as a function of r_0, determined by fitting the theoretically predicted rotation curve (using equation 3) to their observed dark matter rotation curves. Note that for the systems NGC55, NGC300 and NGC1560 (filled circles) the contribution of the visible component had to be subtracted, leading to a new, residual DM rotation curve (Puche & Carignan 1991, Broeils 1992). We find a strong linear correlation between r_0 and $v_0^{1.5}$. Given v_0 amd r_0 one can determine M_0, the total DM mass inside the scale radius, as well as the central DM density ρ_0 as a function of r_0:

$$M_0 = 7.2 \times 10^7 \left(\frac{r_0}{kpc}\right)^{7/3} M_\odot \tag{5}$$
$$\rho_0 = 4.5 \times 10^{-2} \left(\frac{r_0}{kpc}\right)^{-2/3} \frac{M_\odot}{pc^3}$$

Note, that the dark matter density profiles are completely determined by only one free parameter. Given e.g. r_0, equation 4 provides the corresponding ρ_0 which completely specifies the DM density profile through equation 3.

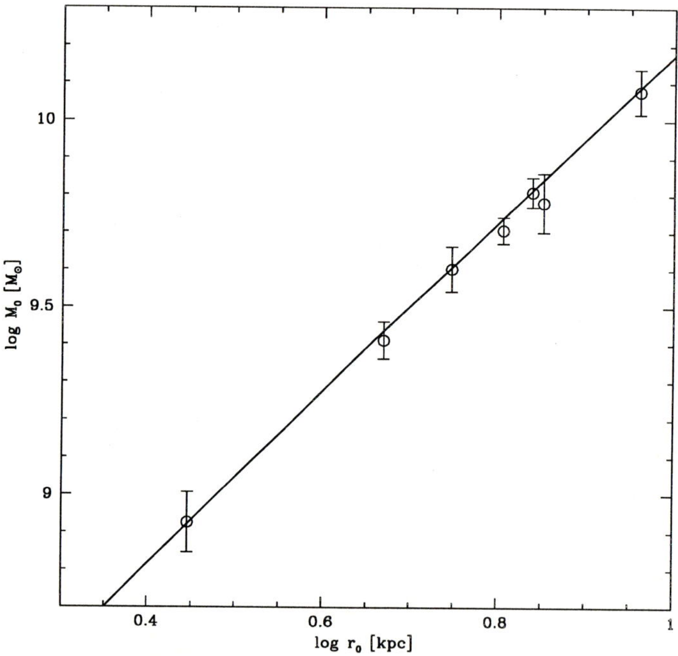

Fig. 3. The observed scale parameters M_0 and r_0 (open circles) are compared with the predictions of a standard CDM model (solid line).

Burkert (1995) demonstrates that there might exist a cosmological explanation for the scaling relations (equation 4). The solid line in figure 3 shows the expected relation between M_0 and r_0 if one assumes that all DM halos formed from primordial cold dark matter density fluctuations with the same initial amplitude. The observations (open triangle) indeed follow very nicely the theoretically predicted mass-radius relation.

5 Conclusions

Cosmological models also predict self-similar dark matter halos, in agreement with the present observations. The observed, universal DM density distribution disagrees however with the results of theoretical models. It has been proposed, that this discrepancy arises from secular energetic processes of the baryonic component which affect the inner DM density distribution through variations of the gravitational potential and which are not considered in the dissipationless cosmological models (Moore 1994). Although this explanation seems at first very attractive, the present results show that considerable fine tuning of these chaotic processes would have to be required in order to lead to self-similar DM halos. One rather would expect to find a second free parameter which is specified by the efficiency with which these secular processes affect the inner DM halo. The observed shallow central density profiles more likely indicate some important features of dark matter which are still missing in current cosmological models.

Figure 3 indicates, that the scaling relations of DM halos around dwarf galaxies can in principle be explained by standard cold dark matter models. It is however puzzling that only fluctuations with a certain fixed primordial amplitude should have managed to form DM dominated dwarf spiral galaxies.

References

Begeman, K.G., Broeils, A.H., and Sanders, R.H. 1991, *Mon. Not. R. Astr. Soc.* **249**, 523

Broeils, A.H. 1990, *Thesis*, Univ. Groningen

Broeils, A.H. 1992, *Astron. Astrophys.* **256**, 19

Burkert, A. 1995, *Astrophys. J. Lett.* **447**, L25

Carignan, C., and Beaulieu, S. 1989, *Astrophys. J.* **347**, 760

Casertano, S., and van Gorkom, J.H. 1991, *Astron. J.* **101**, 1231

Dubinski, J., and Carlberg, R. 1991, *Astrophys. J.* **378**, 496

Flores, R.A., and Primack, J.R. 1994, *Astrophys. J. Lett.* **427**, L1

Frenk, C.S., White, S.D.M., Efstathiou, G.P., and Davis, M. 1985, *Nature* **317**, 595

Lake, G., Schommer, R.A., and van Gorkom, J.H. 1990, *Astron. J.* **99**, 547

Moore, B. 1994, *Nature* **370**, 629

Navarro, J.F., Frenk, C.S., and White, S.D.M. 1995, *Mon. Not. R. Astr. Soc.*, in publication

Puche, D., and Carignan, C. 1991, *Astrophys. J.* **378**, 487

Schramm, D.N. 1992, *Nucl. Phys. B. Proc. Suppl.* **28A**, 243

Spectra of Young Galaxies

Stéphane Charlot

Astronomy Department and Center for Particle Astrophysics,
University of California, Berkeley CA 94720, USA

1 Introduction

The expected widespread population of high-redshift analogs or progenitors of present-day disk galaxies has still not been observed in emission. This leaves large uncertainties on the appearance of young galaxies (see White *et al.* 1989, and Pritchet 1994). Early models predicted that "primeval galaxies" undergoing strong starbursts at redshifts $z \gtrsim 2$ should be detectable at magnitudes $R \sim 21-23$, but deep spectroscopic surveys to $B \lesssim 22.5$, $I \lesssim 22.1$, and $K \lesssim 20$ have not revealed such population of forming galaxies (Partridge & Peebles 1967, Meier 1976, Colless *et al.* 1993, Cowie *et al.* 1994, Lilly *et al.* 1995. Partridge & Peebles (1967) pointed out that Lyα emission could be the most prominent and easily detectable signature of primeval galaxies. The reason for this is that the ionizing radiation from young stars in galaxies should lead to a strong and narrow Lyα line by recombination of the hydrogen in the ambient interstellar medium, which would be more readily visible than continuum radiation against the sky noise. A few galaxy-like objects have been discovered at redshifts $z \lesssim 4$, which occasionally show strong Lyα emission (Pritchet 1994, Spinrad 1989, Macchetto *et al.* 1993, Rowan-Robinson *et al.* 1993). However, most of these objects are peculiar, and their connection to present-day galaxies is not at all clear. In fact, all blank sky searches for Lyα emission from ordinary galactic disks at high redshifts have given null results, suggesting that young galaxies form stars slowly or in large volumes, or that Lyα photons are absorbed by dust (see Baron & White 1987, and references therein).

Independently, much has been learned on the distribution of H I in the universe at redshifts $z \lesssim 3.5$ from absorption-line studies of distant quasars (see for example Petitjean *et al.* 1993). The strongest absorption lines are attributed to the damped Lyα systems, that are generally interpreted as the best candidates for ordinary galactic disks at high redshift (Lanzetta *et al.* 1991). These have observed H I column densities $N_{\rm HI} \gtrsim 2 \times 10^{20}\,{\rm cm}^{-2}$, and their abundances in heavy elements and dust at $z \approx 2.5$ amount to about 10% of the values in the Milky Way (Pei *et al.* 1991, Pettini *et al.* 1994). The damped Lyα systems do not show Lyα emission at the level expected from young, dust-poor disk galaxies (see Pettini *et al.* 1994). Alternatively, the evolution with redshift of the gas density integrated over all H I absorbers, $\Omega_{\rm HI}(z)$, provides some constraint on the global depletion of cold gas though star formation in the universe since $z \approx 3.5$ (Lanzetta *et al.* 1995, Pei & Fall 1995]). Absorption-line systems of distant quasars have also been used successfully as a way to select normal galaxies in emission out to $z \approx 1.6$ against the dominant population of relatively nearby blue galaxies that dominate galaxy number counts at faint magnitudes (Bergeron 1988, Steidel & Dickinson 1995, Steidel *et al.* 1995). These studies have shown that field galaxies with luminosities around L^* exhibit only little evolution in their space density, luminosity, and optical/infrared colors at redshifts $0.2 \lesssim z \lesssim 1.6$.

Recent observations therefore seem to indicate that the formation and evolution of normal disk galaxies has been less spectacular than originally thought. In what follows, we explain how the apparent lack of Lyα emission from young galaxies at high redshift is probably mainly a consequence of the relatively brief periods in which primeval galaxies are dust-free, and hence Lyα-bright. In fact, most present observational constraints on young galaxies appear to be in agreement with the predictions of theories based on hierarchical clustering, in which galaxies form slowly and relatively recently. The very blue, primeval galaxy phase expected at the onset of star formation would then be faint and short-lived. Since at redshifts $z \gtrsim 2$ galaxies are expected to be hard to detect, one may think of using population synthesis models to trace back the early history of star formation from observations at lower redshifts. We also discuss below the limitations of this approach.

2 Lyα Emission from Young Galaxies

The observed Lyα emission from a young galaxy depends on the star formation rate and initial mass function (IMF), but also on several other factors: the contributions by supernova remnants and active galactic nulcei, the orientation of the galaxy, and absorption by dust. The contribution to the Lyα emission by stars can be estimated using stellar population synthesis models. We assume for the moment that circumstellar H II regions are the only sources of Lyα photons and that the column density of the ambient H I is large enough that case B recombination applies ($N_{\rm HI} \gtrsim 10^{17}\,{\rm cm}^{-2}$) but otherwise ignore the effects on the interstellar medium on the transfer of Lyα photons. Under these "minimal" assumptions and using recent population synthesis models, Charlot and Fall have shown that the Lyα emission from a galaxy depends sensitively on the age and IMF slope, even when the star formation rate is constant (Charlot & Fall 1993). The dependence on the IMF upper cutoff and metallicity, on the other hand, are much weaker. Thus, only a rough estimate of the Lyα equivalent width of a young, dust-free galaxy is permitted, about 50 – 120 Å.

We now briefly review the other factors that can affect the observed Lyα emission from a young galaxy (see Charlot & Fall 1993 for more details). Shull & Silk (1979) have computed the time-averaged, Lyα luminosity of a population of Type II supernova remnants using a radiative-shock code with low metallicity. Their results indicate that the contribution to the Lyα emission by supernova remnants is always less than the contribution by stars (typically 10% for a solar-neighborhood IMF) and can therefore be neglected. Active Galactic Nuclei (AGNs) are another potential source of ionizing radiation in a galaxy. We assume for simplicity that the spectrum of an AGN can be approximated by a power law $f_\nu \propto \nu^{-\alpha}$ with an index blueward of Lyα in the range $1 \lesssim \alpha \lesssim 2$. If we also assume that the AGN is completely surrounded by H I, that case B recombination applies, and that absorption by dust is negligible, then the Lyα equivalent width is $827\alpha^{-1}(3/4)^\alpha$ Å, or 600 Å for $\alpha = 1$ and 200 Å for $\alpha = 2$. The fact that most bright quasars have observed Lyα equivalent widths in the range 50 – 150 Å could reflect a partial covering of the AGNs by H I clouds in the broad-line regions (in fact, some ionizing radiation escapes from quasars), attenuation of the Lyα emission by dust, or orientation effects (see below). Thus, in principle, AGNs can produce higher Lyα equivalent widths than stellar populations. However, the presence of an AGN in a galaxy is usually revealed by other readily identifiable signatures: strong emission lines of highly ionized species (C IV, He II, etc.) and broad emission lines with velocity widths several times larger than those expected from the virial motions within galaxies.

The Lyα photons produced in galaxies will suffer a large number of resonant scatterings in the ambient neutral atomic hydrogen. In the absence of dust, this would lead to no net enhancement of the angle-averaged Lyα emission from a galaxy or of the total Lyα emission from a sample of randomly-oriented galaxies. However, since the Lyα line is emitted more isotropically than the continuum, the Lyα equivalent width of an individual galaxy will decrease as it is viewed more nearly edge-on. For example, in the idealized case of a plane-parallel slab, the ratio of the observed to angle-averaged Lyα equivalent width will decrease from 2.3 to 0 for viewing angles to the normal ranging from 0° to 90°. The resonant scattering of Lyα photons by H I also increases enormously their chances of absorption by dust grains. The attenuation is expected to be important when the dimensionless dust-to-gas ratio, defined in terms of the extinction optical depth in the B band by $k \equiv 10^{21}(\tau_B/N_{\rm HI})\,{\rm cm}^{-2}$, exceeds the critical value $k_{\rm crit} \approx 0.01(N_{\rm HI\perp}/10^{21}\,{\rm cm}^{-2})^{-4/3}(\sigma_V/10\,{\rm km\,s}^{-1})^{2/3}$ (see Charlot & Fall 1993). In this expression, $N_{\rm HI\perp}$ and σ_V are the face-on column density and line-of-sight velocity dispersion of HI. For reference, the dust-to-gas ratio in the Milky Way and Large and Small Magellanic Clouds are, respectively, $k \approx 0.8$, $k \approx 0.2$, and $k \approx 0.02$, and the face-on H I column densities within the optically visible regions of most spiral galaxies lie in the range $10^{20} \lesssim N_{\rm HI\perp} \lesssim 10^{21}\,{\rm cm}^{-2}$. Thus, we expect $k \gtrsim k_{\rm crit}$ unless the dust-to-gas ratio is much smaller than the value in the Milky Way. In particular, some attenuation of the Lyα emission by dust is expected in the damped Lyα systems, since $N_{\rm HI} \gtrsim 2 \times 10^{20}\,{\rm cm}^{-2}$ and $k \approx 0.1$ (although there may be a large dispersion around this value; see Pei *et al.* 1991, and Pettini *et al.* 1994). Moreover, the attenuation of Lyα emission by dust depends sensitively on the structure of the interstellar medium in a galaxy. In the case of a multiphase medium, the transfer of Lyα photons will depend largely on the topology of the interfaces between H I and H II regions (see Spitzer, and Neufeld 1991). As a result of these complications, it is nearly impossible to deduce star formation rates from the observed Lyα emission of a galaxy.

Charlot and Fall (see Charlot & Fall 1993) have used the above arguments to interpret the observations of and searches for Lyα emission from nearby star-forming galaxies, damped Lyα systems, blank sky, and the companions of quasars and damped Lyα systems. Their results indicate that, when Lyα emission is weak or absent, as is the case in most star-forming galaxies at low redshifts and in damped Lyα systems at high redshifts, the observed abundance of dust is sufficient to absorb most of the Lyα photons. On the other hand, when Lyα emission is strong, the presence of highly ionized species, large velocity widths, or nearby quasars indicate that much of the ionizing radiation may be supplied by AGNs. The hope has always been that the searches for Lyα emission at high redshifts would reveal a population of primeval galaxies, in which the abundances of heavy elements and hence dust were low enough that most of the Lyα photons could escape. Such a population may exist at some redshifts. However, since the Lyα emission is attenuated when the dust-to-gas ratio exceeds $1 - 10\%$ of the value in the Milky Way, a typical galaxy probably spends only the first few percent of its lifetime in a Lyα-bright phase. We therefore expect primeval galaxies, as defined above, to be relatively rare at most redshifts, consistent with the null results of all searches to date.

3 What Should Young Galaxies Look Like?

If galaxies undergoing their first episodes of star formation in the young universe resemble nearby examples of starburst galaxies, one would expect them to exhibit very blue continua and strong Balmer emission lines. However, the most distant counterparts of nearby disk galaxies found by association with quasar absorption-line systems do not show such extreme signatures (Steidel & Dickinson 1995). Instead, out to $z \approx 1.6$, normal field galaxies appear to display strikingly little evolution in their space density, luminosity, and optical/infrared colors. Other constraints on the history of star formation in galaxies, and hence on their appearance, may be obtained from the evolution of the integrated neutral gas density of the universe. Recent surveys indicate that the total amount of gas in damped Lyα systems at a redshift $z \approx 3.5$ could be nearly as high as the total density of luminous material in present-day galactic disks (Lanzetta *et al.* 1995, Wolfe 1995). This would imply that most stars have formed relatively recently, consistent with the expectation from theories of galaxy formation based on hierarchical clustering (although a more precise interpretation of the observed Ω_{HI} evolution requires including the effect of quasar obscuration by dust in damped Lyα systems; see Pei & Fall 1995).

A natural question to ask, then, is what do models accommodating present observational constraints predict for the appearance of very young galaxies? To answer this question, we have computed the early spectral evolution of a spiral galaxy using a combination of the Kauffmann *et al.* (1993) semi-analytic model of galaxy formation and the Bruzual & Charlot (1993) population synthesis code. The star formation rate can be adjusted to reproduce both the evolution of $\Omega_{\rm HI}(z)$ at $z \lesssim 3.5$ in a standard cold dark matter universe and the spectral energy distribution of a typical nearby spiral galaxy at $z = 0$ (see Kauffmann & Charlot 1994). The early spectral evolution of this model is presented in Figure 1. Figure 1a shows that the galaxy undergoes an extremely blue phase at $z > 3$, during which the spectrum resembles the observed spectrum of the nearby starburst galaxy Mrk 710 with strong H-Balmer and oxygen emission lines (from Vacca & Conti 1992). At this time, the young galaxy could qualify as a "primeval

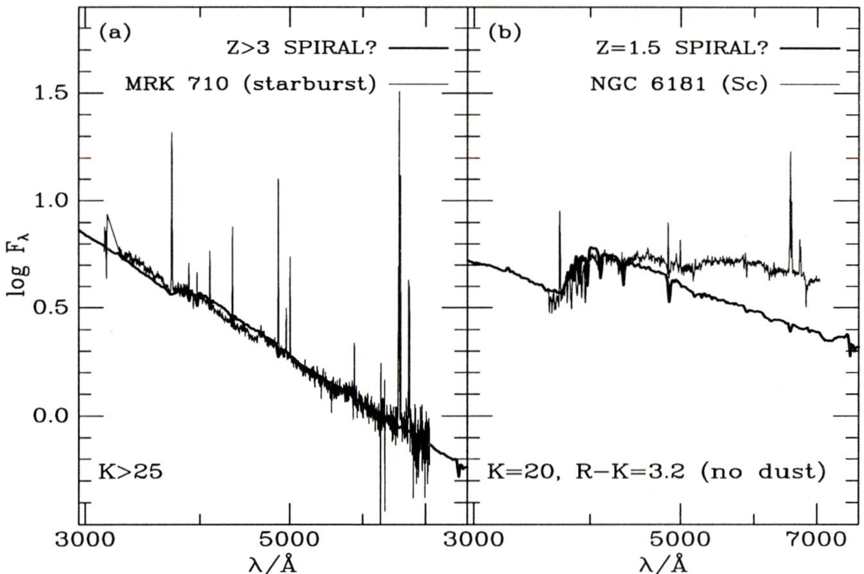

Fig. 1. (a) Possible appearance of a progenitor spiral galaxy at $z > 3$ satisfying the current observational constraints on the evolution of $\Omega_{\rm HI}(z)$ and the observed properties of the Milky Way in a standard cold dark matter universe (thick line) compared to the observed spectrum of the nearby starburst galaxy Mrk 710 (thin line). The model spectrum does not include emission lines (see text for more details). (b) Same model galaxy as in (a) viewed at $z = 1.5$ and compared to the observed spectrum of the nearby spiral galaxy NGC 6181. The spectra in (a) and (b) are in the rest frames of the galaxies, and the predicted apparent K magnitudes (and $R - K$ color at $z = 1.5$) are indicated at the bottom.

galaxy" as defined earlier. However, the phase is short-lived ($\lesssim 10^7$ yr) and very faint ($K > 25$). Then, the onset of evolved supergiant, asymptotic giant branch, and red giant branch stars reddens the spectrum substantially. At $z = 1.5$, twenty percent of the stars present at $z = 0$ have formed, and the model galaxy in Figure 1b has $K \approx 20$ and $R - K \approx 3.2$, i.e., a spectrum only moderately bluer than that of the nearby spiral galaxy NGC 6181 (from Kennicutt 1992).

These predicted colors, which ignore reddening by dust, are interestingly close to the observed $K \approx 19.5$ and $R - K \approx 4$ of galaxies discovered at $z \approx 1.5$ in association with quasar absorption-line systems (Steidel & Dickinson 1995). Hence, the present results would reinforce the suggestion that the spectra of normal disk galaxies have evolved only moderately for much of their lifetime. The extremely blue phase at the onset of star formation, which might coincide with a Lyα-bright phase, is expected to be much fainter and short-lived.

4 Tracing Back the History of Star Formation in Galaxies

Since young galaxies may be hard to detect at redshifts beyond $z \sim 2$, an alternative is to try and trace back the earlier history of star formation from observations at lower redshifts. The conventional approach to this problem is to use stellar population synthesis models and search for the evolution of the star formation rate that will reproduce the observed spectral characteristics of galaxies. We now exemplify the difficulties in this approach using the recent population synthesis models of Bruzual & Charlot (1993). Figure 2a shows the spectral evolution of a burst stellar population computed for a Salpeter IMF with lower and upper cutoffs $0.1\ M_\odot$ and $100\ M_\odot$, respectively. At 10^6 yr, the spectrum is entirely dominated by short-lived, young massive stars on the main sequence. Then, the ultraviolet light declines and the spectrum reddens rapidly as lower-mass stars progressively complete their evolution. The most remarkable feature in Figure 2a is the nearly unevolving shape of the optical to near-infrared spectrum at ages from 4 to 19 Gyr. This "age degeneracy" hampers the dating of passively evolving stellar populations such as elliptical galaxies from the single knowledge of the continuum spectrum, or equivalently, of broad-band colors (see Charlot & Silk 1994, and references therein). The high luminosity of young massive stars also complicates the determination of the past history of star formation in spiral galaxies. This is illustrated in Figure 2b, in which we show the spectral decomposition of a present-day spiral galaxy represented by a model with constant star formation seen at 13.5 Gyr. The ultraviolet to visible spectrum is strongly dominated by massive, main-sequence and supergiant stars. Furthermore, as seen above, the contribution by evolved asymptotic giant branch and red giant branch stars is age-degenerated. Therefore, from the continuum spectrum of a spiral galaxy, one can at best obtain the ratio of the current to past-averaged star formation rate (see for example Kennicutt *et al.* 1994).

Another important limitation in determining the past history of star formation in galaxies is the influence of metallicity. At fixed mass, stars of lower metallicity are hotter and evolve more rapidly than stars of higher metallicity. Thus, the age assigned to a stellar population on the basis of its observed colors is a function of metallicity. Table 1 illustrates the dependence on metallicity of the turnoff age (and mass of the turnoff star) at fixed turnoff temperature for a burst stellar population (after Schaller *et al.* 1992). The assigned age can vary by a factor of up to five when the metallicity changes from 100% to 5% of solar.

This is the "age-metallicity" degeneracy. In reality, the numbers in Table 1 are only roughly indicative of the age-metallicity degeneracy in interpreting

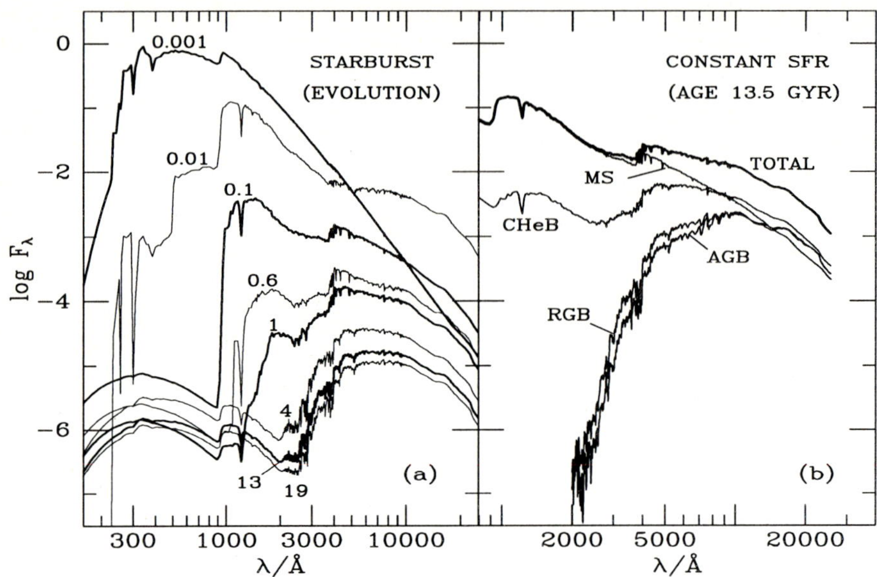

Fig. 2. (a) Spectral evolution of an instantaneous-burst stellar population computed with the population synthesis models of Bruzual & Charlot (1993). The stellar population has a Salpeter IMF with lower and upper cutoffs 0.1 $M_\odot$ and 100 $M_\odot$, respectively. Ages are indicated next to the spectra (in Gyrs). (b) Spectral decomposition of a present-day spiral galaxy represented by a model with constant star formation rate for 13.5 Gyr. The total spectrum is shown, together with the contributions main sequence (MS), supergiant (CHeB), asymptotic giant branch (AGB), red giant branch (RGB) stars. The IMF is the same as in (a).

galaxy continuum spectra because the colors of stars at fixed temperature also change slightly with metallicity. The most recent models of stellar population synthesis now include the effect of metallicity variations on the spectral evolution of galaxies (e.g. Worthey 1994, Bressan *et al.* 1994). These models can illustrate more accurately the age-metallicity degeneracy of spectral fits of galaxies.

Some additional information on the past history of star formation in galaxies may be learned from the stellar absorption lines of hydrogen and of other prominent atoms and molecule such as Mg, Mg_2, Fe, Ca, Na, Sr, and CN. For example, main-sequence A and B stars (that have lifetimes $\lesssim 2\,\mathrm{Gyr}$) are expected to strengthen the H-Balmer series and weaken the prominent metallic lines in the integrated galaxy spectrum. This can be most simply illustrated by considering the case of intermediate-age stellar populations in early-type galaxies. There is growing photometric and dynamical evidence that many E/S0 galaxies have formed stars only a few billion years ago (Pickles 1989, Rose 1985, Schweizer & Seitzer 1992, and references therein). The prototypical example is the dwarf elliptical galaxy M 32, which is believed to have undergone substantial star for-

Table 1. Dependence on metallicity of the turnoff age (and mass of the turnoff star) at fixed turnoff temperature for a burst stellar population (after Schaller *et al.* 1992).

$\log(T_{\rm TO}/{\rm K})$	$Z = Z_\odot$	$Z = 0.05 Z_\odot$
4.0	0.4 Gyr (3.0 $M_\odot$)	1.3 Gyr (1.7 $M_\odot$)
3.8	2.7 Gyr (1.5 $M_\odot$)	15. Gyr (0.8 $M_\odot$)

mation until only about 5 Gyr ago (e.g., O'Connel 1980). Figure 3a illustrates how spectral absorption features such as the Balmer Hδ equivalent width can be used to detect late bursts of star formation in early-type galaxies with colors otherwise typical of old, passively evolving stellar populations. The solid lines correspond to a model elliptical galaxy formed in a major burst at age $t = 0$, on which a new burst involving 10% of the final mass is added at an age of 6 Gyr. At ages $t > 6.5$ Gyr, the $U - B$ and $V - K$ colors of this model differ by less than 0.1 mag from the values in the absence a second burst (shown by the dashed lines). However, the Hδ equivalent width continues to evolve significantly for nearly 1 Gyr because of the presence of A and B stars. Pickles and Rose have shown, using other stellar absorption lines, how similar diagnostics can be used to discriminate between different generations of stars in E/S0 galaxies and, to some extent, to untangle the competing effects of age and metallicity on the spectra (see Pickles 1989 and Rose 1985). Worthey has also recently produced a comprehensive study of the absorption-line characteristics of old stellar populations and their dependence on age and metallicity (Worthey 1994, see also Worthey *et al.* 1993).

We now briefly exemplify some implications of these arguments for the history of star formation in early-type galaxies (see Charlot & Silk 1994 for more details). In the upper panel of Figure 3b we have compiled estimates of the typical fraction of optical light accounted for by intermediate-age stars in normal E/S0 galaxies at redshifts $z \lesssim 0.4$ from several studies of stellar absorption-lines strengths (Rose 1985, Pickles 1985, Couch & Sharples 1987). At $z \gtrsim 0.1$, most galaxies were selected in clusters (the estimates are based on all red galaxies with luminosities $\lesssim L^*$ for which spectra were available at each redshift). In each case, the range of intermediate ages attributed to the stars is indicated in billion years. In the lower two panels of Figure 3b, we have reexpressed these constraints on the ages of stellar populations into constraints on the redshifts of formation for two cosmologies using the Bruzual & Charlot (1993) population synthesis models. A flat universe with $q_0 = 0.5$ and $h = 0.45$, and an open universe with $q_0 = 0.1$ and $h = 0.55$ (where $h = H_0/100\,{\rm km\,s^{-1}Mpc^{-1}}$). Both correspond to a present age of the universe of about 15 Gyr and lead to similar predictions: the mass fraction of stars formed in E/S0 galaxies has decreased smoothly with time, from about 8% at $z \approx 1$ to less than 1% at $z \approx 0$ (see also Pickles 1985 and Schweizer & Seitzer 1992). This evolution of the star formation rate inferred from stellar absorption-

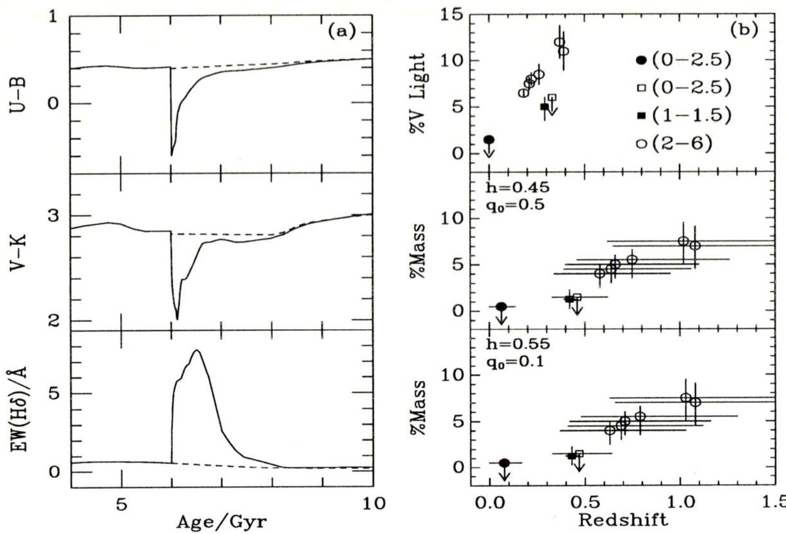

Fig. 3. (a) Evolution of the $U-B$ and $V-K$ colors and Hδ absorption equivalent width of a model elliptical galaxy formed in a single burst at $t = 0$ (dashed line) and of a similar model galaxy on which a new burst involving 10% of the final mass is added at an age of 6 Gyr (solid line). (b) Upper panel: observed contributions by intermediate-age stars to the V luminosity of E/S0 galaxies at $z \approx 0$ and in low-redshift clusters inferred from stellar absorption-line studies (see text for sources). Different symbols correspond to different studies or different ranges of intermediate ages (indicated in Gyrs). Lower panels: mass fraction of stars formed in the progenitors of E/S0 galaxies as a function of redshift derived for two cosmologies from the observations shown in the upper panel. The horizontal error bars follow from the uncertainties on the ages of stars detected in low-redshifts E/S0 galaxies. The vertical error bars follow from the uncertainties on the determination of the contribution by these stars to the mass for the allowed range of ages.

line studies is in reality a mean evolution averaged over large redshift intervals, as the horizontal error bars indicate. It does not imply that E/S0 galaxies should form stars at all times. In fact, the ages of intermediate-age stars estimated from absorption-line strengths in the spectra of galaxies at low redshifts are uncertain by a few billion years. As Figure 3a shows, after a galaxy undergoes a burst of star formation, the colors reach the values characteristic of old, passively evolving stellar populations in less than 1 Gyr. Thus, although galaxies at low redshift may present similar signatures of past star formation, there should be a dispersion in the ages and hence colors of the progenitor galaxies at high redshift around the value corresponding to the mean epoch of star formation estimated in the lower panels of Figure 3b. The relevance of this result for the evolution of galaxies in clusters has been investigated by Charlot & Silk 1994 (see also Belloni *et al.* 1995).

Unfortunately, stellar absorption-line strengths cannot yet be used to trace back the history of star formation in spiral galaxies. The reason for this is that the best-known stellar absorption features arise at ultraviolet and optical wavelengths, where the spectral signatures of old and intermediate-age stars in spiral galaxies are hidden by the strong continuum light of young massive stars (see Fig. 2b). The situation may soon be improved, as substantial progress is underway to understand the infrared spectral signatures of old stars in star-forming galaxies (Lancon & Rocca-Volmerange 1995). However, tracing back the onset of star formation in normal disk galaxies appears to be a long way ahead.

References

Baron, E., White, S.D.M. 1987, *Astrophys. J.* **322**, 585

Belloni, P., Bruzual A., G., Thimm, G.J., Röser, H.-J. 1995, *Astron. Astrophys.* **297**, 61

Bergeron, J. 1988, in *Large Scale Structures in the Universe*, Int. Astron. Union Symp. No. **130**, Reidel, p343

Bressan, A., Chiosi, C., Fagotto, F. 1994, *Astrophys. J. Suppl.* **94**, 63

Bruzual A., G., Charlot, S. 1993, *Astrophys. J.* **405**, 538

Charlot, S., Fall, S.M. 1993, *Astrophys. J.* **415**, 580

Charlot, S., Silk, J. 1994, *Astrophys. J.* **432**, 453

Colless, M., Ellis, R.S., Broadhurst, T.J., Taylor, K., Peterson, B.A. 1993, *Mon. Not. R. Astr. Soc.* **261**, 19

Couch, W.J., Sharples, R.M. 1987, *Mon. Not. R. Astr. Soc.* **229**, 423

Cowie, L.L., Gardner, J.P., Hu, E.M., Songaila, A., Hodapp, K.-W., Wainscoat, R.J. 1994, *Astrophys. J.* **434**, 114

Kauffmann, G., White, S.D.M., Guiderdoni, B. 1993, *Mon. Not. R. Astr. Soc.* **274**, 201

Kauffmann, G., Charlot, S. 1994, *Astrophys. J.* **430**, L97

Kennicutt, R.C. 1992, *Astrophys. J. Suppl.* **79**, 255

Kennicutt, R.C., Tamblyn, P., Congdon, C.W. 1994, *Astrophys. J.* **435**, 22

Lancon, A., Rocca-Volmerange, B. 1995, *Astron. Astrophys.*, preprint

Lanzetta, K.M., Wolfe, A.M., Turnshek, D.A. 1995, *Astrophys. J.* **440**, 435

Lanzetta, K.M., Wolfe, A.M., Turnshek, D.A., Lu, L.M., McMahon, R.C., Hazard, C. 1991, *Astrophys. J. Suppl.* **77**, 1

Lilly, S.J., Tresse, L., Hammer, F., Crampton, D., Le Fevre, O. 1995, **Astrophys. J.**, in press

Macchetto, F., Lipari, S., Giavalisco, M., Turnshek, D.A., Sparks, W.B. 1993, *Astrophys. J.* **404**, 511

Meier, D. 1976, *Astrophys. J.* **207**, 343

Neufeld, D.A. 1991, *Astrophys. J. Lett.* **370**, L85

O'Connell, R.W. 1980, *Astrophys. J.* **236**, 430

Partridge, R.B., Peebles, P.J.E. 1967, *Astrophys. J.* **147**, 868

Pei, Y.C., Fall, S.M. 1995, *Astrophys. J.*, preprint

Pei, Y.C., Fall, S.M., Bechtold, J. 1991, *Astrophys. J.* **378**, 6

Petitjean, P., Webb, J.K., Rauch, M., Carswell, R.F., Lanzetta, K.M. 1993, *Mon. Not. R. Astr. Soc.* **262**, 499

Pettini, M., Smith, L.J., Hunstead, R.W., King, D.L. 1994, *Astrophys. J.* **426**, 79
Pickles, A.J. 1985, *Astrophys. J.* **296**, 340
Pickles, A.J. 1989, in *The Epoch of Galaxy Formation*, eds C.S. Frenk, R.S. Ellis, T. Shanks, A.F. Heavens, J.A. Peacock, Kluwer, p.191
Pritchet, C.J. 1994, *Publ. Astr. Soc. Pac.* **106**, 1052
Rose, J. 1985, *Astron. J.* **90**, 1927
Rowan-Robinson, M., *et al.* 1993, *Mon. Not. R. Astr. Soc.* **261**, 513
Schaller, G., Schaerer, D., Meynet, G., Maeder, A. 1992, *Astron. Astrophys. Suppl.* **96** 269
Schweizer, F., Seitzer, P. 1992, *Astron. J.* **104**, 1039
Shull, J.M., Silk, J. 1979, *Astrophys. J.* **234**, 427
Spinrad, H. 1989, in The Epoch of Galaxy Formation, eds C.S. Frenk, R.S. Ellis, T. Shanks, A.F. Heavens, J.A. Peacock, Kluwer, p.39
Spitzer, L. *Physical Processes in the Interstellar Medium*, Wiley
Steidel, C.C., Dickinson, M. 1995, in *Wide Field Spectroscopy and the Distant Universe*, Cambridge University Press, to appear
Steidel, C.C., Dickinson, M., Persson, S.E. 1995, *Astrophys. J. Lett.* **437**, L75
Vacca, W.D., Conti, P.S. 1992. *Astrophys. J.* **401**, 543
Wolfe, A.M. 1995, in *The Physics of the Interstellar Medium and Intergalactic Medium*, eds A. Ferrara, C. Heiles, C. McKee, P. Shapiro, *Pub. Astr. Soc. Pac. Conf. Ser.*, to appear
Worthey, G., Faber, S.M., Gonzalez, J.J. 1993, *Astrophys. J.* **409**, 530
Worthey, G. 1994, *Astrophys. J. Suppl.* **95**, 107

Pure Luminosity Evolution Models of Faint Galaxy Samples

Gustavo Bruzual A.[1,2] *and Lucia Pozzetti*[3]

[1] Landessternwarte Heidelberg-Königstuhl, 69117 Heidelberg, Germany
[2] C.I.D.A., Apartado Postal 264, Mérida, Venezuela
[3] Dipartimento di Astronomia, Università di Bologna, Via Zamboni 33, Bologna, Italy

1 Introduction

Koo & Kron (1992, hereafter KK92), Koo, Gronwall & Bruzual (1993, hereafter KGB93), and Gronwall & Koo (1995, hereafter GK95) have shown that it is possible to explain most of the observed properties of faint field galaxy samples by means of Pure Luminosity Evolution (PLE) models. KK92 assume that galaxies follow the conventional Schechter (1976) luminosity function (LF) and that their spectral energy distributions (SEDs) evolve according to the Bruzual (1983) models. Instead, KGB93 assume that spectral evolution can be neglected to a first approximation and derive the LF that is required to understand the number counts and the color and redshift distributions. KGB93 derive the K-corrections for the non evolving galaxy SEDs from a set of Bruzual & Charlot (1993, hereafter BC93) models. GK95 extend the KGB93 work to PLE models properly and obtain excellent fits to the data, not far from those of KGB93, but the derived LFs are closer to the conventional ones.

From an independent point of view, Metcalfe *et al.* (1991, 1995a) argue that most of the properties of their galaxy samples in b_j and r_f can be understood by means of PLE models. Their models are somewhat unphysical because the spectral evolution assumed in b_j and r_f is not consistent, and because the (e+K)-corrections used are postulated to remain constant beyond $z = 2$.

In this paper we present a progress report of a large project on the exploration of traditional, simple, and self-consistent PLE models. The details of this work are described elsewhere (Pozzetti, Bruzual, & Zamorani 1995). In our work we assume that *(a)* Galaxies of each type and color class follow the LFs derived locally (Shanks 1990; Efstathiou *et al.* 1988). *(b)* The proportion or mix of galaxies of each type is the local one (Ellis 1983). *(c)* Galaxy spectral evolution is described by the BC93 models. The model representing a given galaxy type is selected from fits to local SEDs. *(d)* The predicted number counts are normalized at $b_j = 19.25$ mag. Metcalfe *et al.* (1995b) address the problem of possible calibration errors and the discrepancy in the number counts at bright magnitudes. *(e)* The geometry of the universe is described by the standard ($\Lambda = 0$) Friedmann cosmology.

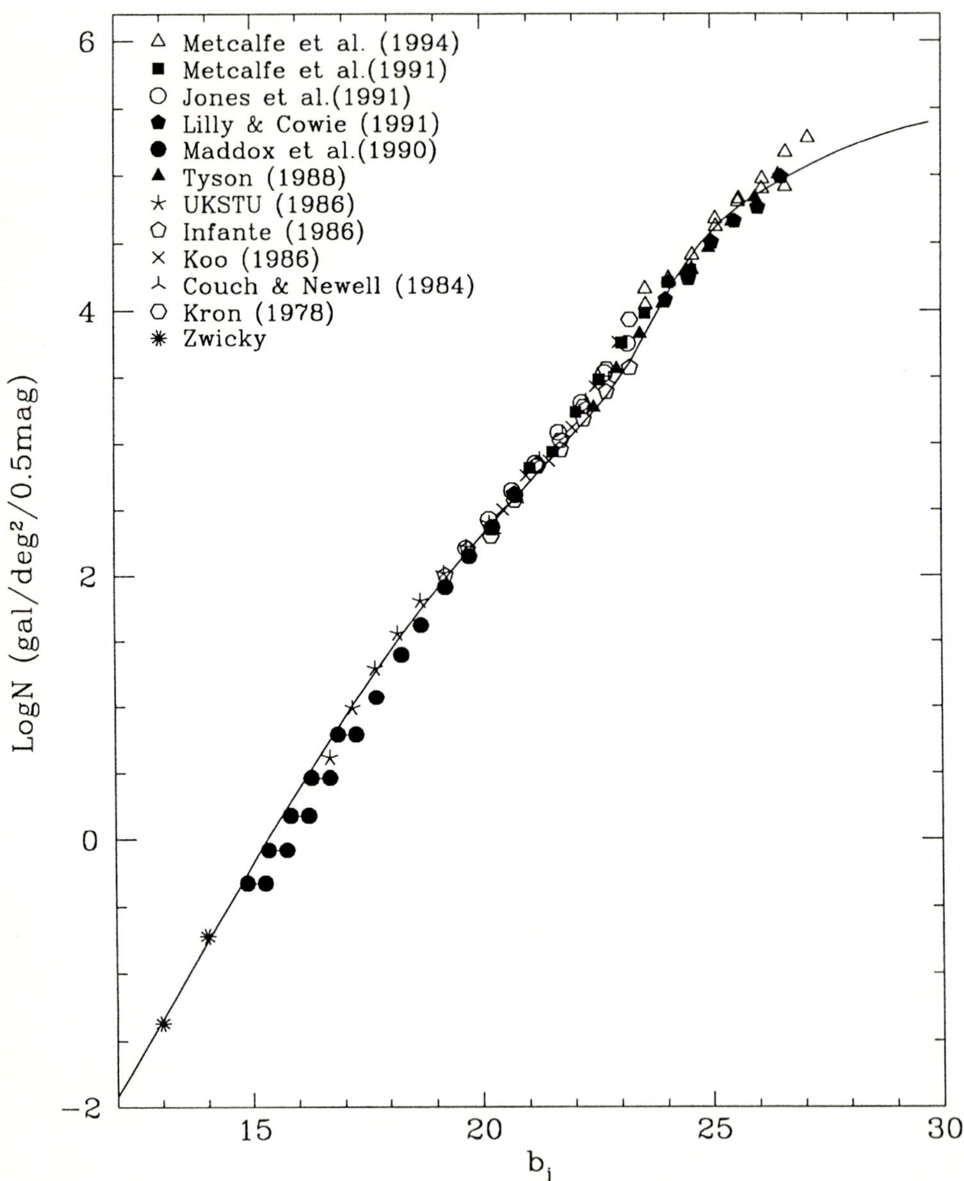

Fig. 1. The predicted number counts in b_j for the standard model described in the text. See Pozzetti, Bruzual & Zamorani (1995) for details.

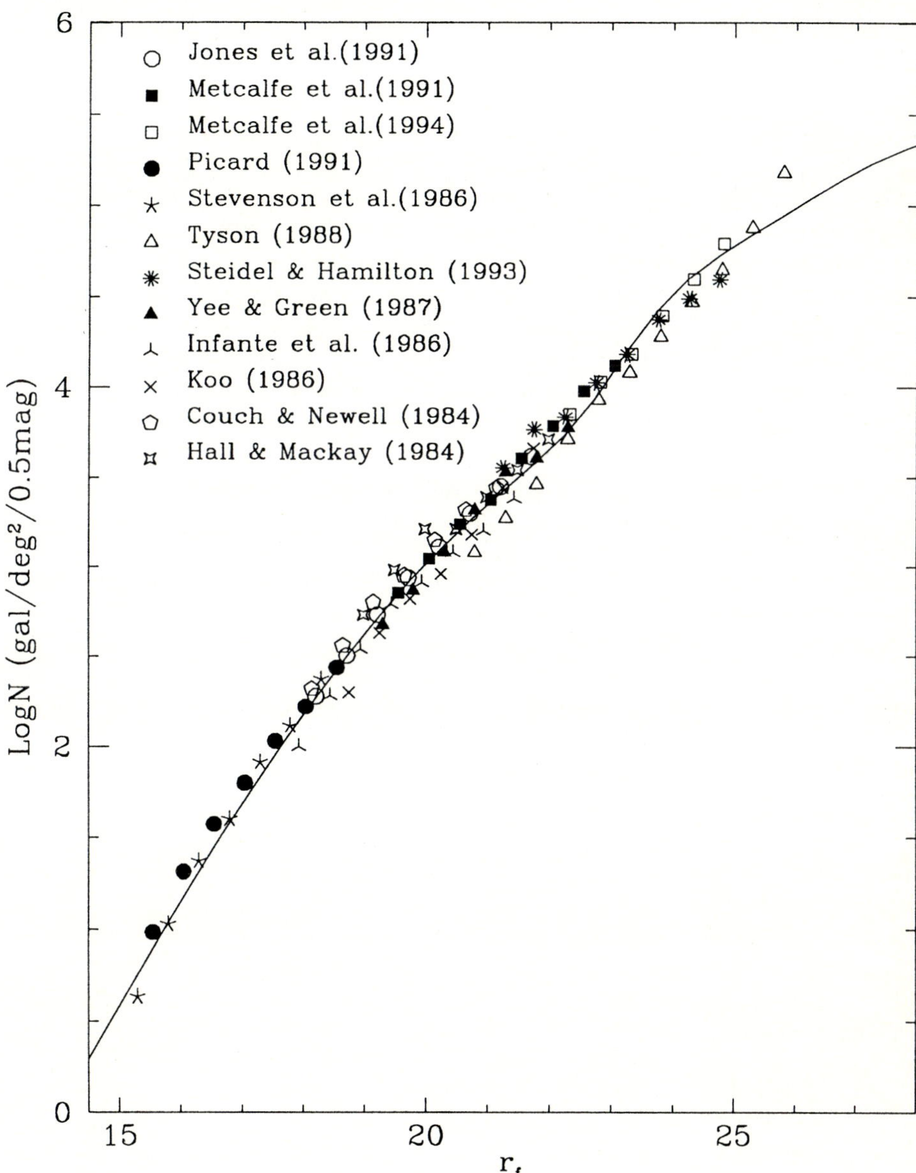

Fig. 2. The predicted number counts in r_f for the standard model described in the text. See Pozzetti, Bruzual & Zamorani (1995) for details.

2 Results

Fig. 1 and Fig. 2 show the predicted number counts in b_j and r_f, respectively, for our standard PLE model. In this model E-S0 galaxies are represented by BC93 models in which star formation decays exponentially with e-folding time $\tau = 1$ and 2 Gyr and follows the Scalo (1986) IMF. Similar models are used for Sab-Sbc galaxies but with $\tau = 8$ Gyr. Scd-Sdm galaxies are represented by a model with constant star formation rate but with the Salpeter IMF. At a galaxy age of $t_g = 16$ Gyr, these models provide excellent fits to the SED's of local galaxies of the indicated type. The number counts shown in Figs. 1 and 2 have been computed for $\Omega = 0$ and $H_o = 50$, which corresponds to a galaxy formation redshift $z_f = 4.5$. If we require all galaxies to follow the Salpeter IMF, then we must choose $t_g \approx 18$ Gyr ($z_f \geq 10$). We mention in passing that GK95 *add dust* to their Salpeter IMF models in order to suppress part of the UV flux and obtain the correct color distributions. The Scalo IMF serves the same purpose.

The results from the simple PLE models shown in Fig. 1 and Fig. 2 provide good fit to the data. The fits to the color and redshift distributions are as good as those of GK95 and are not shown here due to lack of space (see Pozzetti *et al.* 1995). These models show no evidence of an excess in the b_j number counts. The redshift distributions do not show high$-z$ tails, at least up to $b_j = 23$. Models with $\Omega = 1$ have been tested. In this case the fits are not as good as in the $\Omega = 0$ case, but are still feasible and leave some room for galaxy merger events.

The most relevant difference between our models and most of the previous PLE models, which have failed to obtain the right number of galaxy counts, hence the *excess* in the b_j counts, as well as in the shape of the color and z distributions are the (e+K)-corrections. It is thus quite important that our assumed spectral evolution be tested against distant galaxy SED's.

References

Bruzual, G., Charlot, S. 1993, *Astrophys. J.* **405**, 538 (BC93)
Efstathiou, G., Ellis, R.S., Peterson, B.A. 1988, *Mon. Not. R. Astr. Soc.* **232**, 431
Ellis, R.S. 1983, in Jones B.J.T., Jones J.E., eds, The Origin and Evolution of Galaxies. Reidel, Dordrecht, p.255
Gronwall, C., Koo, D.C. 1995, *Astrophys. J. Lett.* **440**, L1 (GK95)
Koo, D.C., Gronwall, C., Bruzual G. 1993, *Astrophys. J. Lett.* **415**, L21 (KGB93)
Koo, D.C., Kron, R.G. 1992, *Ann. Rev. Astr. Astrophys.* **30**, 613 (KK92)
Metcalfe, N., Shanks, T., Fong, R., Jones, L.R. 1991, *Mon. Not. R. Astr. Soc.* **249**, 498
Metcalfe, N., Shanks, T., Fong, R., Roche, N. 1995a, *Mon. Not. R. Astr. Soc.* **273**, 257
Metcalfe, N., Fong, R., Shanks, T. 1995b, *Mon. Not. R. Astr. Soc.* **274**, 769
Pozzetti, L, Bruzual A., G., Zamorani, G. 1995, *Mon. Not. R. Astr. Soc.*, submitted
Scalo, J.M., 1986, Fund Cosmic Phys, 11, 1
Schechter, P. 1976, *Astrophys. J.* **203**, 297

Shanks, T. 1990, in *Galactic and Extragalactic Background Radiation: Optical, Ultraviolet, and Infrared Components*, IAU Symp. **139**. Bowyer S., Leinert Ch., eds, Kluwer, Dordrecht, p.269

Number Density Predictions for Primeval Galaxies

Eduard Thommes and Klaus Meisenheimer

Max-Planck-Institut für Astronomie
Königstuhl 17, D-69117 Heidelberg, Germany

1 Introduction

In this contribution we discuss the question: How many Primeval Galaxies (PGs) can we expect to find in a survey like CADIS (see K. Meisenheimer *et al.* , this volume), which searches for their Ly-α emission ? The answer to this question is very important for the determination of an optimal survey strategy. First, we have to define what we regard as primeval galaxy: The ancestor of a present-day bright galaxy, undergoing its **first** mayor starburst at high redshift, that is in an enviroment of largely primordial matter.

Partridge & Peebles (1967) emphasized for the first time that if young galaxies can be detected al all, large numbers of them should be visible. They estimated the probability that a random line of sight will pass through a highly luminous young galaxy using the present day spatial density of galaxies in the present universe. Repeating the argumentations of Partridge & Peebles with actual values for the local number density of L^* galaxies and assuming a galaxy formation redshift of ≈ 5, Pritchet (1994) predicted a surface density of bright PGs in the order of $10^5 deg^{-2}$.

Detailed predictions of the number density of PGs as a function of their continuum and Ly-α luminosity were given by Baron & White (1987). According to this predictions surveys until now should have detected $10^1 - 10^3$ objects. Yet, no really good candidate has been found. We illustrate this discrepancy between expectations and observations in Fig. 1, where we compare the limits of the Palomar FPI Survey, which searched for Ly-α emission of PGs (Thompson *et al.* 1992) with the numbers of Ly-α emitting PGs one would expect to find in this survey according to the models of Baron & White.
What are the reasons of this discrepancy between models and observations ? Are the assumptions of the models wrong ? Is it possible to learn something about galaxy formation from this discrepancies ?

An often discussed possible explanation for this differences between observations and predictions is the presence of dust , which would especially suppress Ly-α. However, a strong suppression of Ly-α by dust is evidence for a previous generation of stars. In the sense of our above definition, such dusty objects would

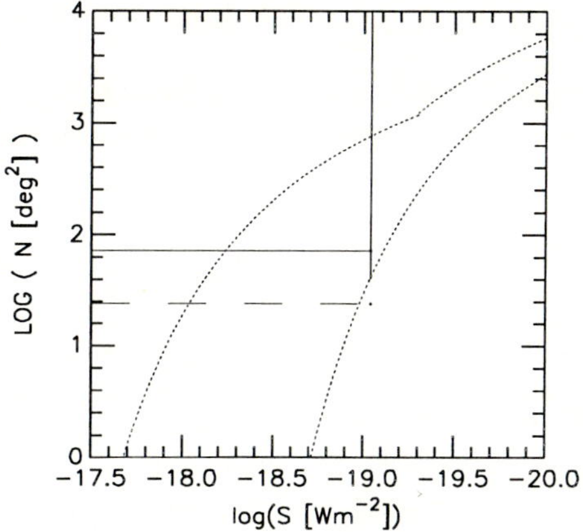

Fig. 1. Range of predictions by Baron & White (1987) (region between dotted curves) for $z = 4.8$ and $\Delta z = 0.15$ ($H_0 = 50 kms^{-1} Mpc^{-1}$) compared with the detection limits of the Palomar FPI survey of Thompson et.al. (1992). The solid box corresponds to an upper limit of ≤ 3 primeval galaxies and the dashed line to ≤ 1 object within the searched area.

not count as genuine PGs. Expressing the problem according to our definition, one has to take into account that the genuine PG phase of a galaxy only lasts as long as the enrichment of the primordial medium with metals (by stellar winds and supernova explosions) is minute. Thus, we can expect the duration of the genuine PG phase to be only some hundred million years[1]. In addition, the models so far always assumed, that all galaxies started at the same time with their first star fomation, which is obviously an over-simplification. The protogalaxies in a certain comoving volume do not need to be simoultaniously in the PG phase. So at a given redshift we expect to see only a fraction of the galaxies in a certain comoving volume to be in the state of detectable Ly-α emission.

Taking these points into account, we show in this contribution that the predictions for example of Baron & White (1987) were over-optimistic and that it was unlikely for previous surveys to find many PGs. On the other hand, we propose here, that it should be possible to limit the parameter space in a sensible way by using constraints from recent observations and theoretical works. This leads to much lower values for the expected number of PGs shining in Ly-α which nevertheless should be in the scope of todays observational techniques.

2 Parametrisation of the Problem

In the following, we assume a search for Ly-α emitting objects using a narrow band filter with a central wavelength λ_0 and a band width of $\Delta\lambda$. The central wavelength λ_0 corresponds to a certain redshift z_0 of the Ly-α line and $\Delta\lambda$ to a

[1] Baron & White (1987) assumed durations of the Ly-α bright PG phase up to 6 Gyrs.

redshift interval Δz. We denote with t_0 the epoch corresponding to the redshift z_0. By z_0,Δz and the solid angle $\delta\Omega$ covered on the sky a certain comoving volume δV_c of the universe is defined. Galaxies in this comoving volume could be detected on the narrow band exposures, provided their Ly-α emission is above the detection limit.

We assume, that a PG starts to shine in Ly-α as soon as the first star formation sets in and that the Ly-α luminosity is proportional to the star formation rate (SFR) as long as dust absorption is negligible. Furthermore, it is reasonable to assume that the SFR of the first star burst is proportional to the mass of the PG. Thus, the Ly-α luminosity of a PG should be proportional to their mass. If the SFR is not constant but changes with time, so does the Ly-α luminosity. However, after some time other effects like dust absorption may quench the Ly-α luminosity of the PGs, so the proportionality to the SFR is no longer valid. We describe this time dependence of the Ly-α luminosity with a function $f(t - t_s)$ where t_s is the epoch, at which the galaxy starts to shine in Ly-α:

$$L^{(m)}_{Ly_\alpha}(t - t_s) = k \ m \ f(t - t_s) \quad \text{with} \quad f(t - t_s) = \begin{cases} 0 \ \text{if} \ t < t_s \\ > 0 \ \text{if} \ t > t_s \end{cases} \tag{1}$$

Because of the assumed proportionality between Ly-α flux and the PG mass m , there exists a lower limit m_{min} for the PGs, which are luminous enough that their Ly-α at z_0 can be detected. PGs with $m < m_{min}$ remain always too faint in Ly-α, to be detectable above the surveys flux limit S_{lim}. The maximal number of PGs in the volume δV_c which we can expect to be bright enough in Ly-α is then given by the integral over the mass function of PGs with lower integration limit m_{min} multiplied by δV_c: $\int_{m_{min}}^{\infty} \Phi(z_0, m) dm \, \delta V_c$ where the mass function $\Phi(z_0, m)dm$ gives the comoving density of objects with mass $m \in [m, m + dm]$ at the observed redshift.

But not all of these galaxies are in the PG phase at the observed epoch t_0. Some of them already left the PG phase or others enter this phase at a later epoch, so that they are dormant at t_0. To take this into account, we introduce a function $P_m(t)dt$, which gives the fraction of objects with mass m in the volume δV_c starting their first star formation during the time intervall $[t, t + dt]$. With this function we get for the number of PGs per solid angle $\delta\Omega$ with a detectable Ly-α flux above S_{lim}:

$$N_{det}(z_0) = \int_{m_{min}(z_0)}^{\infty} \left\{ \Phi(z_0, m) \int_{t_2(m)}^{t_1(m)} P_m(t) dt \right\} dm \, \delta V_c(z_0) \tag{2}$$

This expression is very universal but note, that the integration limits $t_1(m)$ and $t_2(m)$ are determined by the duration of the phase, in which the PGs of a certain mass are bright enough in Ly-α to be detectable. They depend strongly on the shape of the function $f(t - t_s)$ and the distance of the object.

3 Some Special Cases

In order to make expression (2) more transparent, we choose some special functions for $P_m(t)$ and $L^{(m)}_{Ly-\alpha}(t-t_s)$. For $P_m(t)$ we take a delta-function $P_m(t) = \delta(t-t_{in})$. That is, all galaxies have the same ignition epoch $t_s = t_{in}$ and shine in Ly-α simultaneously. If we assume that a fraction ϵ of the stars of a galaxy are born in the first starburst with a constant star formation rate (SFR) over a period Δt (that is $SFR = \epsilon m/\Delta t$, where m is the mass of the galaxy) and that the Ly-α luminosity is proportional to the SFR, we get for the Ly-α luminosity

$$L^{(m)}_{Ly-\alpha}(t-t_s) = \begin{cases} k\frac{\epsilon m}{\Delta t} & \text{if } t \in [t_s, t_s + \Delta t] \\ 0 & \text{otherwise} \end{cases} \tag{3}$$

With this choice for $P_m(t)$ and $L^{(m)}_{Ly-\alpha}(t-t_s)$ (2) gives immediatley:

$$N_{det}(z_0) = \begin{cases} \int_{m_{min}(z_0)}^{\infty} \Phi(z_0,m)dm\ \delta V_c(z_0) & \text{if } t_0 \in [t_{in}, t_{in}+\Delta t] \\ 0 & \text{otherwise} \end{cases} \tag{4}$$

With the additional assumption $\Phi(z_0,m) = \Phi(0,m)$ (*i.e.* the mass function of the PGs is perserved into the mass function of present-day galaxies), this is the formula Baron & White (1987) used for their calculations. They took $t_{in} = \frac{1}{5}t_{coll}$ and $\Delta t = \frac{4}{5}t_{coll}$, where t_{coll} is the collapse time associated with a uniform spherical perturbation of the same initial mean density as the protogalaxy[2] They varied z_{coll} between 6 and 1.5 and took $q_0 = 0.5$ and $q_0 = 0.05$ resulting in the range of predictions we displayed in Fig. 1. This corresponds to a variation of Δt between 0.6 and 6 Gyrs.

But now let us take into account, that the galaxies in the volume δV_c do not at the same time start their Ly-α bright PG phase. A simple way to approximate this is to assume that the ignition times t_s of the galaxies are destributed equally over a certain time intervall. So the delta function for $P_m(t)$ has to be replaced by the function

$$P_m(t) = \begin{cases} \frac{1}{t_{out}-t_{in}} =: \frac{1}{\Delta t_P} & \text{if } t \in [t_{in}, t_{out}] \\ 0 & \text{otherwise} \end{cases} \tag{5}$$

Furthermore, if we take into account, that the genuine Ly-α bright PG phase only lasts for a limited period, e.g. $\Delta t_{Ly-\alpha} = 0.1\,$Gyr as discussed above, we get immediately from (2):

$$N_{det}(z_0) \leq \frac{\Delta t_{Ly-\alpha}}{\Delta t_P} \int_{m_{min}(z_0)}^{\infty} \Phi(z_0,m)dm\ \delta V_c \tag{6}$$

Thus, in comparison with (4) the numbers are at least reduced by the factor $\Delta t_{Ly-\alpha}/\Delta t_P$. In Table 1 we show this factor for different z_{in}, z_{out} (z_{in} and z_{out} are the redshifts, corresponding to t_{in} and t_{out} in (5) and $q_0 = 0.5$,0.1 and a $\Delta t_{Ly-\alpha}$ of 10^8 yr. Obviously the factors by which the numbers are reduced can be quite large.

[2] Note, that the collapse of a spherical perturbation starts at the redshift $z_{coll} = z(t_{coll})$.

Table 1. Factors $\Delta t_{Ly-\alpha}/\Delta t_P$ for different z_{in} and z_{out} and $\Delta t_{Ly-\alpha} = 10^8 yr$.

z_{in}	z_{out}	q_0	$\Delta t_{Ly-\alpha}/\Delta t_P$
20	5	0.5 0.1	1/8 1/14
20	2	0.5 0.1	1/24 1/40

4 Results of More Detailed Calculations

The assumption of a constant starfomation rate and Ly-α luminosity of a PG as in (3) is a crude simplification. Detailed numerical simulations concerning the formation of galaxies (Steinmetz 1993, 1994) show, that the SFR of PGs start with very low values, increase rapidly and reach a peak after $\approx 0.5\,Gyr$. The metallicity of the early bulge increases also rapidly in this phase and reaches values around $0.1 Z_\odot$ during the first $0.2\,Gyr$ after the onset of star formation. According to our definition this limits the genuine PG phase to only about $0.2\,Gyrs$, as already assumed above. After this time the metallicity reaches values which may lead to significant dust formation combined with an absorption of the Ly-α flux. Thus, the Ly-α flux of the PG first increases proportional to the increasing SFR. After $0.2\,Gyrs$, it is no longer proportional to the SFR and decreases rapidly. We approximate this behaviour by a Gaussian for $f(t - t_s)$ with a $\sigma_{Ly-\alpha} = 0.1\,Gyr$. Steinmetz (1993), simulated a Milky-Way like galaxy with a baryonic mass of $8 \times 10^{10} M_\odot$. Half of this mass is contained in the bulge of the galaxy. In this simulations the SFR reaches values around $10 M_\odot/yr$ when the metallicity reaches values around $0.1 Z_\odot$. Because only the bulge stars form during the first star burst, it seems to be more appropriate to scale the SFR of the PGs only with their bulge mass instead with their total mass as we did in section 3. This gives us the normalization that a PG with a bulge of $4 \times 10^{10} M_\odot$ reaches a SFR of $10 M_\odot/yr$ during the PG phase. We converted the SFR in a Ly-α flux using the constant of proportionality of $10^{35} W m^{-2} M_\odot^{-1} yr$ (Kennicutt 1983) which is determined by observations of nearby galaxies. By using this value, we implicitly assume that the initial mass function (IMF) of the early star formation phase is the same as in todays galaxies. All this fixes the function $L^{(m)}_{Ly-\alpha}(t - t_s)$ of the PGs which in the end determines the integration limits $t_1(m)$ and $t_2(m)$ in (2).

Now we turn to the mass function of the PGs. Because we scale the SFR of the PGs with their bulge mass, we need the mass function of the bulges at z_0. A straight forward procedure to get an estimate for this, is to neglect any merging or massloss process of galaxies between the epoch of PGs and today. With this assumtion a good approximation for the mass function of the bulges of the PGs should be the mass function of the bulges of the galaxies we see today (locally). We derive the local bulge mass function from the type dependent luminosity functions determined from the CFA data by Marzke et. al. (1994) together with values for the type dependent ratio of the bulge to disk luminosity from Simien & De Vaucouleurs (1986) and a constant mass to light ration for spheroids of $10 M_\odot / L_\odot$.

More realistic for the function $P_m(t)$ than (5) is a function which increases smoothly up to a maximum and then decreases again. A suitable function, which can be motivated in some way by the Press-Schechter formalism (see e.g. Efstathiou 1990), is

$$P(t) = \frac{4}{3 \Delta t_P \Gamma(\frac{3}{2})} \left(\frac{t - t_{in}}{\Delta t_P} \right)^{-3} \exp \left\{ - \left(\frac{\Delta t_P}{t - t_{in}} \right)^{\frac{4}{3}} \right\} \tag{7}$$

In (7) we neglect a possible mass dependence of $P_m(t)$. A reasonable value for t_{in} is given by $z_{in} \approx 20$ (see e.g. Peebles 1990). If we fix z_{in} at this value, the only free parameters left are h_0, q_0 and Δt_p in (7), which is in same sense a measure of the age spread of the galaxies.

In Fig. 2 we present the result of our calculations of $N_{det}(z_0)$ as a function of the detectable flux limit S_{lim} for $z_0 = 4.8$ and $\Delta z = 0.15$, i.e. the values used in Fig. 1. The comparison with the Baron & White predictions shows that we indeed expect significantly fewer PGs than they did. Accordingly there is no conflict between our predictions and the limits of the Palomar FPI survey of Thompson *et al.* (1992).

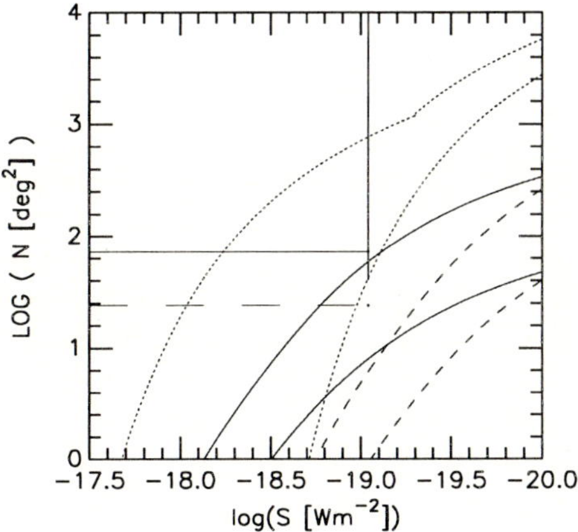

Fig. 2. Predictions of Baron & White (1987) for $z = 4.8, \Delta z = 0.15$ (region between the dotted curves) compared with the results of our calculations and the limits of the survey of Thompson *et al.* (1992). The region between the solid curves show the results of our calculations for $q_0 = 0.5$ and the region between the dashed curves for $q_0 = 0.1$. The two curves correspond respectively to $\Delta t_P = \Delta t_P^{opt}$ and $1/5 \Delta t_P^{opt}$.

The mean slope of our predicted number counts *versus* flux limit reveals that two approaches in spending more observing time are virtually equivalent: Either one could decrease the flux limit by a factor of 3 in observing a field 10 times longer or one could cover a tenfold increased field. Both would enhance the chance of finding a PG by a factor of 10. Due to the large field of view of our new focal reducers the new Calar Alto Deep Imaging Survey (CADIS, see Meisenheimer *et al.* , this volume) will improve in both directions: The survey will cover about 6 times the area of the Palomar FPI survey and should reach a 3 times fainter detection limit S_{lim}. In Fig. 3 we display our predicted $N_{det}(z_0)$ above the detectable flux limit S_{lim} for the redshifts $z_0 = 4.75, 5.75$ and 6.55 with $\Delta z = 0.1$. These are the redshift intervalls CADIS searches for Ly-α emission of PGs. We chose $h_0 = 0.5$ and did the calculations for $q_0 = 0.5$ (solid lines) and $q_0 = 0.1$ (dashed lines). We varied Δt_P with all the other parameters fixed. We found, that for each combination of the other parameters, there exists an "optimal" Δt_P^{opt}, at which N_{det} as a function of Δt_P has a maximum. In Fig. 3 the results for $\Delta t_P = \Delta t_P^{opt}$, $\Delta t_P = \frac{1}{2}\Delta t_P^{opt}$ and $\frac{1}{5}\Delta t_P^{opt}$ are shown.

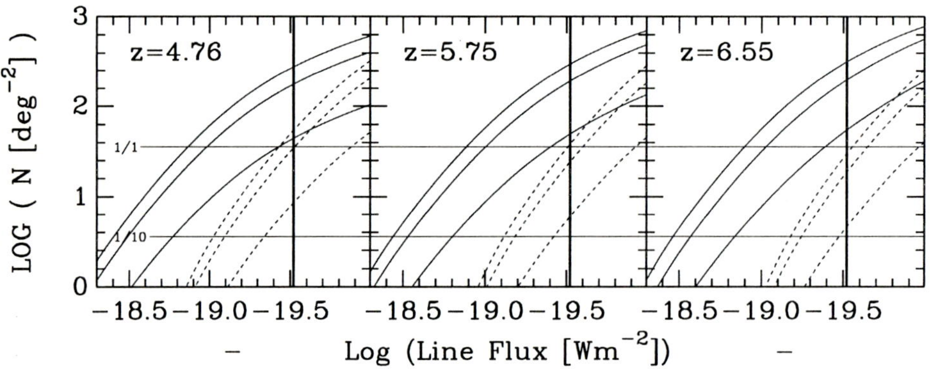

Fig. 3. Results of the calculations for the three redshift intervals where CADIS searches for Ly-α emission of PGs. The solid curves are for $q_0 = 0.5$, dotted curves for $q_0 = 0.1$. We show the results for $\Delta t_P = \Delta t_P^{opt}, \frac{1}{2}\Delta t_P^{opt}$ and $\frac{1}{5}\Delta t_P^{opt}$. For $z = 4.75$ we got $\Delta t_P^{opt} = 1.8 Gyr$ ($q_0 = 0.1$), $0.8 Gyr$($q_0 = 0.5$), for $z = 5.75$ $\Delta t_P^{opt} = 1.3 Gyr$ ($q_0 = 0.1$), $0.5 Gyr$($q_0 = 0.5$) and for $z = 6.55$ we got $\Delta t_P^{opt} = 1.0 Gyr$ ($q_0 = 0.1$), $0.3 Gyr$($q_0 = 0.5$). A thick vertical line indicates the limiting flux of CADIS. Horizontal lines are drawn at those surface densities which correspond to the detection of 1 object per $10' \times 10'$ field (1/1) and 1 object in 10 fields (1/10).

Unless our assumption $z_{in} = 20$ is utterly wrong or the age spread $P_m(t)$ has an unexpected narrow distribution (that is almost all galaxies have left their Ly-α bright phase already at the observed epochs) these lines should embrace the most likely scenarios. So we expect to detect at least a few and perhaps up to a hundred PGs by the huge effort which will be undertaken with CADIS.

References

Partrige & Peebles, 1967, *Astroph. J.* **147**, 868

Pritchet, C.J. 1994, *Proc. Astr. Soc. Pac.* **106**, 1052

Baron, E., White, S.D.W. 1987, *Astron. J.* **322**, 585

Thompson, D., Djorgovski, S.,Trauger, J. 1992, in *Cosmology and Large-Scale Structure in the Universe*, ed. R. de Carvalho, *ASPCS* **24**, 147

Steinmetz, M. 1993, *Thesis, Technische Universität München*

Steinmetz, M. 1994, *Astron. Astrophys.* **281**, L97

Kennicutt, R.C. 1983, *Astron. J.* **272**, 54

Felten, J.E. 1998, *Comm. Ap. Space Phys.* **11**, 53

Simien, F, De Vaucouleurs G. 1986, *Astron. J.* **302**, 564

Marzke, R.O., Geller , M.J., Huchra, J.P. 1994, *Astron. J.* **108**, 437

Efstathiou, G 1990, *Physics of The Early Universe*, Proceedings of the Thirty-Sixth Scottish Universities Summer School in Physics, eds J.A. Peacock, A.F. Heavens, A.T. Davies

Peebles, P.J.E. 1993, *Principles of Physical Cosmology*, Princeton University Press, Princeton, New Jersey

Self-Regulation Models for Star Formation

Joseph Silk

Departments of Astronomy and Physics, and Center for Particle Astrophysics, University of California, Berkeley, CA 94720

1 Introduction

Despite intensive study at optical, infrared, x-ray submillimeter and radio wavelengths, star formation remains a topic that has proven remarkably elusive. We observe many of the ingredients that are necessary for a successful theory, yet no such theory has been developed that can predict such critical ingredients as the star formation rate and the initial mass function. Galaxy formation, arguably the most important problem in cosmology, remains as elusive as ever because of our lack of knowledge of the details of star formation. If we cannot account for star formation in the intensively observed Orion nebula and associated molecular clouds, we cannot hope to incorporate star formation into realistic predictions of the properties of forming galaxies.

The brute force approach of computational power is not going to elucidate the innards of star formation, at least in the foreseeable future. There are simply too many variables, including magnetic fields, ionization, dust, molecular chemistry, molecular cooling, turbulence and environment. Processes that play a role include accretion, fragmentation, coagulation of fragments, Alfven wave generation, momentum-driven flows, and binary formation. In a situation of such complexity, it is important to search for some overall guiding principle that may help elucidate the underlying physics and provide some predictive power. This approach has worked well in diverse areas of physics, including the theories of gravitation, of fundamental interactions and of elementary particles. Astrophysics is a far dirtier domain, but it is useful nevertheless to seek some unifying idea that can be applied to star formation physics.

The principle that I wish to pursue here is that of self-regulation. I will show that the global properties of galactic disks can be understood if star formation self-regulates, and I will provide a plausible physical argument that self-regulation occurs and accounts for the global inefficiency of star formation. On a finer scale, I will argue that feedback from newly formed protostars explains not only why star formation is locally so inefficient, but can also lead to an understanding of the initial mass function. All of this is intended to provide a stepping stone towards predicting the observable characteristics of forming galaxies.

2 Star Formation in Galactic Disks

A successful theory of star formation in galactic disks must account for the longevity of star formation and for its relative constancy with time. A major clue is provided by the observations of Kennicutt (1989), which demonstrate that the gas surface density in nearby spirals remains close to the critical value above which the gas is gravitationally unstable.

The star formation theshold was identified by Kennicutt with the Toomre criterion for gravitational instability applied to the gas (HI and H_2) disk. A more sophisticated model developed by Wang & Silk (1994) incorporated the stellar disk and generalized theshold into an expression for the star formation rate given by

$$\mathrm{SFR} = \epsilon \Sigma_{\mathrm{gas}} \Omega \frac{(1-\mathrm{Q})}{\mathrm{Q}}; \ \mathrm{Q} \leq 1.$$

Here, Q is the Toomre parameter, given by

$$Q = \frac{\kappa \sigma_g}{f G \sigma_{gas}},$$

$\kappa \simeq \sqrt{2}\Omega$ is the epicyclic frequency, σ is the velocity dispersion of the gas, $f(<1)$ corrects for the stellar component in the disk, and ϵ is the star formation efficiency. This star formation rate was applied to compute the star and gas surface density and abundances, as a function of radius. SN II yields were used to compare with galactic H II region oxygen abundances. Gas recycling was included, and gas infall was not necessary to obtain sufficient longevity of the star formation rate in the disk. The various observational properties of the Milky Way disk including the G dwarf problem were explained, provided an initial metallicity was adopted.

Prantzos & Aubert (1995) have noted two difficulties in the model of Wang and Silk. The small but nonzero number of low metallicity disk stars may require infall of some unenriched gas. Recent data on oxygen abundances in disks suggests that outside the central few kpc, there is no abundance gradient. An improved model is given by

$$\mathrm{SFR} = \epsilon \Sigma_{\mathrm{gas}} (\Omega(\mathrm{r}) - \Omega_{\mathrm{p}})$$

where Ω_p is the spiral arm pattern speed. This is the model for the star formation rate originally proposed by Wyse & Silk (1989). The gravitational instability of the disk is now implicit in writing the SFR as proportional to $\Omega - \Omega_p$: wherever there are spiral arms, the disk is unstable to star formation. Moreover at large galactocentric radii, the star formation rate is suppressed, and at small radii it is large.

The star formation efficiency may be estimated as follows. Assume that heating from supernova remnants (and H II regions) provides momentum input to the interstellar medium, essentially accelerating the gas. The interstellar medium is very dissipative, and one may therefore estimate the efficiency of momentum input, identified with the star formation inefficiency, as

$$\epsilon = \sigma_g / p_{SN},$$

where

$$p_{SN} = 2E_{SN}v_{cool}^{-1}m_{SN}^{-1}$$

is the specific momentum injected, that for simplicity I have identified with supernova injection. E_{SN} is the initial supernova remnant energy, m_{SN} is the mass found in stars per supernovae, and v_{cool} is the supernova remnant velocity at the onset of the approximately momentum-conserving snowplow phase. Typical numbers are $E_{SN} = 10^{51}$ ergs, $m_{SN} = 100\,\mathrm{M}_\odot$, $v_{cool} \approx 300\mathrm{km\,s^{-1}}$, and $\sigma_g \approx 10\,\mathrm{km\,s^{-1}}$, so that $p_{SN} \approx 1000\,\mathrm{km\,s^{-1}}$ and $\epsilon \approx 0.01$. Of course, this is more or less what is observed in giant molecular clouds, and also not coincidentally what is required to infer the observed galactic star formation rate from the galactic molecular cloud mass that would otherwise collapse over a cloud free-fall time.

One can in fact do a little better than extract the star formation rate. I shall argue that the Tully-Fisher relation is a natural consequence of the adopted star formation law. I approximate the star formation rate by

$$SFR = \epsilon \Sigma_{gas} \Omega,$$

and identify luminosity with star formation rate in spirals. I now can write the luminosity of a disk galaxy as

$$L = \beta\epsilon M_{gas}\Omega = \beta\epsilon \frac{v_{rot}^3}{G}\frac{M_{gas}}{M}, \quad \beta \equiv L/SFR,$$

after applying the virial theorem, $M = Rv_{rot}^2 G^{-1}$. The derived value for ϵ gives the correct normalization for the Tully-Fisher relation:

$$L = 10^{10}\left(\frac{\epsilon}{0.01}\right)\left(\frac{v_{rot}}{220\mathrm{km\,s^{-1}}}\right)^3\left(\frac{10M_{gas}}{M}\right)\left(\frac{\beta}{5\times 10^9\mathrm{L_\odot yr M_\odot^{-1}}}\right)\mathrm{L_\odot}$$

as well as the star formation rate, to within a factor of 2,

$$SFR = 2\left(\frac{\epsilon}{0.01}\right)\left(\frac{4\times 10^7\mathrm{yr}}{\Omega^{-1}}\right)\left(\frac{M_{gas}}{6\times 10^9 M_\odot}\right)\mathrm{M_\odot\,yr^{-1}}.$$

Note that $L \propto v_{rot}^3$ fits the observed Tully-Fisher relation (in B and R bands), whereas the naive virial theorem plus constant surface density and mass-to-light ratio assumptions would yield the incorrect result that $L \propto v_{rot}^4$. The dependence on efficiency and gas fraction, constant within a Hubble type, means that environment should influence the normalization, but not the slope, of the Tully-Fisher relation.

The Tully-Fisher relation applies not just in the B and R bands, but also at longer wavelengths. It appears to steepen slightly in the I and in the H bands (Burstein 1995). The fact that a universal relation is seen in the different bands means that the Tully-Fisher correlation is satisfied for both the young and the old stellar populations in spiral galaxies. One can understand the coincidence in a phenomenological way as follows. Dynamically hot stellar populations are found to lie in the fundamental plane of parameter space, defined by Bender *et al.* (1992)

$$r_e \propto v^{1.4} \Sigma_L^{-0.85},$$

where r_e is effective radius, v is velocity dispersion and Σ_L is surface brightness. Applying the virial theorem, $v^2 = GM/r_e$, one can rewrite this as

$$L \propto v^{10/3} \Sigma^{-5/6}.$$

Comparison of the latter with the predicted Tully-Fisher relation shows remarkable concordance at fixed Σ_{gas}. One expects the self-regulation theshold argument to specify Σ_{gas}, if $Q \sim 1$, to be

$$\Sigma_{gas} \approx \Sigma_{crit} \approx \kappa \sigma_g / \pi G.$$

Hence at fixed Hubble type, for which one can plausibly take the rotation curve, and even more plausibly the differential rotation rate, as a dominant characterizing parameter, both bulges and disks satisfy a similar Tully-Fisher relation. The Σ_{gas}/Σ dependence implies that low surface brightness galaxies satisfy an identical Tully-Fisher relation, at the same Σ_{gas}/Σ ratio. If star formation is regulated by disk instability, one would also expect this ratio not to vary greatly in star-forming disks, since gas self-gravity is ultimately the cause of the star formation.

The fundamental plane relation is equivalent to

$$M/L \propto L^{1/5}.$$

There is no simple explanation for this relation. It may be due to baryonic dissipation being more efficient at lower galaxy mass, to star formation efficiency decreasing with increasing galaxy mass, or even to a systematic variation in the IMF. Hints in support of the latter, possibly most radical, alternative come from several directions. There are indications of gradients in the ratio of α-nuclei to iron in the central regions of ellipticals (Worthey *et al.* 1992), and the low dispersion in the ratio of α-nuclei to iron abundances in old disk stars (Edvardsson *et al.* 1993) may be suggestive of a common origin for these elements. The high mass of iron in intracluster gas was presumably ejected from cluster galaxies but is not compatible with the standard IMF yields that are predominantly of a Type I supernova origin, these supernovae being the major present-day contributors of iron (Arnaud *et al.* 1992). Most recently, there are hints of a predominanly Type II supernova yield characterizing the heavy element abundance ratios in intracluster gas (Tanaka 1995). All of these observations could plausibly be more easily understood if the initial mass function favored somewhat more massive stars at earlier epochs, and thereby had both a higher nucleosynthetic yield and one more characteristic of Type II supernovae.

For extreme dwarf galaxies, the relation changes dramatically. Dwarf spheroids have high M/L values, and one finds that $M/L \propto L^{-1/3}$ for these very low mass systems. The simplest explanation for this appeals to wind-driven mass loss, something most easily accomplished from shallower potential wells (Dekel

& Silk 1986). One expects that the retained baryonic fraction should be proportional to v/v_{SN}, and confirmation that is likely to be correct comes from abundance studies. For dynamically hot galaxies, one finds that $Z \propto v$, in a relation that has surprisingly low dispersion (Bender 1992). This is precisely what one expects if the mass loss driven by winds interrupts chemical enrichment before it saturates at the nucleosynthesis yield of about $2Z_\odot$, appropriate to a closed box model, as found for luminous elliptical galaxy cores and inner disks.

Consider the implications of the adopted global star formation law for the mass-to-light ratio of disks. The derived expression for L can be rewritten as

$$\frac{M}{L} = (\epsilon\Omega)^{-1}\frac{M}{M_{gas}} = \epsilon^{-2/3}G^{-1/3}\Sigma_L^{-1/2}L^{1/6}\left(\frac{M}{M_{gas}}\right)^{2/3}.$$

Evidently for fixed $\epsilon M_{gas}/M$ and surface brightness, $\frac{M}{L}$ increases slowly with L. Indeed, along the disk sequence, earlier Hubble types, which satisfy a trend in the mean of slowly increasing luminosity, have slightly higher $\frac{M}{L}$ than later Hubble types, an effect that can be interpreted as due to the increasing dominance of the older bulge component. The observed slow increase of $\frac{M}{L}$ with L, together with the fact that Σ_L is approximately constant, implies that $\epsilon M_{gas}/M$ must also be more or less constant with Hubble type. While M_{gas} decreases towards earlier Hubble types, the star formation efficiency parameter must increase. Star formation is necessarily more efficient in bulges than in disks.

One recovers almost the identical weak $\frac{M}{L}$ dependence on L in disks at specified Hubble type or gas fraction to that found in the fundamental plane for ellipticals and for bulges. This is perhaps a hint that protodisks may have been the fundamental building blocks from which ellipticals and even bulges formed. The fact that dwarf ellipticals deviate from the fundamental plane suggests a different origin, as indeed is consistent with theoretical expectation. The surface brightness and $\frac{M}{L}$ of dwarfs is most likely a consequence of wind-driven mass loss.

3 The Initial Mass Function

The key to understanding star formation lies in the theory that determines the initial mass function of stars. Unfortunately there is no consensus on such a theory, nor indeed even as to its most important ingredients. These may include processes of accretion and fragmentation as modified by magnetic field coupling; Alfvenic and hydrodynamic turbulence; photo-dissociation; cosmic ray pressure, heating and ionization; depletion onto grains; feedback from protostars via ionization, jets or outflows. It has been argued that with so many physical effects occurring that are of comparable importance, a log normal distribution of protostar masses may be the natural outcome (Zinnecker 1984). Indeed, the observed initial mass function is, within the errors, indistinguishable from a log normal distribution (Scalo 1986).

However I prefer to believe that there is more likely to be a dominant process that helps determine the IMF. Hints of variations in the IMF with environment in regions of extreme star formation (*i.e.* luminous starbursts) tend to support this thesis. The process that I will advocate is that of protostellar feedback. My argument is motivated by the idea that one has to commence with an understanding of what determines the mass of a star, and that the competition between gravitational accretion and protostellar outflows is central to such an understanding.

Theoretical arguments have suggested that fragmentation of a cold, dark collapsing cloud is likely to lead to formation of very low mass opaque cores. Estimated masses are in the range $0.001 - 0.01 M_\odot$. However these have little in common with the final stellar masses. Inclusion of more realistic physical processes, including accretion and coagulation of fragments, presumably results in the formation of stellar mass objects.

The physics of the evolution of an individual accreting protostellar core is better understood. Collapse of an isothermal sphere onto a core results in the accretion rate $\Delta V^3/G$, where ΔV is the generalized sound speed, defined to include both turbulent and magnetohydrodynamic contributions (Shu 1977). This assumes that the collapsing sphere exceeds the Jeans or magnetic Jeans mass, a situation that should be relevant for the formation of massive stars. In the case of low mass star formation, the accreting gas sphere may be of subcritical mass, relative to the magnetic Jeans scale. However once the magnetic field decouples from the neutral gas, on an ambipolar timescale that is controlled by the local ionization fraction, the accretion rate should again control the rate of protostellar core growth.

What then limits the stellar mass? The most logical mechanism comes from feedback by the forming protostars. Intermediate and massive protostars develop vigorous bipolar outflows that have sufficient momentum to halt the accretion. These occur in the premain sequence, even in the pre-T-Tauri phase, when Kelvin-Helmholtz contraction provides the primary energy source from the protostar. Low mass protostars are found to be relatively prolific x-ray sources, with $L_x/L_{bol} \sim 10^{-3}$. The measured x-ray ionization rate in, for example, the ρ Oph molecular cloud from young stellar objects and T-Tauri stars dominates the ionization of the molecular gas, and thereby controls the ambipolar diffusion rate, $t_{amb}^{-1} = 10^{-14} x^{-1} \mathrm{yr}^{-1}$. Recoupling to the field will effectively halt accretion. The x–rays are chromospheric in origin in a convective protostar, but as with the bipolar flows the ultimate energy source must be Kelvin-Helmholtz contraction.

A simple model which explicitly gives the mass of a protostar may be constructed as follows (Silk 1995). Note first that there is adequate energy release from accretion to generate the outflow. The bolometric luminosity of the central source exceeds the outflow energy by some two orders of magnitude. However the detailed coupling between accretion and outflow is unknown. I shall take the Kelvin-Helmholtz luminosity of the protostar, which is ultimately responsible for the outflow, to be equal to the rate of release of gravitational energy at the accretion shock. This shock occurs at the edge of the accretion disk, whose radius is inferred to be at the point where angular momentum conservation strongly modifies the infall geometry,

$$r_c = \frac{G^3 M^3 \Omega^2}{16 \Delta V^8}.$$

In order of magnitude, the accretion luminosity is

$$L_{acc} \sim GM_* \frac{\dot{M}}{r_c}, \quad \text{where } \dot{M} \approx \Delta v^3 G^{-1}, \quad \text{or } L_{acc} \sim \frac{\Delta v^{11}}{G^3 M_*^2 \Omega^2},$$

if the disk mass is comparable to the protostellar mass. The Kelvin-Helmholtz luminosity of the protostellar core, of mass M_*, is of order

$$L_{KH} \sim \sigma T_*^4 R_* / \kappa\rho, \quad \text{where } R_* \sim GM_*\mu/kT_*, \quad \rho \sim M_* R_*^{-3},$$

or

$$L_{KH} \sim \sigma \left(\frac{G\mu}{k}\right)^4 M_*^3 \kappa^{-1}.$$

Equating the two rates yields

$$M_* \sim \Delta v^{11/5} \left(\frac{k}{\sigma}\right)^{1/5} \Omega^{-2/5} G^{-7/5} (\frac{k}{\mu})^{4/5}.$$

A further simplification is that the rotation rate of molecular cloud cores scales as $\Omega \propto R^{-0.4}$ and in addition $\Delta v \propto R^{0.4}$, so that

$$M_* \propto (\Delta V)^{13/5} \kappa^{1/5}.$$

This is remarkably close to the correlation observed for the most luminous embedded IRAS source within dense molecular cloud cores (Myers & Fuller 1992):

$$M_* = 1.8(\pm 0.1)\, (\Delta V / 1 \mathrm{kms}^{-1})^{2.38(\pm 0.17)} \mathrm{M}_\odot,$$

observed to hold over $0.3 \lesssim M_* \lesssim 30 M_\odot$. The interpretation of this result is that the turbulent velocity in molecular cloud cores is responsible for limiting stellar masses via outflows which limit accretion. Lower mass stars can also form: the feedback determines the maximum mass that can grow by accretion.

One can now apply this result to generate a stellar initial mass function as follows. I assume that molecular cloud support is sustained by outflow-driven turbulence. Linewidths are supersonic, although probably sub-Alfvenic, and the observed momentum input from bipolar flows in nearby clouds such as Orion appears to be capable of accounting for the observed linewidths. Suppose further that stars form at a constant rate within a cloud. Each protostar is assumed to generate an outflow, which in the following schematic model I shall assume to be identical. I model the cloud turbulence as being driven by a network of interacting bipolar outflows. Each sweeps up a bubble of molecular gas, that is driven into the cloud at a speed

$$\Delta V \equiv \frac{dR}{dt} \equiv A t^{-1/2}$$

for a momentum-conserving shell. Hence there are a few shells of high velocity, but many shells, and more generically, pieces of shells that have collided with other shells and broken up, or become Rayleigh-Taylor unstable after being halted by the ambient pressure. Indeed, the ambient pressure is due to the accumulation of the momentum input from the many shreds of old shells. The velocity turbulence distribution in the cloud can be estimated as follows

$$N(>\Delta V) \sim R^3(t).t \sim A^8(\Delta V)^{-5}.$$

The resulting mass function is determined by the distribution of ΔV :

$$\frac{dN}{dM} = \frac{dN(>\Delta V)}{d\Delta V}\frac{d\Delta V}{dM} \propto M^{-1-x},$$

where I have set $R \propto t^\delta$, $M \propto (\Delta V)^\epsilon$, and $\gamma = \frac{3\delta+1}{\epsilon(1-\delta)}$. With $\epsilon = 2.4$, as observed, then $0.3 \lesssim \delta \lesssim 0.5$ (corresponding to the expected range of δ for generalized outflows) implies $1.1 \lesssim x \lesssim 2.1$. For a strongly time-dependent flow, such as a blast wave, $\delta = 0.3$, while for a wind, $\delta = 0.5$. This spans precisely the observed range of IMF slope over $0.3 \lesssim M \lesssim 100 M_\odot$. A density gradient steepens the slope: with $\rho \propto r^{-k}$, one finds that $\delta = 2/(4-k)$. The uniform core of a cloud, where the lowest mass stars are likely to form, should have a flatter IMF than the outer regions.

The IMF in a large molecular cloud complex arises from the addition of stars produced in regions with varying degrees of turbulence. Quiescent regions dominate, so the net effect is generically an IMF dominated by low mass stars. Do observations of cloud cores indeed confirm that stellar energy input drives the linewidths? A comparison of cores with and without embedded cores reveals that the linewidth-size relation is almost independent of the presence of a core (Fuller & Myers 1992). However, study of the proximity of young star clusters to dense cores (Caselli & Myers 1995) shows that the linewidths are strongly correlated with projected distance to the clusters. These observations are consistent with the hypothesis that the collective effect of young stars, whether via winds or indirectly via magnetic coupling through x-ray ionization, controls core linewidths. In turn, the linewidths influence the resulting IMF to form in the cores.

A further application is to the mass spectrum of cloud cores. These are supported by turbulence, and gravitational support requires that

$$M_{core} \propto \Delta V^4 p^{-1/2}.$$

Observed scaling laws for cloud cores yield $p \approx$ constant, the usual interpretation appealing to magnetic support with $p = B^2/8\pi$ and $B \approx 15 - 40\mu G$ (Myers & Goodman 1988). In this case, the turbulence is likely to be Alfvenic. However the stirring by bipolar flows still generates the turbulence, and the preceeding derivation of the mass function now yields

$$x = \frac{3\delta+1}{4(1-\delta)} \approx 0.7 \quad \text{for } \delta = 0.3.$$

One indeed measures $x \approx 0.6$ for molecular cloud cores, with x steepening to ~ 1 for very low mass cores.

4 Star Formation in More Exotic Environments

Formation of the stellar mass distribution may be controlled by the local turbulence according to the feedback argument that I have developed. However one can imagine situations where the feedback is modified. For example, in a merger between two galaxies, the highly non-circular cloud velocities induced by tidal torques will result in a greatly enhanced frequency of cloud collisions, which in turn are likely to raise the level of internal cloud turbulence. It would be a reasonable inference that there is then a significant shortfall in low values of ΔV, resulting in an IMF that is top-heavy, or biased towards massive star formation.

Conversly, consider a region of highly quiescent star formation where only stars of mass comparable to or below the nuclear burning limit are formed. There may be no bipolar flows to induce turbulence in such a region. Only low mass stars would form. One could argue that accretion is ineffective in producing more massive stars if there is sufficient ionization from chromospheric activity to couple neutral gas and magnetic fields. Eventually collapse along field lines will result in massive star formation, but this is controlled by the magnetic Jeans mass. If, in addition, magnetic fields are small, as may be appropriate for pregalactic star formation, one could imagine forming exclusively low mass stars in exceptionally quiescent regions. In this case, the accretion energy deposition rate ($\propto \Delta V^{11.8} M_*^{-2}$) is completely quenched by the Kelvin-Helmholtz energy release ($\propto M_*^3$), and accretion should be ineffective. In such an environment, one could conceivably imagine forming prolific numbers of stars near the hydrogen-burning limit.

5 Concluding Remarks

An understanding of star formation has proven to be remarkably elusive, despite the proliferation of data, both galactic and extragalactic. I have argued that local self-regulation may help explain the initial mass function of stars, and global self-regulation in disks can account for the rate of star formation. The next challenge is to apply these arguments to more diverse and remote regions. Perhaps the early IMF was very different, either top-heavy when ellipticals and bulges formed or bottom-heavy when galaxy halos condensed. If this were the case, one might have explanations for the heavy element abundances in intracluster gas in terms of enhanced enrichment by massive star ejecta and for the dark matter in galaxy halos as a population of brown dwarfs.

The star formation rate in disks provides clues to the origin of the Tully-Fisher relation, and suggests that environmental variations may be a common phenomenon. The gas content and rotation curves of disks may vary with environment, and these are crucial ingredients in determining star formation rates and histories. The formation of bulges and ellipticals remains more obscure insofar as we have no understanding of star formation in dynamically hot systems. Did stars form in massive dense clouds? Or did star formation occur in protodisks that subsequently merged?

The fundamental plane relation provides clues to the star formation history both with regard to the implicit M(L) dependence and the recently recognized fundamental plane substructure that seems to have a dissipative signature. The intense star formation rates inferred in starbursts are intriguingly similar to the star formation rates that once occurred in protobulges, and the underlying density profiles of starburst nuclei can be fit with de Vaucouleurs' profiles. Here too may be yet another, relatively nearby, clue to the formation of bulges and ellipticals. Whether mergers, tidal interactions or global instabilities provide the initial trigger is but one of the challenges yet to be elucidated.

References

Arnaud, M. *et al.* 1992, *Astron. Astrophys.* **254**, 49

Bender, R. 1992, in *The Stellar Populations of Galaxies*, eds. B. Barbuy and R. Renzini (Kluwer: Dordrecht), p.267

Bender, R., Burstein, D. and Faber, S.M. 1992, *Astrophys. J.* **399**, 462

Burstein, D. 1995, in *The Opacity of Spiral Disks*, ed. J.I. Davies (Kluwer Academic, Dordrecht), in press

Caselli, P. & Myers, P.C. 1995, *Astrophys. J.*, in press.

Dekel, A. and Silk, J. 1986, *Astrophys. J.* **303**, 39

Edvardsson, B. *et al.* 1993, *Astron. Astrophys.*, **275**, 101

Fuller, G. & Myers, P.C. 1992, *Astrophys. J.* **384**, 523

Kennicutt, R. 1989, *Astrophys. J.* **344**, 685

Myers, P.C. & Fuller, G. 1992, *Astrophys. J.* **396**, 631

Myers, P.C. & Goodman, A.A. 1988, *Astrophys. J.* **329**, 392

Prantzos, N. & Aubert, O. 1995, preprint

Scalo, J.M. 1986, *Fundam. Cosmic Phys.* **11**, 1

Shu, F.H. 1977, *Astrophys. J.* **214**, 488

Silk, J. 1995, *Astrophys. J. Lett.* **438**, L41

Tanaka, Y. 1995,, in Proc. 17th Texas Symposium, N.Y. Acad. Sci., in press

Wang, B, & Silk, J. 1994, *Astrophys. J.* **427**, 759

Worthey,, G., Faber, S.M., & Gonzales, J.J. 1992, *Astrophys. J.* **398**, 69

Wyse, R.F.G. and Silk, J. 1989, *Astrophys. J.* **339**, 700

Zinnecker, H. 1984, *Mon. Not. R. Astr. Soc.* **210**, 43

The Sloan Digital Sky Survey

Richard G. Kron

Fermilab, PO Box 500, Batavia, IL 60510, USA

1 The Survey

The Sloan Digital Sky Survey (SDSS; see Kron 1992 and Gunn & Knapp 1993) represents an effort to map 10,000 square degrees of the Northern Galactic Cap photometrically in five wavebands and spectroscopically to V $\sim$ 18 for extragalactic objects identified in the photometric map, yielding one million redshifts. The top-priority goals are large-scale-structure studies and related topics in extragalactic astronomy, but the data products will also support many programs in Galactic astronomy. In a number of respects the SDSS can be regarded as a next-generation Palomar Sky Survey.

2 Hardware

The nature of the undertaking has been based on something like the following logic: 1) To make significant progress in large-scale- structure studies, we need a much larger set of redshifts than currently exists, and the galaxies and quasars must be selected in a carefully controlled way. 2) Such a program requires a high rate of obtaining redshifts, a level of performance that can be provided (affordably) by building hardware dedicated to and designed for the task. 3) To select properly the galaxies and quasars for which spectra will be obtained, we require multi-band imaging and good angular sampling, which means a new imaging survey must be undertaken prior to, or in parallel with, the spectroscopic survey. 4) The same telescope can be used for both the imaging and the spectroscopic parts of the survey. Given the desiderata of a focal length that would yield 0.4-arcsec pixels on CCD's with 24-micron pixels, and a focal ratio of around f/5 in order to feed optical fibers with good performance in terms of focal ratio degradation, we arrive at a telescope aperture of 2.5 meters. 5) Fiber diameters of around 3 arcsec are desired for a number of reasons, for example minimizing losses from atmospheric dispersion. At this diameter, the flux from the night sky background is about equal to the flux from a V $\sim$ 18 galaxy (after accounting for light not included in the fiber). This defines an effective limiting flux for a wide-field survey. 6) To avoid strong interference from Galactic absorption and star crowding, we choose to survey above a Galactic latitude of approximately 30 degrees. In this 10,000 square degree area, there are about one million galaxies brighter than V $\sim$ 18. If the spectroscopic exposure times are approximately 1 hour long, and if a survey duration of about 5 years is considered, then about 600 redshifts must be obtained for each pointing of the telescope, and it follows that the field of view of the telescope must be about 3 degrees. 7) An optical design of adequate performance can be achieved in a modified Ritchey-Chrètien configuration - there are two large correcting lenses, one about 1 meter in front of the focal plane, and the other immediately in front of the focal plane.

Not required by this logic are some of the details of the camera and the spectrograph. The most distinctive aspect of the camera - aside from its wide field of view - is the array of 5 filters spanning the full optical range. The two extremes, a so-called u' band centered at 340 nm and a z' band centered at 910 nm, have been included on the basis of detecting quasars over the full range of redshifts, but both of these bands will likely prove to be of essential value for the galaxy program as well. For a given source, all five bands are obtained within an interval of about 8 minutes. The observations are obtained by clocking the charge on the CCD's at the same rate the sky is scanned across the focal plane. The field is so large that it is obligatory to scan nearly on great circles (and the optical distortion across the field must be small). It turns out that if the scanning rate is the same as the sidereal rate on the celestial equator, then the imaging can be completed in about 100 photometric nights - about 5 years of observing if our estimate of the frequency of photometric conditions is accurate. (The spectroscopic survey, which takes most of the telescope time, will be conducted in parallel with the imaging survey, using the remainder of the astronomically useful time.) This rate yields an exposure time per source and per filter of 55 seconds. The signal-to-noise ratio at the V $\sim$ 18 limit is more than adequate for selection of spectroscopic targets.

One dimension of the camera area is occupied by five CCD's, each with its own filter, as described above. The other dimension (that is, perpendicular to the scanning dimension) is taken up by six such arrays in order to cover area adequately quickly. The gaps between the scan lines are filled in later by taking another scan that is offset with respect to the first. Including the necessary overlaps due to this and to the convergence of the system of great circles, a significant fraction of the sky will be scanned more than once.

The scientific requirements include high-precision photometry, but most of the sky is scanned only once. Consequently, a separate smaller telescope called the Monitor Telescope is dedicated to measuring the transparency of the atmosphere and the brightnesses of calibrating stars within the field of the 2.5-m telescope, while the latter telescope is scanning.

In order to obtain 600 spectra per exposure, two fiber-fed spectrographs are required. Each is fed by 320 fibers (the extra being for sky), and each fiber maps onto three pixels perpendicular to the dispersion. With some space between the spectra, this format fills up a 2048-pixel CCD. Each of the two spectrographs is itself double, that is, there is a red and a blue channel split by a dichroic filter, yielding a net spectral range from 390 nm to 910 nm at a resolution of about

1800. This relatively large spectral range and high resolution is anticipated to enhance the yield of redshifts. As designed, the spectrographs provide much more than just accurate redshifts: in particular, line widths can be measured for the brighter, more massive galaxies.

The spectrographs are mounted permanently on the back of the telescope primary mirror support structure, surrounding the 3-degree focal plane. The fibers are positioned in the focal plane via drilled plates that are plugged with fiber harnesses during the daytime. A battery of such fiber assemblies can be swapped into the telescope rapidly during the night. When photometric conditions permit, the imaging camera replaces the fiber plug plate assembly, a swap that is also designed to be fast.

3 Data System

The data system features minimal on-line processing. The plan is to be able to submit the data tapes to the off-line main reduction system such that they are required to be read only once (besides speeding up the operation, this helps to automate the processing). To accomplish that, the data from the CCD's need to be re-formatted prior to being written to tape, the positions of the brighter stars need to be known, and information enabling flat-field vectors needs to be computed. Thus, these quantities are determined on-line by the data-acquisition system. The only other significant on-the-mountain processing relates to quality control aspects, since the observers must be able to decide if conditions are good enough to support imaging. The principal job of the off-line processing is to identify the galaxy and quasar spectroscopic targets from the imaging data and to undertake astrometry at the precision required to place the fibers. Moreover, we need to solve for the pattern of overlaps of the spectroscopic fields such that regions of high galaxy surface density are still fully sampled. The goal for this level of data reduction is a turn-around time of one month. If this can be achieved, it will ensure that the spectroscopic plan can be adaptively fit to the available imaging data, thus minimizing the duration of the survey.

The amount of data in the imaging survey is 8.4 Terabytes (10,000 square degrees sampled by 0.4-arcsec pixels in five separate bands). This number is a lower limit, since it does not count the overlaps, and necessarily we will take significantly more data than what ultimately pass the quality-assurance tests. The intent is to make the calibrations sufficiently reliable, and the detection of sources sufficiently complete, that recourse to the raw data will not be required on a routine basis.

Not counting the raw data, the data products from the survey will include calibrated spectra and redshifts for one million galaxies and quasars; calibrated pixel maps for these same sources in the five bands; similar pixel maps for sources that were too faint to be selected as spectroscopic targets; image parameters derived from the pixel maps; and various kinds of bookkeeping information. The intent is to make these data available soon after the quality of the calibrations can be verified.

4 Project Status

The project has obtained major funding from the Sloan Foundation and the National Science Foundation. In addition, the survey is supported by the participating groups and institutions (Fermilab, The University of Chicago, The Johns Hopkins University, Princeton University, The Institute for Advanced Study, the U.S. Naval Observatory, the University of Washington, and the Japanese Promotion Group).

The site is Apache Point Observatory, near Sunspot, New Mexico (latitude 32.8 degrees, elevation 2800 meters). The 2.5-m telescope and Monitor Telescope are adjacent to the ARC 3.5-m telescope, which was built and is operated by many of the same institutions; we can therefore draw on an existing infrastructure at the site.

The major construction elements of the project include the site development; the 2.5-m telescope mechanical assembly; the main optics (primary and secondary mirrors, and the corrector lenses); the Monitor Telescope; the instruments (the imaging camera and the two spectrographs); the procurement of the CCD's; the data acquisition system; the fiber cartridges; the software supporting target selection and database access; and the computers for the main processing system. All of these are in advanced stages of work, and some are complete. The Monitor Telescope has obtained its first images, and is now undergoing upgrades to function in survey mode. The 2.5-m telescope mechanical assembly is to be installed by July 1995, and the main optics by the end of 1995. After integrating all of the delivered components, there will be an engineering commissioning phase, followed by a "scientific commissioning" phase where we experiment with the details of the galaxy and quasar selection based on actual photometric measurements (as opposed to the simulations we are currently using to test the software). Having made decisions about the selection parameters, we intend to conduct the entire survey with the same selection criteria.

5 Applications to Young Galaxies

What will the Sloan Digital Sky Survey do to advance the field of distant galaxies? At the very least, it will provide a definitive census of the Universe to a median redshift of 0.1 that will form the basis for comparison with statistical samples of much more distant volumes. In order to model the observables for galaxies at high redshift, we would like to specify dependencies of galaxy type on environment, galaxy luminosity on spectral class, galaxy profile type on spectral class and on luminosity, etc. Having a million galaxies provides a unique capability to subdivide the parent sample into many subsamples, to look for weak correlations, and to define the tails of distribution functions.

The high resolution and wide wavelength interval of the spectra will greatly and generally advance studies of the integrated light in galaxies, which will constrain evolutionary models. In particular, the line-width measurements will vastly augment the currently available data, and correlations of visible properties to the underlying mass distribution will be correspondingly better defined.

For a million or so galaxies, we will have both broad-band photometry through five filters at high signal-to-noise ratio, and the spectra. These in combination will provide an unprecedented training sample to enable approximate redshifts to be derived from colors alone for the many millions of galaxies that are fainter than the spectroscopic limit. The redshift distribution is sensitive to both the cosmological model and to luminosity and color evolution, and even coarse redshift information may suffice to constrain these quantities. Moreover, the angular correlation function for galaxies can be computed according to their photometric redshifts, enabling a look at large-scale structure and its evolution at high redshift.

The description of the SDSS given above is for the Spring observing season. In the Autumn, we have an opportunity to adopt a different observing strategy that would complement the survey of the North Galactic Cap. Our plan for the South is to scan an equatorial stripe many times (in fact as often as possible given competing demands) in order to search for variable objects and in order to gain signal-to-noise ratio. This deep stripe could be used for many faint-galaxy programs, including for example a search for distant clusters.

References

Kron, R.G. 1992, in *Progress in Telescope and Instrumentation Technologies*, ESO Conference and Workshop Proceedings **42**, ed. M.-H. Ulrich (Garching:ESO), p.635

Gunn, J.E. & Knapp, G.R. 1993, in *Sky Surveys: Protostars to Protogalaxies*, A.S.P. Conference Series Vol **43**, ed. B.T. Soifer (San Francisco:ASP), p. 267

The Nature of Faint Galaxies from the Medium Deep Survey and Other Deep HST Images

R. A. Windhorst[1], *S. P. Driver*[1], *E. J. Ostrander*[1], *S. B. Mutz*[1], *P. C. Schmidtke*[1], *R. E. Griffiths*[2], *K. U. Ratnatunga*[2], *S. Casertano*[2], *M. Im*[2], *L. W. Neuschaefer*[2], *R. S. Ellis*[3], *G. F. Gilmore*[3], *R. A. W. Elson*[3], *K. Glazebrook*[3], *B. Santiago*[3], *W. C. Keel*[4], *D. C. Koo*[5], *G. D. Illingworth*[5], *D. A. Forbes*[5], *A. C. Phillips*[5], *R. F. Green*[6], *J. P. Huchra*[7], *and A. J. Tyson*[8]

[1] ASU, Dept. of Physics and Astronomy, Tempe, AZ 85287-1504, USA
[2] The Johns Hopkins Univ., Baltimore, MD, USA
[3] Institute of Astronomy, U. Cambridge, UK
[4] Univ. of Alabama, Tuscaloosa, AL, USA
[5] Lick Obs., UCSC, Santa Cruz, USA
[6] NOAO, Tucson, AZ, USA
[7] Harvard-Smithsonian CfA, Cambridge, MA, USA
[8] AT&T Bell Labs., NJ, USA

1 Abstract

High-resolution HST images with WF/PC and WFPC2 have allowed the morphological classification of faint field galaxies to $I_{814} \sim 24.25$ mag ($B_J \sim 26.0$). We find that the differential galaxy counts at these faint magnitudes are dominated by late-type spiral/Irregular galaxies. The elliptical and early-type spirals remain roughly consistent with the *no-evolution* predictions for these types, and firmly place the epoch of giant galaxy formation to beyond $z \sim 0.8$. Close examination of the late-type spiral/Irregular population supports a picture in which recent star-formation has occurred in a significant fraction (35 %) of this population, and that the remainder are lower redshift objects under-represented in the local surveys.

2 Introduction

The Medium-Deep Survey (MDS) is a Key Project with the Hubble Space Telescope (HST) which relies exclusively on parallel observations of random fields taken with the Wide Field and Planetary Cameras. A major aim of this HST project is to resolve one of the fundamental problems in observational cosmology: the nature of the numerous "faint blue field galaxies" ("FBG's", for a review, see *e.g.*, Koo & Kron 1992). Observations from the ground have been unable to solve this problem, or to distinguish between the numerous models proposed in the literature (e.g. *fading dwarfs* - Broadhurst, Ellis & Shanks 1988; Cowie, Songaila & Hu 1991; Babul & Rees 1992; Phillipps & Driver 1995; *merging* - Rocca-Volmerange & Guiderdoni 1990; Broadhurst, Ellis & Glazebrook 1992; *dwarf-rich* - Koo, Gronwall & Bruzual 1993; Driver *et al.* 1995a).

Here we report on recent results from the MDS, which shed new light on the FBG problem and has advanced our understanding of the faint blue field galaxies significantly. These results are due to the unique combination of HST's superb resolution (0.1″ FWHM for WFPC2) and the acquisition of a very large database of faint field galaxy images from which statistical studies can be made. The Cycle 1–3 WF/PC MDS data set contains 13,500 profile-fitted galaxies covering $18 < m_{I_{F785LP}} < 21$ mag (Casertano *et al.* 1995). This database has been used to study the angular sizes, bulge-to-disk ratio's, and color gradients of the field galaxy population down to $I_{785} \sim 21.0$ mag.

The Cycle 4 WFPC2 database, obtained after HST's e+a successful refurbishment, allowed the morphological classification of field galaxies to $I_{814} \gtrsim 22$–23 mag (Griffiths *et al.* 1994a; Glazebrook *et al.* 1995; Driver, Windhorst & Griffiths 1994a, Phillips *et al.* 1995b). In addition to the MDS WFPC2 database, an ultra-deep pointed WFPC2 field allowed to extend the morphological classification of field galaxies to $I_{814} \sim 24.25$ mag (Driver *et al.* 1994b). The color plate on the frontispiece shows the 5.7hr I_{814} and 5.7hr V_{606} WFPC2 images surrounding the weak radio galaxy 53W002 at $z = 2.390$ (Windhorst & Keel 1995). Images in V_{606} , $(V + I)/2$ and I_{814} are shown in the blue, green and red channels, respectively. A total of 600 objects in the I_{814} and 850 objects in the V_{606} frames have been detected in 0.00136 deg^2 down to $m_I \simeq 25.5$ and $m_V \simeq 26.75$ mag, implying a surface density of 4.4×10^5 and 6.2×10^5 deg^{-2} in I_{814} and V_{606} , respectively. The I_{814} and V_{606} image stacks have a 6.72 degree difference in HST roll angle, leading to apparent extreme object colors at the chip edges.

The determination of the morphological mix of field galaxies to such faint magnitudes represents a significant observational breakthrough against which the proposed faint field galaxy models can be tested. Both the WF/PC and WFPC2 datasets point to a population of low luminosity systems as being responsible for the faint blue excess. The morphological studies also suggest that a significant fraction of this population (35 %) has undergone — or is undergoing — violent and rapid star-formation. The remaining inert population suggests a severe underestimation of the *local* space-density of late-type spirals/Irregular systems. Here we briefly summarize the main observational results, our interpretation of the data, and the continuing work underway.

3 Morphological Studies with WF/PC and WFPC2

1) The WF/PC MDS database contains 13,500 galaxies taken from 112 MDS fields in Cycle 1–3. Their scale-lengths, total magnitudes, colors, and crude classifications have been determined via two-dimensional model fitting of *undeconvolved* images to a limiting magnitude of $I_{785} \sim 21$–22 mag (Casertano *et al.* 1995). Down to this limit, most galaxies observed before refurbishment have profiles which are rather regular, smooth, and consistent with those of local galaxies. However, at the faintest limit achievable before refurbishment — $I_{785} \sim 22.0$ mag — the mean angular size is somewhat lower than expected, suggesting

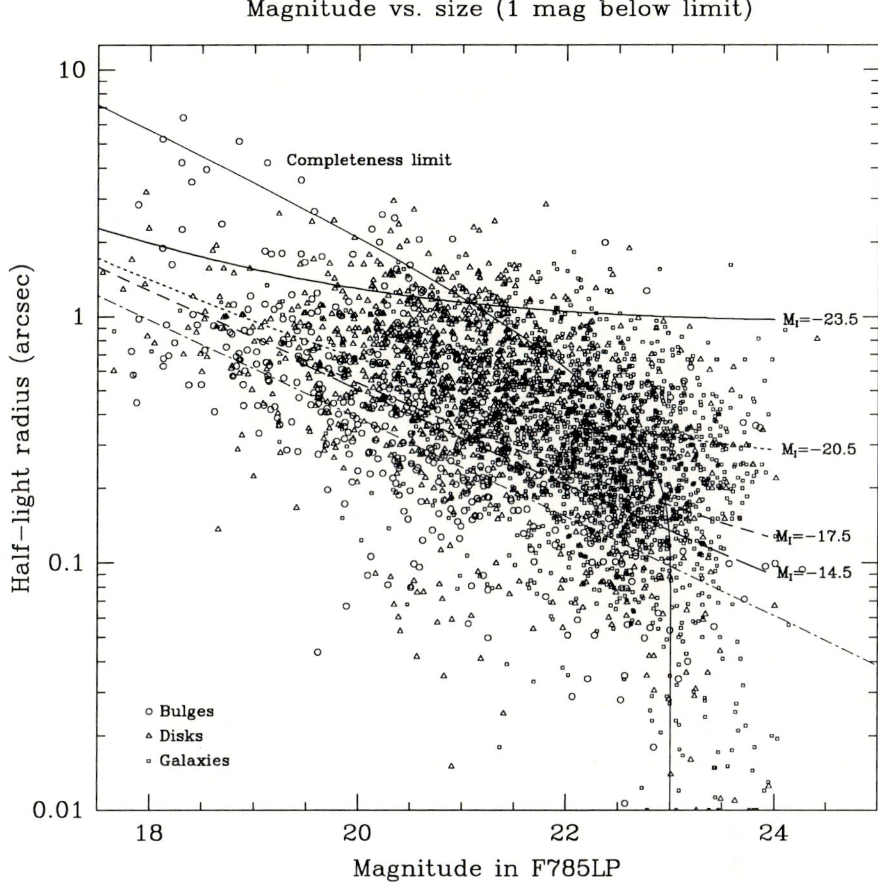

Fig. 1. Half-light radius vs. I_{785} magnitude for all MDS galaxies observed with WF/PC in Cycle 1–3 (Casertano *et al.* 1995). The lines indicate the expected relation for different values of M_I. Most galaxies are fainter — and smaller ! — than L^* galaxies ($M_I^* \sim -22.9$ mag). The straight line is the Euclidean extrapolation of the median for low-redshift galaxies with $M_I \sim -20.5$ mag ($\sim 0.1L^*$).

a higher proportion of compact systems. Figure 1 shows the size distribution as a function of apparent I_{785}-magnitude from Casertano *et al.* (1995), along with the Euclidean extrapolation for a galaxy with $M_I = -20.5$ mag (for $H_o = 50$ $\mathrm{km\,s^{-1}Mpc^{-1}}$). Note that most MDS field galaxies are fainter and smaller than L^* galaxies ($M_I^* \sim -22.9$ mag). The best fit median occurs for $M_I \sim -20.5$ mag (about a $\sim 0.1L^*$ galaxy), and the median half-light radius for all MDS galaxies down to $I_{785} = 22$ mag is 0.″3-0.″4 (Griffiths *et al.* 1994b, Im *et al.* 1995). Since at least a few independent beams are required across an object for accurate scale-length determination, size-estimates from ground-based images — even in the best possible seeing — will be biased for a good fraction of faint field galaxies (scale-length determination below 0.″1 is even difficult to do with HST's PC).

The HST number counts in the range $18 < I_{785} < 22$ mag exceed the ground-based numbers by about 50%, a range over which many ground-based objects may have been mis-classified as stars. Ellipticals are observed to become redder for $I_{785} > 20$ mag, but spirals show no large trend of bluer color with fainter magnitude, as expected from their K-corrections (based on local SED's in the near UV).

2) Studies of the intrinsic properties of MDS galaxies in Cycle 1–3 — *i.e.* those for which spectra have already been obtained — show little change in linear size versus redshift out to $z \sim 0.8$ (Mutz *et al.* 1994), and little change in structural parameters (Windhorst *et al.* 1994a, 1994b; Phillips *et al.* 1995a). However, it should be noted that the number of Cycle 1–3 galaxies for which spectra exists is still relatively small. Even so, a clear case of a strongly evolving luminous elliptical or early-type spiral galaxy is yet to be observed.

3) The clustering properties (*i.e.* the angular two-point correlation function) of elliptical galaxies in Cycle 1–3 are found to be stronger than for spirals (Neuschaefer *et al.* 1995). This is consistent with the idea that ellipticals are more prevalent in high density environments, but also consistent with a picture in which merging is more frequent for ellipticals. Studies of the clustering properties as a function of color also shows a *higher* correlation amplitude for bluer objects, suggesting that they are at lower redshifts, and/or more likely to be mergers when selected as blue (Neuschaefer *et al.* 1995). The deep HST images of Burkey *et al.* (1994) suggest an increase in the galaxy pair fraction towards fainter magnitudes, thereby constraining the epoch dependent merger rate to $\sim\propto (1+z)^3$. This is currently being checked with a deeper and much larger HST sample in Cycle 4.

4) With the refurbished WFPC2 MDS images, morphological classification into Hubble types can now be made down to $I_{814} \gtrsim$ 22-23 mag (Griffiths *et al.* 1994a, Driver, Windhorst, & Griffiths 1994a, Forbes *et al.* 1995, Glazebrook *et al.* 1995). Magnitude limited WFPC2 studies by Glazebrook *et al.* (1995, 301 galaxies) and Driver *et al.* (1994a, 144 galaxies) both independently find that beyond $I_{814} \sim 21.0$ mag ($\equiv I_{785} \sim 20.44^{\dagger}$) the observed counts are primarily made up of Irregular type galaxies, *i.e.* galaxies with low mean surface brightness ("SB"), irregular profiles and multiple-core structures. (For a typical galaxy in our sample, the Kron-Cousins I and the WFPC2 I_{814} filters are related by $I_{814} - I = 0.1$ mag (Holtzman *et al.* 1994). For comparison with the redshift sample of Lilly (1993), $I_{814} = I_{AB} - 0.38$ mag).

Without a large number of spectra for the Cycle 4 object available as yet, we cannot say if these systems are local dwarf-like systems, or the progenitors of today's giant spirals undergoing a major evolutionary process. The inference from the images is that, while many of these Irregulars show multiple cores and close companions (35 %), the majority (65 %) are of low mean and central SB, show no spiral structure, no central cores and have non-uniform asymmetrical outer profiles consistent with local Irregulars (e.g. LMC, SMC). However, a significant fraction (35 %) of this irregular population show signs of ongoing star-formation, and so have likely undergone more rapid evolution since $z \lesssim 0.5$–0.8 ($I \lesssim 22$-23 mag).

5) A single deep pointed 12-hr WFPC2 field (6-hr in both V_{606} and I_{814} ; Windhorst & Keel 1995) is shown in the frontispiece color plate and has extended the morphological study of field galaxies to $I_{814} \sim 24.25$ mag (Driver *et al.* 1994b, 227 galaxies). Although this ultra-deep HST WFPC2 image only constitutes a single line-of-sight — and therefore is susceptible to field-to-field fluctuations — the results from this field and the shallower MDS studies agree quite well in the magnitude range where they overlap ($21 \lesssim I_{814} \lesssim 22$ mag, see Figure 2). The rising trend of Irregular galaxies continues down to the *classification* limit of the deepest HST images ($I_{814} \sim 24.25$ mag; the 6-hr *detection* limit for compact galaxies is another $\sim$2 mag fainter: $I_{814} \sim 26.5$), with a smaller fraction of elliptical and early-type spirals at fainter magnitudes. Part of this may be due to the more irregular appearance of even nearby luminous galaxies in the restframe UV (Hill *et al.* 1992), compared to the optical. However, the current HST samples were selected in the I-band, and at $I_{814} \sim 22$ mag, the median redshift $z_{med} \sim 0.5$ (Lilly 1993), while at $I_{814} \sim 24$ mag, it may not exceed $z_{med} \sim 1$, according to spectral evolution models (*e.g.*, Carlberg & Charlot 1992). Hence, the restframe sampled by the I_{814} filter is likely to lie near or longward of the 4000 Å break.

4 Implications for Faint Galaxy Models

Im *et al.* (1995) compared the observed WF/PC size distributions of the Cycle 1–3 MDS database to three faint galaxy models: (1) no-evolution; (2) merger-driven evolution; and (3) dwarf-enriched no-evolution models. The mild-evolution or no-evolution models predict luminous galaxies of large angular size at high redshift: such objects are not seen in a majority of the MDS images, and are instead consistent with the dwarf-rich models and/or a local luminosity function with a steep faint-end slope. The MDS results rule out models in which the faint excess number counts are caused by large, low-SB galaxies (e.g. Ferguson & McGaugh 1995), since we do not see the predicted numbers of large, low-SB objects (HST selects against the largest low-SB galaxies, but this has been taken into account; Neuschaefer *et al.* 1995).

Figure 2 shows the breakdown of the differential I_{814} -band galaxy counts into ellipticals, early-type spirals and late-type spirals/Irregulars from the morphological HST studies summarized above (the Cycle 1–4 MDS and the ultradeep Cycle 4 WFPC field around 53W002 at $z = 2.390$). The model lines represent the no-evolution predictions for ellipticals and early-type spirals based on the observed local luminosity functions for these populations (c.f. Binggeli, Sandage & Tammann 1988). The normalization of the models to the data was done at $I_{814} \simeq 18.0$ mag, assuming a standard Einstein-de-Sitter model. The good fit

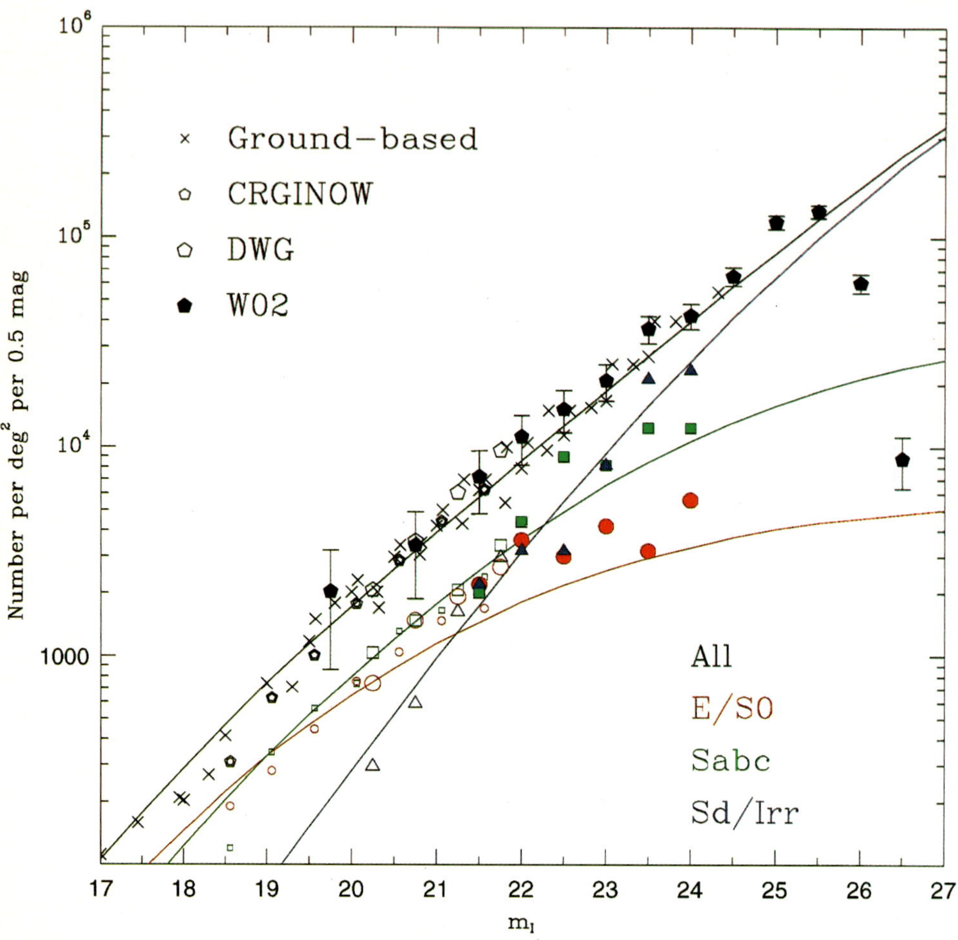

Fig. 2. The differential I_{814}-band galaxy number counts are shown for all HST galaxy types, as well as separately for HST ellipticals (red), early-type spirals (green), and late-type spirals/Irregulars (blue). The data come from Casertano *et al.* (1995, CRGINOW, 13,500 galaxies with $18.0 < I_{785} < 22$ mag), Driver, Windhorst & Griffiths (1994a, DWG, 144 galaxies with $20.0 < I_{814} < 22.0$ mag), and Driver *et al.* (1994b, W02, 227 galaxies with $22.0 < I_{814} < 24.25$ mag). The solid lines show the best fit models to the HST data for these three galaxy classes, derived from non-evolving elliptical and spiral LF's and from a dwarf-rich no-evolution model (c.f. Driver *et al.* 1994, 1994b), respectively, and present a possible solution to the FBG problem.

between the models and the data leaves little room for strong luminosity evolution in the elliptical or early-type spiral population. We conclude that neither have undergone major evolution down to $I_{814} \sim 24.25$ mag (Driver *et al.* 1994b). This places the epoch of galaxy formation (or at least the last major epoch of star-formation) for these giant types at $z_{form} \gtrsim 0.8$. Additional evidence comes from the lack of significant change in linear size and structural parameters for the elliptical population (Mutz *et al.* 1994; Windhorst *et al.* 1994a, b; Casertano *et al.* 1995; Phillips *et al.* 1995a, b). Note that the faintest HST ellipticals are slightly above the non-evolving model predictions, which could reflect some epoch dependence in the merger rate (*e.g.*, Burkey *et al.* 1994).

The observed late-type spiral/Irregular population that dominates the galaxy counts beyond $I_{814} \gtrsim 22.0$ mag appears inconsistent with a pure no-evolution extrapolation of the conventional local luminosity function for these types. In fact, at $I_{814} \sim 22.0$ mag the late-type spiral/Irregular population has a factor of 5-10$\times$ higher space density than expected (Driver *et al.* 1994b). The population responsible for the FBG problem has therefore been isolated in the deep HST images. Detailed inspection of their WFPC2 light-profiles and V_{606} & I_{814} contours suggest a population of Irregular galaxies, which should have a high *local* space density, or they would not explain the entire Sd/Irr count as observed by HST without drastic evolution. Only about a third of the late type galaxies seen by HST have a morphology consistent with recent evolution via both mergers and/or spontaneous starbursts. Hence, we conclude that a hybrid evolving dwarf-rich model (Phillipps & Driver 1995) provides a consistent solution to the long-standing FBG problem. A systematic spectroscopic follow-up is needed to confirm this picture beyond doubt, and this is currently underway with the Multiple Mirror Telescope at Mt Hopkins, the William Herschel at La Palma, the Kitt Peak Mayall, and the Keck telescopes.

5 Discussion

Parallel HST operations have steadily improved in efficiency through Cycle 1–4. Systematic parallel observations with WF/PC and WFPC2 resulted in a very large, uniform HST database, which allowed us to address the goals of the Medium Deep Survey: (1) measurement of field galaxy properties beyond $I \sim 22$ mag; (2) number counts as a function of galaxy type; (3) constrain the evolution of field galaxies out to $z \gtrsim 0.5$–0.8; and (4) study of galaxy morphology as a function of environment. These morphological studies also include: (a) the search for compact and multiple galaxy nuclei; (b) the frequency of galaxy mergers, interactions, and groups with cosmic time; and (c) the frequency of irregular morphology, starburst knots, bars, arms, and rings, etc.

A picture is emerging from the deep refurbished HST images in which the familiar excess number counts of FBG's – known for almost two decades from the ground – is being resolved by HST in terms of some evolution in a fraction of a *local* dwarf-rich galaxy population. This is manifested by a large number of irregular and peculiar galaxies, some of which seem to be merger remnants.

We thank the STScI staff for their help in obtaining the MDS and other HST images. We acknowledge support from HST grants GO.2684.0*.93A (to the MDS team) and GO.5308.0*.93A (to RAW and WCK).

References

Babul, A., & Rees, M.J. 1992, *Mon. Not. R. Astr. Soc.* **255**, 346
Binggeli, B., Sandage, A., & Tammann, G.A. 1988, *Ann. Rev. Astr. Astrophys.* **26**, 509
Broadhurst, T.J., Ellis, R.S., & Shanks, T. 1988, *Mon. Not. R. Astr. Soc.* **235**, 827
Broadhurst, T.J., Ellis, R.S., & Glazebrook K. 1992, *Nature* **355**, 827
Burkey, J.M., *et al.* 1994, *Astrophys. J. Lett.* **429**, L13
Carlberg, R.G., & Charlot, S. 1992, *Astrophys. J.* 397, 5
Casertano, S., Ratnatunga, K.U., Griffiths, R.E., Im, M., Neuschaefer, L.W., Ostrander, E.J., & Windhorst, R.A. 1995, *Astrophys. J.*, in press (CRGINOW)
Cowie, L.L., Songalia, A., & Hu, E.M. 1991, *Nature* **354**, 460
Driver, S.P., *et al.* 1994, *Mon. Not. R. Astr. Soc.* **266**, 155
Driver, S.P., Windhorst, R.A., & Griffiths R.E. 1995a, *Astrophys. J.*, submitted (DWG)
Driver, S. P., *et al.* 1995b, *Astrophys. J. Lett.* **449**, L23 (W02)
Ferguson, H. & McGaugh 1995, *Astrophys. J.*, in press
Forbes, D.A., Elson, R.A. W., Phillips, A.C., Illingworth, G.D., & Koo, D.C. 1995, *Astrophys. J. Lett.*, in press
Glazebrook, *et al.* 1994, *Nature*, submitted
Griffiths, R.E., *et al.* 1994a, *Astrophys. J.* **437**, 67
Griffiths, R.E., *et al.* 1994b, *Astrophys. J. Lett.* **435**, L49
Hill, J.K., *et al.* 1992, *Astrophys. J.* **395**, L37
Holtzman, J.A., *et al.* 1994, *Proc. Astr. Soc. Pac.*, in press
Im, M., Casertano, S., Griffiths, R.E., Ratnatunga, K.U., & Tyson, J.A. 1995a, *Astrophys. J.*, in press
Koo, D.C., & Kron, R.G. 1992, *Ann. Rev. Astr. Astrophys.* **30**, 613
Koo, D.C., Gronwall, C., & Bruzual, G. 1993, *Astrophys. J. Lett.* **415**, L21
Mutz, S.B., *et al.* 1994, *Astrophys. J. Lett.* **434**, L55
Neuschaefer, L.W., Casertano, S., Griffiths, R.E., & Ratnatunga, K.U. 1995, *Astrophys. J.*, submitted
Phillipps, S., & Driver, S.P. 1995, *Mon. Not. R. Astr. Soc.*, in press
Phillips, A.C., *et al.* 1995a *Astrophys. J.* 443, in press
Phillips, A. ., *et al.* 1995b *Astrophys. J. Lett.*, submitted
Rocca-Volmerange, B., & Guiderdoni, B. 1990, *Mon. Not. R. Astr. Soc.* **247**, 166
Windhorst, R.A., *et al.* 1994a, *Astron. J.* **107**, 930
Windhorst, R.A., *et al.* 1994b, *Astrophys. J.* **435**, 577
Windhorst, R.A., *et al.* 1995, *Natyure*, in press
Windhorst, R.A., & Keel, W.C. 1995, *Astrophys. J.*, in prep.

The Calar Alto Deep Imaging Survey

Klaus Meisenheimer, Steven Beckwith, Josef Fried, Hans Hippelein, Ulrich Hopp, Christoph Leinert, Hermann-Josef Röser, Eduard Thommes

Max-Planck-Institut für Astronomie
Königstuhl 17, D–69117 Heidelberg, Germany

The Quest for Primeval Galaxies

One of the great mysteries of the young universe is how matter turned its apparently smooth distribution (at the epoch of recombination) into the extremely structured form we observe today. Although we believe to know the driving force – gravity – it is by no means clear how the process worked in detail and under which circumstances the first stars have been formed. But certainly their occurrence completely changed the state of matter by providing a new source of energy – thermonuclear reactions – and enriched its chemical composition with heavy elements. Thus matter quickly developed into its current mature form with all its effective coolants like metal lines and molecular transitions.

With the technical revolution of observational astrophysics during the last two decades it seemed more and more feasible that one could observe those early stages of galaxy formation directly by looking back at high redshifts (Partridge & Peebles 1967). This very prospect initially provoked a rather optimistic view how these first or "primeval" galaxies may have formed: Having crossed the Jeans limit the radial collapse of a spherical gas cloud should develop a core dense enough to form the first stars within a free-fall time of several 10^8 years. The back-reaction of these seed stars onto the collapsing cloud would cause a violent "primeval burst of star formation". Under the assumption that a large fraction of the old stellar population in spherical systems (*i.e.* ellipticals, bulges of spirals) have been formed in this initial burst, Meier (1976) predicted that (a) the starburst should be very strong ($\geq 1000 M_{\odot}$/yr) and (b) primeval galaxies should be rather common ($> 1000/\Box^{\circ}$) in order to explain present numbers of galaxies. The most spectacular sign of these primeval starbursts should be the Ly-α emission from the ultraluminous H II region formed near the center of the collapsing cloud. Assuming that primeval galaxy formation took place around $z_f = 5$ one could expect a substantial number of detectable bright Ly-α galaxies between $z = 2$ and $z = 5$. Although the brightest of these objects should be within the reach of photographic plates the prospects of finding primeval galaxies became very realistic indeed with the advent of optical CCDs which pushed the detectable flux limit a factor of 10 fainter — albeit initially on a dramatic expense of the field of view.

However, early attempts to find primeval galaxies have not been successful (see review by Koo 1986). In the mid-eighties Meier's estimate was replaced by more detailed models which tried to tie the luminosity function of primeval galaxies to that of present-day galaxies (Baron & White 1987). This showed that most primeval galaxies are expected to be of lower luminosity than the "model galaxies" assumed by Meier. But sensitivity and collecting area of the CCDs still kept rising thus balancing this loss of optimism. Likewise the range of observing modes broadened from medium-band imaging through interference filters to narrow-band imaging through Fabry-Pérot-interferometers (FPI), longslit spectroscopy and broad-band searches including the near infrared wavelength region (see reviews by Djorgovski 1992, Djorgovski & Thompson 1992).

Despite considerable efforts a genuine primeval galaxy remains to be found. Thus the initial optimism has ceased and most discussions of primeval galaxy surveys focus on the question how primeval galaxies might avoid detection. We regard it premature to claim that the non-detection is already in clear conflict with robust theoretical expectations (see Thommes, these proceedings). Moreover, since we consider observational studies of the epoch of galaxy formation as one of the key projects of modern day astrophysics, we have initiated a large survey project — the Calar Alto Deep Imaging Survey (CADIS) — which is specifically designed to be effective in detecting primeval galaxies but will in addition produce a large data base for investigations of faint galaxies and quasars. So the effort will not be lost if no primeval galaxies are found. Nevertheless, in this contribution we will concentrate on the primeval galaxy search with CADIS and mention other aims of the survey only in passing.

Promising Search Strategy for Primeval Galaxies

First, one needs to guess in which redshift range the genuine primeval galaxy phase – *i.e.* the first burst of star formation – is most likely to occur. Although Cold Dark Matter models tend to place the main epoch of the formation of large galaxies rather late (at $2 \lesssim z \lesssim 4$) there are several reasons to believe that **the most promising redshift range lies beyond $z \simeq 5$:**

- Quasars and radio galaxies have been found out to $z \gtrsim 4$ and show strong emission lines of C,N,O ... Fe, indicating that the first generation of stars preceeded the observed epoch.
- Likewise, damped Ly-α systems which are regarded as progenitors of disk galaxies are detected out to $z \simeq 4$. They also contain a considerable amount of heavy elements, again pointing to an earlier epoch of star formation.
- The old stellar population of elliptical galaxies at intermediate redshifts seems to have been formed within a short period at very high redshift $z > 3$ (*e.g.*, Bender, these proceedings).
- Previous surveys which searched the range $2.5 < z < 4.8$ have not been successful in finding primeval galaxies.

Second, one has to decide whether to look for line or for continuum emission. We much favour an **emission line search** since the large equivalent widths expected from the H II region around a primeval starburst (emitted $W_\lambda \geq 10\,\text{nm}$, that is observed $W_\lambda \geq 50\,\text{nm}$ at $z \geq 4$) will make the detection of lines comparatively easy even in those objects which hardly have a detectable continuum ($R > 25^{\text{m}}$). But which line is to be searched for? Although strong suppression of Ly-α by even a small amount of dust currently is one of the favoured explanations for the low success of previous searches, we are still convinced that there is an overwhelmingly **strong case for Ly-α**:

- In the targeted redshift range $z > 4$ all other strong hydrogen lines lie in the nasty wavelength regime $\lambda > 2.4\,\mu\text{m}$ where groundbased observations are severely hampered by the strong thermal emission of the earth's atmosphere.
- In the primeval starburst all heavy element lines (like CIV $\lambda 154.9$ or [O II]$\lambda 372.7$) should be intrinsically weak.
- Likewise, a strong suppression of Ly-α by dust is already evidence for a previous generation of stars. So in a sense, the presence of Ly-α might be regarded as the defining property of a genuine primeval galaxy!

Finally, one has to choose between a search in the field or around density peaks which are marked by a known radio galaxy or quasar. We opt for **an unbiased field search** because

- biased searches are subject to completely unknown selection effects which may even lower the success rate (for instance if the presence of a quasar would indicate that the density peak has experienced its main galaxy formation phase already some time ago),
- in order to get astrophysical insight into the galaxy formation process we ultimately need to know the unbiased statistics of primeval galaxies.

The CADIS Concept

Any search for emission line objects of unknown redshift requires a trade-off between spectral coverage and field of view. In principle, a spectroscopic survey with a longslit[1] of width δy covering N_λ spectral resolution elements is equivalent to narrow-band imaging with the same spectral resolution $\delta\lambda$ since the number of exposures to cover the entire field $\Delta x \times \Delta y$ in the former case is $\Delta y/\delta y \simeq N_\lambda$, while in the latter case N_λ wavelength settings are required to cover the full wavelength range $\Delta\lambda = N_\lambda \cdot \delta\lambda$. In practice, however, narrow-band imaging has several advantages:

(1) The optimum sampling aperture can be chosen for each object individually when analysing the exposures.
(2) In the targeted wavelength region $\lambda > 700\,\text{nm}$ the night sky emission is strongly wavelength dependent so that an optimum search should be confined to some rather narrow wavelength ranges (Fig. 1).
(3) Background subtraction is much more accurate on 2-dimensional (imaging) data. This is especially important for primeval galaxies which we expect to be 10 to 100 times fainter than the sky, even in the Ly-α line.

[1] Due to the high surface density of objects at $R > 25^{\text{m}}$ and the need of blank sky areas any arrangement of spectroscopic apertures is equivalent.

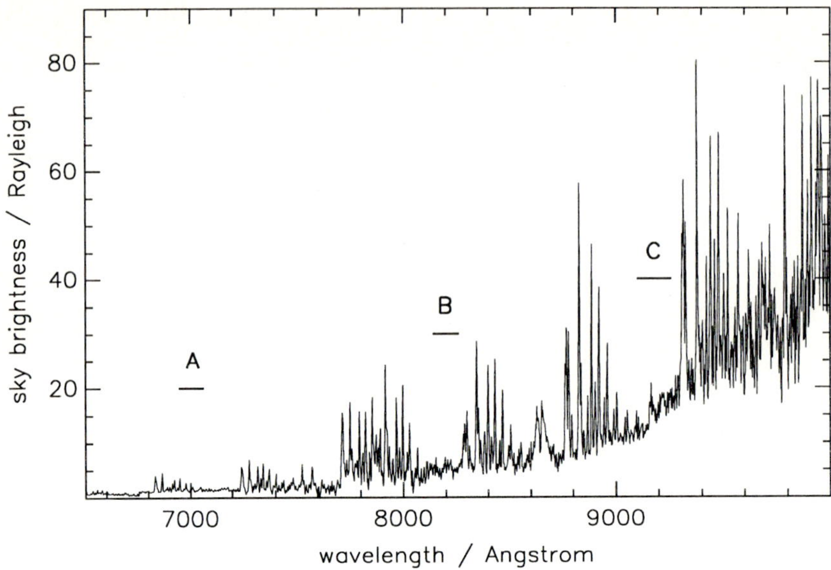

Fig. 1. Spectrum of the night sky emission. CADIS will search in the windows A, B, C for emission line galaxies.

The Calar Alto Deep Imaging Survey will therefore search in three narrow wavelength intervals around $\lambda = 700,\ 820,\ 918\,\mathrm{nm}$ (see Fig. 1 and Tab. 1) for primeval galaxies by employing an imaging Fabry-Pérot-Interferometer with $\delta\lambda \gtrsim 1.5\,\mathrm{nm}$. Each interval will be covered by about 10 wavelength settings. The attempted limiting flux (5σ) of $3 \times 10^{-20}\,\mathrm{Wm}^{-2}$ will be reached after an integration of 15 ksec with the focal reducer CAFOS 2.2 at the 2.2 m telescope (or after 6 ksec at the 3.5 m-telescope). Again, formally there is no advantage in using a narrow spectral resolution $\delta\lambda \simeq 1.5\,\mathrm{nm}$ since the gain in S/N ratio of an emission line is exactly balanced by the longer total time which is necessary to cover the wavelength range $\Delta\lambda$ (or depth in redshift space) of a broader filter. But the narrow-band observations directly provide accurate redshifts for every emission line object and, due to their higher contrast, background subtraction is less affected by instrumental imperfections.

Nevertheless, the limited spectral coverage of the narrow-band imaging approach has three main disadvantages in comparison to spectroscopy:

(4) Less accurate determination of continuum level underneath an emission line,
(5) no measurement of the global spectral energy distribution (SED) of detected objects, and
(6) detection of one emission line only. So the redshift of an object cannot be determined unambiguously.

The latter two points are of vital relevance for the primeval galaxy search, since due to the large number of faint blue galaxies (most are at intermediate redshift) we expect to find 10...100 foreground emission line galaxies for every

Table 1. Emission line bands to be observed with the Fabry-Pérot interferometer.

Line	z range	Δz	N_{Gal}	λ of other lines
		(A) 694.5 to 706.5 nm		
Lyman-α	4.713 ... 4.813	0.100	3	[O II] at 2130 ... 2167 nm
[O II]λ3727	0.863 ... 0.895	0.032	5	H_β at 905.5 ... 921.1 nm
H_β	0.428 ... 0.453	0.025	41	[O II] at 532.5 ... 541.8 nm
[O III]λ5007	0.387 ... 0.411	0.024	42	[O II] at 516.9 ... 526.0 nm
H_α	0.058 ... 0.076	0.018	20	H_β at 514.3 ... 521.0 nm
		(B) 814.0 to 826.0 nm		
Lyman-α	5.696 ... 5.795	0.099	2	
[O II]λ3727	1.184 ... 1.216	0.032	1	
H_β	0.675 ... 0.700	0.025	21	[O II] at 624.4 ... 633.8 nm
[O III]λ5007	0.626 ... 0.650	0.024	25	[O II] at 606.0 ... 615.1 nm
H_α	0.240 ... 0.259	0.018	36	[O II] at 462.4 ... 469.2 nm
		(C) 910 to 926 nm		
Lyman-α	6.485 ... 6.617	0.132	2	
H_β	0.872 ... 0.905	0.033	12	[O II] at 697.9 ... 710.2 nm
[O III]λ5007	0.818 ... 0.849	0.032	16	[O II] at 677.8 ... 689.3 nm
H_α	0.387 ... 0.411	0.024	42	[O III] at 694.3 ... 706.5 nm

"true" primeval galaxy[2]. CADIS will deal with these problems in the following way:

(5^+) The global SED of every object in the field and the continuum underneath the emission line will be measured by photometry in a set of broad- and medium-band filters. The relative calibration of this photometry will be tied to several brighter field stars, the SED of which will be determined accurately by spectrophotometric observations.

(6^+) For every FPI range (A,B,C see Tab. 1) we will carry out additional narrow-band observations at those wavelengths where one expects to find a second line for the most likely (low redshift) identifications of emission lines detected in the FPI intervals: H_α, [O III]λ500.7, H_β, [O II]λ373.7. The appropriate wavelengths of these so-called *veto* filters (because a signal in them excludes Ly-α) can be read off plots like the one displayed in Fig. 2.

The wavelength coverage of CADIS is summarized in Tab. 2. Note the rather good spectral resolution at $\lambda > 560$ nm and the inclusion of a deep K-band observation. They will allow an unequivocal identification of late type dwarf stars and a good redshift measurement ($\sigma(1+z) \lesssim 0.02$) for early type galaxies

[2] This large amount of contamination which had to be sorted out by extremely time-consuming spectroscopic follow-up, mainly has brought the previous primeval galaxy searches to a halt.

between $z = 0.5$ and 1.2 using the multi-color method developed by Belloni & Röser (these proceedings, 153).

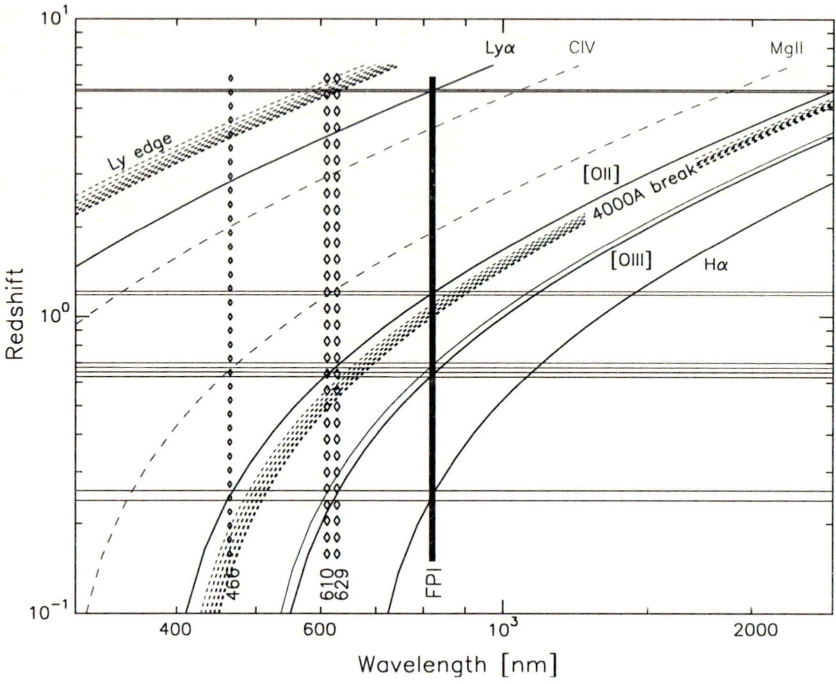

Fig. 2. Emission line redshifts and *veto*-filters for FPI interval (B) around $\lambda = 820$ nm. For instance, H_α enters the window for $0.24 < z < 0.26$ and the wavelength of other prominent spectral features can be read off the diagram by following this redshift interval horizontally towards the right. Three *veto* filters are indicated by vertical chains of diamonds.

CADIS will be carried out with the new focal reducers at the Calar Alto 2.2 m and 3.5 m telescopes. With the LORAL 2048 × 2048 CCD the pixel scale will be 0.″32 and the field of view 10 × 10 □′ (even after allowing for the necessary dithering of the images by some 10″ to get good sky flatfields). A mosaic of 2 × 2 exposures with the wide field near-infrared camera OMEGA will cover the same field. We plan to observe 10 fields distributed over the northern sky in order to assure that at least one field is at low zenith distance at any time. So the total field is 1000 □′ = 0.28 □°. Fields have been selected on the IRAS maps and the Palomar Sky Survey plates to minimize galactic extinction and avoid bright stars (R< 16.ᵐ).

In the entire survey we expect to discover 10 ...100 primeval galaxies (see Thommes, these proc.), > 1000 emission line galaxies in several narrow redshift bins between $z = 0.24$ and $z = 0.90$, at least 500 early type galaxies (with

Table 2. Broad and medium band filters and limiting magnitude for CADIS.

Filter	λ [nm]	$\Delta\lambda$ [nm]	$F_\nu(20^m)$ [μJy]	$m_{lim}(10\sigma)$ [mag]	F_{lim} [μJy]
Broad:					
BV′	465	110	44	$24^m_{\cdot}0$	1.3
R-2	650	150	32	$23^m_{\cdot}5$	1.3
K′	2150	300	6.6	$20^m_{\cdot}5$	4.2
Narrow:					
F 572/21	527	21	37	$23^m_{\cdot}6$	1.3
F 614/28	614	28	34	$23^m_{\cdot}5$	1.3
F 664/28	664	28	31	$23^m_{\cdot}5$	1.3
F 703/21	703	21	30	$23^m_{\cdot}3$	1.5
F 755/30	755	30	28	$23^m_{\cdot}2$	1.6
F 812/17	812	17	27	$23^m_{\cdot}0$	1.7
F 850/30	861	25	26	$22^m_{\cdot}9$	1.7
F 915/35	915	35	24	$22^m_{\cdot}8$	1.8

redshifts determined in the range $0.5 < z < 1.2$), and several hundred faint quasars, most of them at redshifts $z > 3$. So CADIS will not only probe the space density of primeval galaxies to unprecedented depth and redshift but also provide a unique data base for studies of the luminosity function and clustering properties of faint field galaxies and the faint end of the quasar luminosity function.

First Results

In November 1993 we carried out a pilot study with the new focal reducer CAFOS 2.2 at the 2.2 m telescope on Calar Alto. The aim of this study was to determine the limiting magnitudes which can be reach in practice and to test the data reduction tools which are currently developed for the survey. Since neither the FPI nor the $2k \times 2k$ CCDs were available at this early stage we had to carry out the emission line search with a medium-band filter $\lambda/\Delta\lambda = 812/17$ nm and using a $1k \times 1k$ Tektronix CCD (pixel scale 24 μm *i.e.* $0\farcs51$ on the sky, field of view $9 \times 9\,\square'$). Two survey field have been observed. In addition to the $\lambda 812/17$ exposures (integration time $t_{int} = 7500$ sec) we took deep images through broad-band B,R,I-filters ($t_{int} = 8000, 3500, 7000$ sec resp.). From the results of this pilot study we determined the required integration times for CADIS as given in Tab. 2.

Here we would like to present the most spectacular emission line objects in the field we analysed first (centered at $\alpha_{1950} = 9^h\ 10^m_{\cdot}5$, $\delta = +46^\circ\ 25'$) in order to exemplify our search method and demonstrate that we indeed find the kind of objects we are looking for.

In the first step, we search for objects *independently* on all four images B, R, I, $\lambda 812/17$ by using FOCAS (Jarvis & Tyson 1981). All objects above a 3σ-level are regarded as "candidate objects" and the four object lists are merged (assuming that "objects" found within $1''$ of each other on different frames are the same and their "true" position is given by the average of the FOCAS positions). In a second step, we employ our CCD photometry software (*cf.* Röser & Meisenheimer 1991) to calculate accurate colours of every object detected (referring to exactly the "true" position and a common beam width on every frame). In the third step, objects with a significant excess at $\lambda 812/17$ are identified on a two dimensional diagram in which the excess flux $F_{812} - F_{cont}$ is plotted versus $-2.5 \log F_{812}$. We find > 10 emission line candidates with flux above the 5σ-limit of $F_{lim} = 10^{-19}\,\mathrm{Wm}^{-2}$. The spectra of the best 6 of them are displayed in Fig. 3

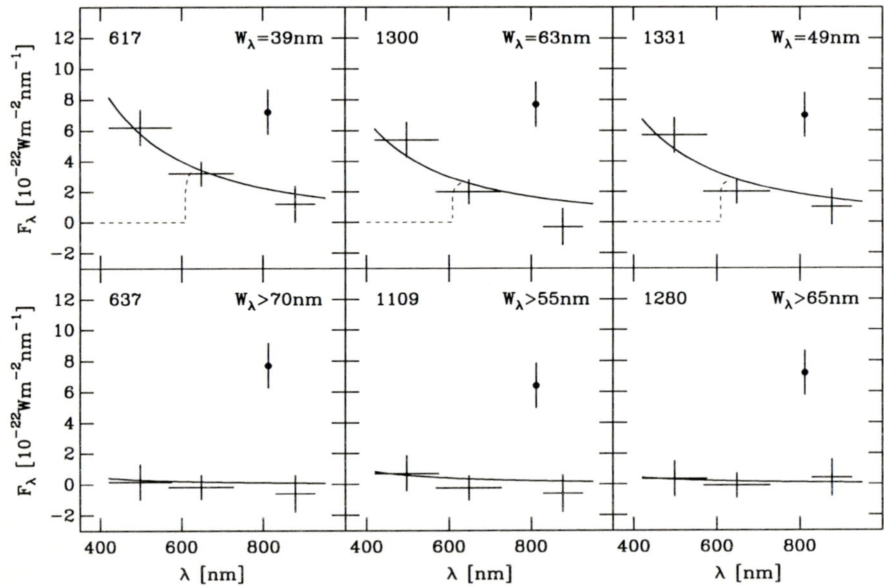

Fig. 3. Spectral energy distribution (SED) of emission line objects in 9h field. The SED is fitted by a flat continuum $F_\nu = constant$. The upper row of the plot shows objects with detected blue continuum. This excludes to identify the emission line with Lyman-α which would imply a Lyman edge at 610 nm. The objects in the lower row have no detectable continuum and their redshift remains undetermined. The lower limit to the (observed) equivalent width W_λ refers to the 3σ upper limit in the R-band.

and show that we indeed detected emission line objects with those spectacular equivalent widths $W_\lambda(obs) > 50\,\mathrm{nm}$ which are expected for primeval galaxies. However, from the very crude measurement of the SED with only three broad-band filters we can already exclude 3 of them being Ly-α galaxies since their detected continuum shows no indication of a break at the predicted Lyman limit.

At first sight the non-detection of any continuum makes the other three objects prime candidates for primeval galaxies. However, two facts are arguing against the impression that our survey might have been successful before it fully started:

- Two of the "no continuum" objects (#1109, #1280) lie within $1'$ of several emission line objects (including #1300, #1331) which are obviously at low redshift. They may thus be members of the same low redshift association.
- Even under optimistic assumptions we expect < 1 primeval galaxy down to our limit of $10^{-19}\,\mathrm{Wm}^{-2}$ in our field (see Thommes, these proceedings). So detecting three of them is much too fortunate to be true.

On the other hand, the very existence of these spectacular emission line galaxies (with *intrinsic* $W_\lambda > 40\,\mathrm{nm}$ if we adopt the most likely redshift $z = 0.24$, *i.e.* an identification with H_α) which have been overlooked in previous surveys strongly supports our concept of employing *veto*-filters: In the present case we have to search for the [O II]λ372.7nm line around $\lambda = 812 \cdot \frac{327.7}{656.3} = 460\,\mathrm{nm}$ to test the hypothesis that the detected line is H_α, and around $\lambda = 605\,\mathrm{nm}$ to verify a possible identification with [O III]λ500.7nm. So far bad weather and the lack of suitable filters prohibited these *veto*-observations and so our best candidate (#637) which is detached spatially from the above mentioned group may still turn out to be a genuine primeval galaxy.

Project Summary and Current Status

The Calar Alto Deep Imaging Survey (CADIS) will combine a very deep emission line survey (limiting flux for 5σ-detection: $3 \times 10^{-20}\,\mathrm{Wm}^{-2}$) with deep broad- and medium-band photometry between 450 nm and 2200 nm (limiting magnitude for 10% photometry: $R \simeq 23\overset{\mathrm{m}}{.}5$, $K \simeq 20\overset{\mathrm{m}}{.}5$). Object selection will be performed on all frames, thus providing a complete list of detected objects without colour selection. We plan to observe ten fields of $100\,\square'$ each, distributed over the northern sky. Although the survey method is tailored to a search for primeval galaxies, the multi-colour part of CADIS will provide a classification and rough redshift measurements of objects with strong continua (stars, quasars and early type galaxies). CADIS should detect and classify several 100 faint field stars, several 100 faint quasars at redshifts $z \gtrsim 3$ and > 500 galaxies of Hubble type earlier then Sb with redshifts determined ($\sigma(z) \lesssim 0.02$) in the range $0.5 < z < 1.2$.

The vast majority of the objects detected (> 1000) will be faint emission line galaxies in several intermediate redshift bins distributed over the range $0.2 < z < 0.9$ which (together with the early type galaxies detected in a much larger volume) will provide an ideal data base to study the evolution of the luminosity function and clustering properties of field galaxies in the above redshift range. CADIS will be complementary to ongoing or previous deep spectroscopic surveys, since emission line objects with faint continua (*e.g.*, H II galaxies) do not escape undetected.

The total integration time per field will be 130 hours (at the 2.2 m telescope, the 3.5 m is 2.5× faster). So we need about 160 clear nights distributed among both telescopes for the project. The current status (as of March 1995) is as follows: The directors of the MPIA have agreed that CADIS will be a major part of our institute's research in the years 1995–98. The necessary telescope time will be allocated to the project. This plan has also been supported by the time allocation committee for Calar Alto. Due to unforeseen delays in the delivery of thinned, science grade LORAL CCDs and the refurbishment of the telescope drives at the 2.2 m (October 1995 to March 1996) the full data production cannot commence before spring 1996. Fortunately, the telescope time allocated for 1995 should enable us to collect a meaningful data set for two fields before fall of this year.

References

Baron, E. & White, S.D.M. 1987, *Astrophys.J.* **322**, 585
Belloni, P. & Röser, H.-J. 1995, *these proceedings,* p.153
Bender, R. 1995, *these proceedings,* p.169
Djorgovski, S. 1992, in *Towards Understanding Galaxies at large redshift*, R.G.Kron & A. Renzini (eds.), p.259
Djorgovski, S. & Thompson, D. 1992, in *The stellar population in galaxies*, A.Renzini & B.Barbury (eds.), IAU Symp. No. **149**, p.337
Jarvis, J.F. & Tyson, J.A. 1981, *Astron. J.* **86**, 476
Koo, D.C. 1986, in *Spectral Evolution of Galaxies*, A. Chiosi & A. Renzini (eds.), p. 419
Meier, D. 1976, *Astrophys. J.* **207**, 343
Partridge, R.B. & Peebles, P.J.E. 1967, *Astrophys. J.* **147** 868
Röser, H.-J. & Meisenheimer, K. 1991, *Astron. Astrophys* **252**, 458
Thommes, E. 1995, *these proceedings,* p.242

An Infrared Search for Primeval Galaxies

Steven V. W. Beckwith[1], *Filippo Mannucci*[2], *and David Thompson*[1]

[1] Max-Planck-Institut für Astronomie, Heidelberg, Germany
[2] Osservatorio di Arcetri, Florence, Italy

1 Motivation

The importance of finding the earliest generation of stars has been stressed by many of the speakers in this volume as well as in the literature (Djorgovski, Piotto, & Capaccioli 1993; Thompson & Djorgovski 1995). The primary benefits are an understanding of the manner by which elliptical galaxies are born, the conditions in the early universe, constraints on cosmological parameters when coupled to models of structure formation, and the possibility to study star formation in different environments. A common view is that the oldest stars are in elliptical galaxies, and the "primeval galaxies" will be young ellipticals in the first stages of star formation.

Although it is possible to produce plausible models for primeval galaxies, the uncertainties in almost all important parameters provide relatively little constraint on the imagination of the modeller. We have few good constraints on the rate of star formation – whether it occurs in many short bursts, for example, or is almost continuous – the concentration of matter during the time of star formation, the times or, equivalently, redshifts at which star formation begins, the shape of the initial mass function, and the rate at which enrichment of interstellar material feeds back on the observational characteristics of the luminosity – through the generation of dust, for example. Nevertheless, we expect that the onset of star formation occured at redshifts higher than about 2, and probably higher than about 4 (Thompson & Djorgovski 1995), that it contained a large number of low-mass stars to give rise to the present populations of old stars, that the interstellar gas was metal poor, and that most of the energy was liberated rather quickly, within a Gyr or so of the onset.

Under a variety of different assumptions, these characteristics should give rise to light which can be observed today (Partridge & Peebles 1967; Tinsley 1977, 1978; Baron & White 1987). To date, there has been no unambiguous observation of a primeval galaxy – a young elliptical – despite a number of sensitive searches. The principle searches have been for rest frame UV radiation in the form of Lyα or continuum radiation redshifted into the optical passband (cf. recent review by Pritchet, C. J. 1994, and references therein) and the study of very high redshift galaxies found through radio surveys (e.g. Meisenheimer 1994). The upper limits to the UV light from the first generation of stars has already put useful constraints on even the conservative models (Thompson & Djorgovski 1995; Thommes 1995).

Light from the first generation of stars might well be observable at present-day infrared wavelengths, even if the optical light is undetectable (Boughn, Saulson, & Uson 1986; Parkes, Collins, & Joseph 1994; Mannucci & Beckwith 1995). If most star formation took place at redshifts higher than about 7, placing the Lyα and/or Lyman limit radiation beyond $1\,\mu$m, the luminosity would be in the near and thermal infrared. If even small amounts of dust were created during the first generation of stars, it would be possible to obscure most of the present-day optical light and shift the luminosity to the near infrared and submillimeter wavelengths. Since both of these conditions are plausible and perhaps even likely, it is highly desireable to extend the search for primeval galaxies to infrared wavelengths.

2 Method

We have undertaken a search for infrared light from high redshift galaxies using a wide-field infrared camera, the MAGIC camera used at the German-Spanish Calar Alto Observatory 3.5 m telescope (Herbst *et al.* 1994). The wide-field optics give a square FOV 207" on each side in a 256×256 pixel format (0".81 pixels). The HgCdTe NICMOS3 detector is sensitive between 1 and $2.5\,\mu$m, and the camera is background limited through all filters with a total efficiency of about 30%. Background-limited observations at J, H, and K (or K$'$) reach $5\,\sigma$ point source limiting magnitudes of 22.5, 21.5, and 21.1, respectively, in 1 hour of integration time.

The approach is to search for emission line objects by taking images through narrow band filters – typically $\Delta\lambda/\lambda \approx 0.01$ – and comparing them with images taken through broad band filters. It is possible to normalize the narrow and broad band images on a bright continuum source in the field and subtract the images to look for objects with an excess in the narrow band.

Fields are chosen to contain a known high redshift object, usually a quasar, and the redshift of this object is such that a prominent emission line should appear in the narrow passband. For example, the Hα line ($0.6562\,\mu$m) appears in the $2.12\,\mu$m filter – the filter commonly used to study the molecular hydrogren $v = 1 - 0$ S(1) line – at a redshift of 2.23; [O II] ($0.3727\,\mu$m) at $z = 4.69$ appears in this filter, and at $z = 5.32$ appears in the filter at $2.355\,\mu$m. MAGIC contains 15 narrow filters between 1.083 and $2.2483\,\mu$m that can be used in this way.

Deep images are made by the usual technique of combining many short-exposure (1 minute) images, each offset from the others by approximately 10" , using averages of the offset images for sky subtraction. A total of 22 fields have been searched amounting to a total of $240^{\square'}$ or $\sim 0.07^{\square\circ}$. The average exposure time per field is $2\frac{1}{2}$ hours. Typical limiting magnitudes for point sources are greater than 19^m at K$'$ and $3 \times 10^{-16}\,\mathrm{erg\,cm^{-2}\,s^{-1}}$ in a narrow band filter.

The relatively large sky coverage is considerably larger than previous searches using similar techniques (Cowie, Songaila, & Hu 1991; Parkes, Collins, & Joseph 1994), and although the limiting sensitivity is brighter than that possible using larger telescopes (Pahre and Djorgovski 1995), the volume of universe sampled is considerably larger, owing to the large field of view.

3 Preliminary Results

Figure 1 shows the field near the quasar Q1159+123 taken through broad and narrow band filters. Three candidates for emission line objects were seen in the narrow band image; they are marked in the figure. These candidates were discovered independently by two different methods. The first was to "blink" the two images rapidly and search for candidates by eye. The second was to use an automatic photometry program, FOCAS (Jarvis & Tyson 1981), to search for objects in each frame and plot the relative brightnesses at the two different wavelengths. Figure 2 is a plot of the relative brightnesses of the objects discovered in the FOCAS search for this field.

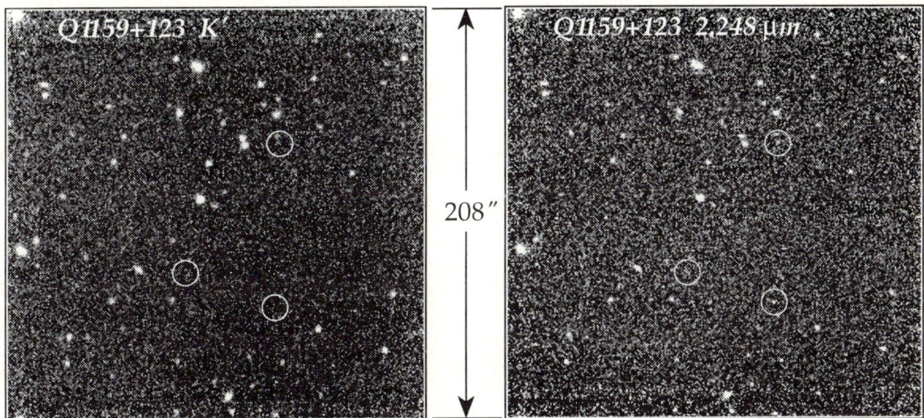

Fig. 1. a: An image of the field near the quasar Q1159+123 taken through a K′ filter. The limiting magnitude for point sources is approximately 19^m. The quasar itself (14.4^m at K′) is near the center of the field. **b:** An image of the Q1159+123 field taken through a narrow band filter at 2.248 μm. The limiting flux is about 3×10^{-16} erg cm^{-2} s^{-1}. Three objects marked in the frame appear weakly or not at all in the broad band image of the same field.

A total of eight candidates have been found in three of the fields using these methods. At the time of this writing, we have been able to repeat the observations of the Q1159+123 field, and none of these candidates appear in the second set of narrowband observations. It is clear that the five remaining candidates, all of which are near the limiting sensitivity of the observations, must then be considered suspect. We henceforth assume the observations have produced only upper limits to the number of bright, emission line objects in the search fields.

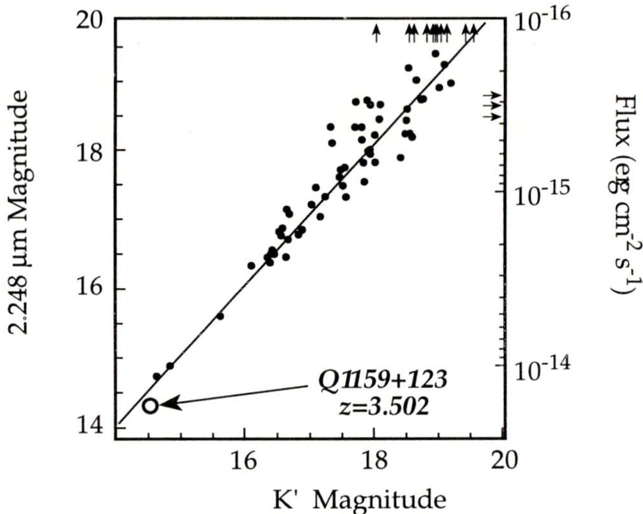

Fig. 2. This figure compares the narrow and broad band magnitudes of the objects discovered by the FOCAS search of the field around Q1159+123. The vertically pointed arrows represent objects seen in the broad band image for which only upper limits are obtained in the narrow band image. The horizontally pointed arrows in the upper right hand part indicate the three objects which are seen in the narrow band image but not in the broad band image. These objects are candidates for emission-line sources.

The upper limits to the searched area may be compared to estimates of the surface density of primeval galaxies as a function of their brightness. Figure 3 is a plot of the surface density of galaxies on the sky as a function of line flux for lines shifted into the K′ band taken from Mannucci & Beckwith (1995). The upper limits from four different searches including the present work are shown in the figure.

4 Summary

We have carried out a preliminary search for primeval galaxies at infrared wavelengths, using a narrowband imaging technique. A total area of $\sim 240^{\square'}$ has been surveyed to a limiting emission line flux of $\sim 4 \times 10^{-16}$ erg cm^{-2} s^{-1}. Although several candidate emission-line objects have been identified, all lie close to the limit of the narrowband data. A second set of narrowband data on one field failed to confirm the original three candidates, thus all candidates are currently considered to be only tentative.

Unlike surveys based on the Lyα emission line, surveys such as this one, targetting restframe optical emission lines, should be much less sensitive to the effects of dust in the PGs. They also offer the possibility of detecting restframe Lyα emission from objects at redshifts of 7 or more.

It is evident from the calculations that the large-area, low sensitivity searches possible with 3–4m class telescopes are entirely complementary to the narrow field, deep searches possible with the new 8–10m class telescopes. The calculations leave considerable room for a population of primeval galaxies that would have escaped observation in the infrared; the lower part of the hatched region in the figure allows a population of galaxies whose surface densities are too small at all limiting line fluxes to have been picked up by now. Increasing the searched area by an order of magnitude or decreasing the limiting flux by half an order of magnitude would begin to seriously challenge the model calculations.

The next generation of large format infrared detectors should make it possible to increase the search areas by approximately an order of magnitude within a similar investment of telescope time, approximately 10 nights. It will be more difficult to decrease the limiting flux without substantially increasing the time allocated to such search projects. Nevertheless, it appears that the detection of very young galaxies should be within our reach, if the calculations are a reasonable indicator of emission from the first generations of stars.

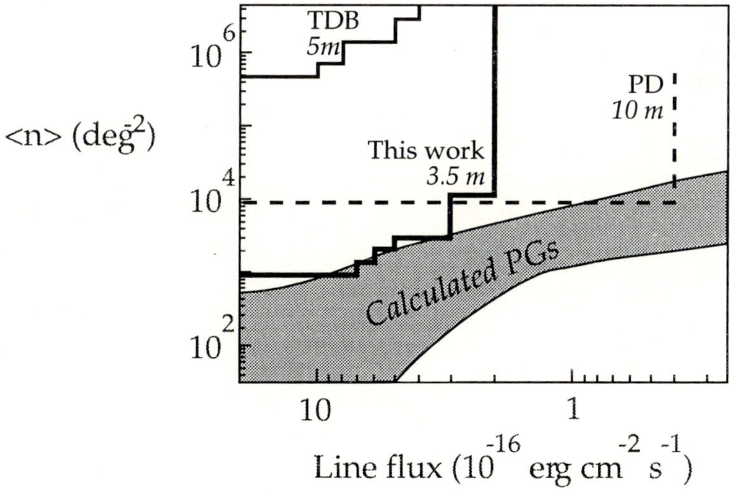

Fig. 3. This figure plots the predictions for a large grid of models of primeval galaxies as surface density vs. apparent line flux. The shaded region shows the predictions of the calculations. The solid and dashed lines show the upper limits established by Thompson, Djorgovski, & Beckwith (TDB, 1994); Pahre and Djorgovski (PD, 1995); and this work.

References

Baron, E., White, S.D.M. 1987, *Astrophys. J.* **322**, 585

Boughn, S.P., Saulson, P.R., Uson, J.M. 1986, *Astrophys. J.* **301**, 17

Collins, C.A., Joseph, R.D. 1988, *Mon. Not. R. Astr. Soc.* **235**, 209

Cowie, L.L., Songaila, A., and Hu, E.M. 1991, *Nature*, **354**, 460

Djorgovski, S., Piotto, G., Capaccioli, M. 1993, *Astron. J.* **105**, 2148

Djorgovski, S. and Thompson, D.J. 1992, in *The Stellar Population of Galaxies*, eds. B. Barbury and A. Renzini, p.337

Herbst, T.M., Beckwith, S.V.W., Birk, C., Hippler, S., McCaughrean, M.J., Mannucci, F. and Wolf, J. in *Infrared Detectors and Instrumentation*, ed A.M. Fowler, SPIE Conference 1946.

Jarvis, J.F. & Tyson, J.A. 1981, *Astron. J.* **86**, 476

Mannucci, F. & Beckwith, S.V.W. 1995, *Astrophys. J.* **442**, 569

Meisenheimer, K. 1994, in *A Panchromatic View of Galaxies*, eds. G. Hensler & Ch. Theis, Edition Frontiere.

Parkes, I.M., Collins, C.A., and Joseph, R.D. 1994, *Mon. Not. R. Astr. Soc.* **266**, 983

Partridge, R.B., Peebles, P.J.E. 1967, *Astrophys. J.* **147**, 868

Pritchet, C.J. 1994, *Proc. Astr. Soc. Pac.*, **106**, 1052

Pahre, M. A. & Djorgovski, S. 1995, *Astrophys. J. Lett.*, in press

Thommes, E. 1995, in preparation.

Thompson, D. & Djorgovski, S. 1995, *Astrophys. J.*, in press.

Thompson, D., Djorgovski, S., Trauger, J. 1992, in *Cosmology and the Large Scale Structure of the Universe*, ed. R. de Carvalho, ASPCS **24**, p.147

Thompson, D., Djorgovski, S., Beckwith, S.V.W. 1994, *Astron. J.* **107**, 1

Tinsley, B.M. 1977, *Astrophys. J.* **211**, 621

Tinsley, B.M. 1978, *Astrophys. J.* **220**, 816

Radio Surveys

James J. Condon

National Radio Astronomy Observatory*
520 Edgemont Road, Charlottesville, VA 22903, USA

1 Introduction

Nearly all galaxies with measured redshifts $z > 1$ have been found because they contain strong radio sources. While such luminous radio galaxies are essential for our current understanding the young universe, they have two drawbacks:
(1) They are intrinsically rare. There will never be enough in the whole sky to answer many statistical questions.
(2) They contain active galactic nuclei (AGN) whose radio luminosities exceed the bolometric luminosities of "normal" galaxies of stars. Their interstellar gas and stellar populations are probably altered by the AGN, through jet-induced star formation for example. Many could even be obscured and misaligned quasars.
A promising way to construct larger and more representative samples of galaxies in the young universe is to find high-redshift galaxies with significantly lower radio luminosities; that is, galaxies identified with faint radio sources. The nearby radio galaxy M87 is a good benchmark. Its bolometric luminosity is dominated by starlight, and the AGN does not seem to have affected its stellar population significantly. If moved to $z \approx 1$, M87 would have a flux density $S \approx 10\,\mathrm{mJy}$ at $\nu \approx 1\,\mathrm{GHz}$.

A new generation of radio surveys will soon cover the entire sky with greatly improved sensitivity, resolution, and position accuracy. These surveys should discover more than 10^6 high-redshift galaxies containing low-luminosity AGN plus nearly 10^5 "starburst" galaxies whose dominant energy sources are young stars and supernova remnants, not AGN. All-sky radio surveys already exist, as do very sensitive small-scale ($\Omega \ll 1\,\mathrm{sr}$) surveys, so we know what to expect from sensitive all-sky surveys. As I review the relevant characteristics of the extragalactic radio source populations and suggest some applications of deep all-sky surveys to studies of galaxies in the young universe, you should be asking yourself how you might use the new radio data for your own research. You don't have to wait until the observers have already exploited the data because the

* The National Radio Astronomomy Observatory is operated by Associated Universities, Inc., under cooperative agreement with the National Science Foundation.

largest of these surveys, the NRAO VLA Sky Survey (NVSS), is *your* survey. It is being made as a service to the international astronomical community. All of the (u, v)-data, sky images, and source catalogs are being released electronically via anonymous ftp as soon as they are produced. The NVSS group (J.J. Broderick, J.J. Condon, W.D. Cotton, E.W. Greisen, R.A. Perley, and Q.F. Yin) reserves *no* proprietary rights to these data products. To guarantee equal access for everyone, we use *only* the data in this open directory for our own research.

2 Extragalactic Radio Source Populations

Nearly all discrete radio sources more than a few degrees from the Galactic plane are extragalactic. The two energy sources responsible for their radio emission are true AGN and stars plus stellar remnants. Over 99% of sources strong enough to appear in published large-scale surveys are the classical radio galaxies and quasars containing AGN. The remaining strong sources, and the majority of sources fainter than about 1 mJy at $\nu = 1.4\,\mathrm{GHz}$, can be identified with star-forming galaxies. The contributions of these two populations to the radio source counts are shown in Fig. 1. As the new surveys approach $S \approx 1\,\mathrm{mJy}$ sensitivity, they will be the first to detect significant numbers of starburst galaxies at radio wavelengths. The tight far-infared (FIR) – radio correlation obeyed by "normal" and starburst spiral galaxies ensures that radio- and FIR-selected samples are almost identical. The median FIR/radio luminosity ratio of galaxies selected at $\lambda = 60\,\mu\mathrm{m}$ is $\log(S_{60\,\mu\mathrm{m}}/S_{1.4\,\mathrm{GHz}}) \approx 2.15$, so the NVSS should detect most of the galaxies above the $S_{60\,\mu\mathrm{m}} = 0.28\,\mathrm{Jy}$ completeness limit of the *IRAS* Faint Source Catalog, Version 2 (Moshir *et al.* 1992).

Cosmological evolution is so strong that the median redshift of radio AGN is $\langle z \rangle \approx 0.8$ for all $S < 1\,\mathrm{Jy}$. Going to lower flux densities primarily selects AGN with lower radio luminosities, not greater distances. Luminous radio sources such as Cyg A and 3C 273 would have to be at redshifts $z > 10$ to be as faint as 1 mJy at 1.4 GHz, and even M87 at $z \approx 2$ would be a 1 mJy source. The population of "starburst" galaxies also evolves, but these radio sources have lower luminosities and their redshifts do increase at lower flux densities (Fig. 2). Identifications of radio sources with $S \approx 1$ mJy should contain some of the most luminous starburst galaxies at cosmological redshifts. For example, the nearby starburst/Seyfert galaxy NGC 1068 would be 1 mJy source if moved to $z \approx 0.3$.

The two goals of detecting weak extragalactic sources and optically identifying them with faint galaxies at high redshifts impose conflicting requirements on radio surveys. The sensitivity limit of any radio survey is actually a *surface brightness* (mJy beam^{-1}), not a flux density (mJy). This distinction becomes important if the survey beam is not much larger than the sources being sought. The cumulative fraction $f(< \phi_\mathrm{M})$ of radio sources with angular diameters smaller than ϕ_M (Windhorst, Mathis, & Neuschaefer 1990) is shown in Fig. 3. Any survey made with a beam much smaller than the $\theta = 45''$ beam of the VLA D configuration at $\nu = 1.4\,\mathrm{GHz}$ will miss those extended sources whose integrated flux densities are above the survey limit but whose peak flux densities are not.

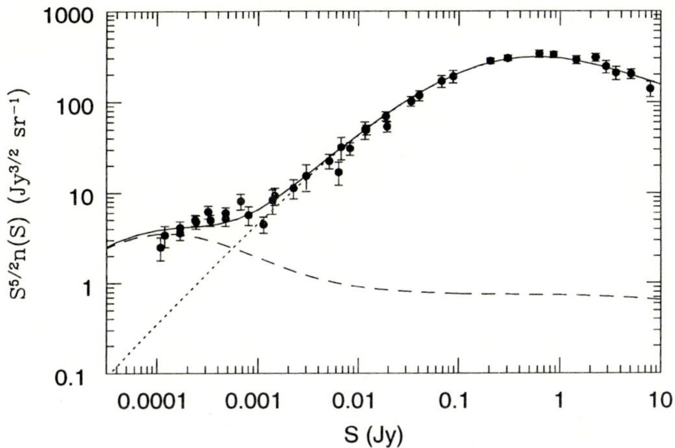

Fig. 1. The observed source counts (data points) at $\nu = 1.4\,\mathrm{GHz}$ are the sums of contributions from AGN (dashes) and star-forming galaxies (dots). *Abscissa:* 1.4 GHz flux density S (Jy). *Ordinate:* Differential number $n(S)$ of sources per steradian weighted by $S^{5/2}$ so that counts in a static Euclidean universe would lie on a horizontal line.

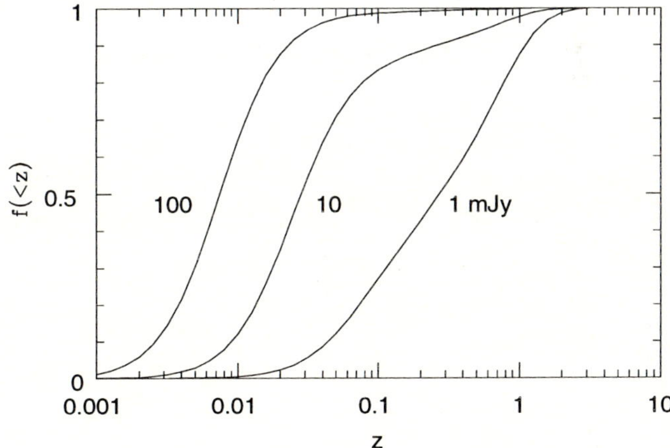

Fig. 2. The cumulative fraction $f(< z)$ of radio sources with redshifts $< z$ and powered by young stars (Condon 1984). The redshift range increases rapidly as the 1.4 GHz flux density decreases from 100 mJy (left curve) to 10 mJy (middle curve) and to 1 mJy (right curve). The new radio surveys should detect a significant number of luminous starburst galaxies at cosmological redshifts.

On the other hand, high resolution is required to produce the accurate radio positions necessary to identify radio sources with very faint galaxies. The rms uncertainty in each fitted coordinate of a weak radio source is limited by noise to $\sigma_p \approx \theta\sigma/(2S)$, where θ is the FHWM resolution and S/σ is the signal-to-noise ratio. The completeness C and reliability R of identifications of faint radio sources with optical objects brighter than $J \approx 22.5$ (the limit of the UK Schmidt and POSS 2 plates) vary with σ_P as shown in Fig. 4. The $\theta = 45''$ resolution needed for source detection is not sufficient for making complete and reliable identifications near the survey limit $S/\sigma \approx 5$. For this reason, *two* 1.4 GHz VLA sky surveys are being made, one with $45''$ resolution and the other with $5''$ resolution.

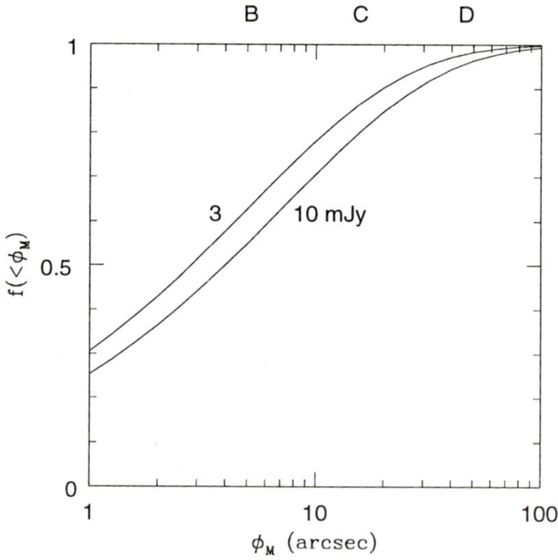

Fig. 3. The cumulative fraction $f(< \phi_M)$ of radio sources with angular size $< \phi_M$ is plotted for 1.4 GHz flux densities 3 and 10 mJy. The resolving powers of the VLA B-, C-, and D-configurations are indicated at the top.

3 A Survey of Surveys

3.1 Recent Large-Scale Surveys

Some of the largest radio surveys already in wide use are:

38 MHz: The Cambridge 8C survey covered $\Omega \approx 1$ sr north of $\delta = +60°$ with $270'' \times 270''$ cosecδ resolution and detected about 5×10^3 sources stronger than $S \approx 1$ Jy (Rees 1990).

151 MHz: The Cambridge 6C surveys have about $250''\times250''$ cosecδ resolution and cover most of the sky north of $\delta = +30°$, detecting sources as faint as 200 mJy (Baldwin *et al.* 1985; Hales *et al.* 1988, 1990, 1991, 1993).

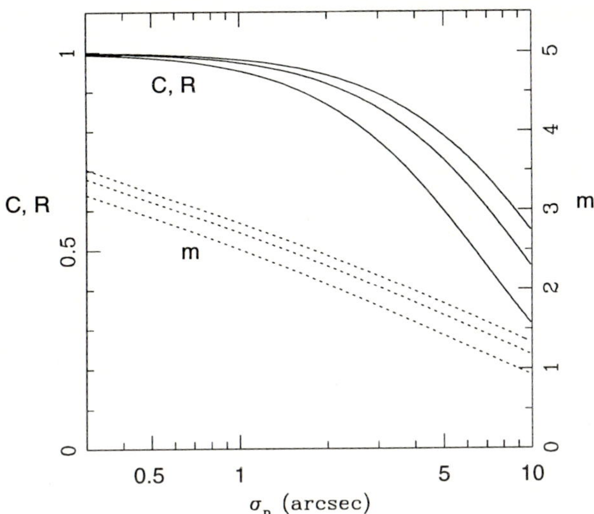

Fig. 4. The completeness C and reliability R of radio identifications with objects visible on the IIIaJ surveys ($J \leq 22.5$) are equal (solid curves) if the normalized search radius $m \equiv r_s/\sigma_p$ is varied as shown by the three dotted lines corresponding to $|b| = 90°$ (top), $|b| = 30°$, $|l| = 90°$ (middle), and $|b| = 30°$, $|l| = 45°$ (bottom). *Abscissa:* rms position uncertainty in each coordinate (arcsec). *Left ordinate:* Identification completeness and reliability. *Right ordinate:* Optimum ratio of search circle radius to rms position uncertainty (dimensionless).

151 MHz: The Cambridge 7C surveys have $70'' \times 70''$ cosecδ resolution and $\sigma \approx 20$ mJy rms noise (McGilchrist *et al.* 1990). This ongoing survey (cf. Visser *et al.* 1995) should eventually cover most of the sky north of $\delta = +20°$ and detect $\approx 10^5$ sources.

365 MHz: The UTRAO survey covers $\delta > -30°$ with a detection limit $S \approx 0.25$ Jy (Bash & Douglas 1974). The large unfilled aperture used to make this survey provides very accurate positions but misses some extended sources. Unpublished data from this survey may be obtained directly from the authors.

408 MHz: The Bologna B3 survey of the declination band $+37°15' < \delta < +47°37'$ was made with about $160'' \times 300''$ resolution and detected over 1.3×10^4 sources stronger than 0.1 Jy (Ficarra, Grueff, & Tomassetti 1985).

1.4 GHz: The Green Bank sky maps cover $-5° < \delta < +82°$ with $\approx 12'$ resolution (Condon & Broderick 1985, 1986). They are confusion limited, with rms confusion $\sigma \approx 25$ mJy. A catalog of $\approx 3 \times 10^4$ sources has been produced from these images (White & Becker 1992).

4.85 GHz: The NRAO seven-beam receiver was used on the (former) Green Bank 91 m telescope and on the Parkes 64 m telescope to survey most of the northern and southern hemispheres. Sky maps based on the 1987 Green Bank data cover $0° < \delta < +75°$ with $\approx 210''$ resolution and rms noise $\sigma \approx 5$ mJy (Condon, Broderick, & Seielstad 1989). Two lists of $\approx 5 \times 10^4$ sources were extracted from these images (Becker, White, & Edwards 1991; Gregory & Condon 1991). Improved images with $\sigma < 4$ mJy noise were made from the combined 1986 and 1987 Green Bank data (Condon *et al.* 1994a). Images with $5'$ resolution covering $-88° < \delta < -37°$ (Condon, Griffith, & Wright 1993) and $-29° < \delta < -9°$ (Tasker *et al.* 1994) have been published, and those for the $-10° < \delta < +10°$ zone are in preparation. Source lists have been extracted from the Parkes images (Gregory *et al.* 1994) and directly from the scan data (Griffith *et al.* 1994; Wright *et al.* 1994).

3.2 The New Surveys

The largest of the surveys planned or in progess are:

325 and 608 MHz: The two-part Westerbork Northern Sky Survey (WENSS) is in progess and will cover $\delta > +30°$ at $\lambda = 92$ cm with $55'' \times 55''$ cosecδ resolution and some portion of $\delta > +30°$, $b > 30°$ at $\lambda = 49$ cm with $30'' \times 30''$ cosecδ resolution (de Bruyn *et al.* 1994; Röttgering 1995). The 5σ detection limit is $S \approx 15$ mJy at both wavelengths. All four Stokes parameters (I, Q, U, and V) are being mapped. The scientific goals of the WENSS include finding sources with very steep spectra (distant radio galaxies, sources in rich clusters of galaxies, and pulsars), discovering new gravitational lenses, and studying peaked-spectrum sources.

843 MHz: A planned survey south of $\delta = -30°$ with the Molonglo Observatory Synthesis Telescope (MOST) will have $43'' \times 43''$ cosecδ resolution and detect about 4×10^5 sources with $S \geq 5$ mJy at 843 MHz (Large *et al.* 1994). It will complement the NVSS, completing the sky coverage with comparable sensitivity and resolution.

1.4 GHz: The VLA Faint Images of the Radio Sky at Twenty cm (FIRST) survey is mapping the north Galactic polar cap in total intensity (Becker *et al.* 1994). The $5''$ FWHM resolution of this B-configuration survey ensures that even the weakest detectable sources, $S \approx 1$ mJy beam^{-1}, will have position uncertainties $< 1''$, small enough for reliable optical identifications with the faintest galaxies detectable by the Sloan Digital Sky Survey (Gunn & Knapp 1993).

1.4 GHz: Observations for the NRAO VLA Sky Survey (NVSS) began in 1993 September and will cover the entire sky north of $\delta = -40°$ with $45''$ resolution by October 1996 (Condon *et al.* 1994b). The compact DnC and D configurations of the VLA are being used to image the sky in three Stokes parameters with maximum surface-brightness sensitivity ($\sigma \approx 0.45$ mJy beam$^{-1} \approx 0.14$ K for Stokes I and $\sigma \approx 0.29$ mJy beam$^{-1} \approx 0.09$ K for Stokes Q and U) and high photometric accuracy. The principal data product will be a set of 2326 $4° \times 4°$ continuum map "cubes" plus lists of discrete sources. The I polarization plane from the $4° \times 4°$ map cube centered on J2000 $\alpha = 22^{\rm h}\,30^{\rm m}$, $\delta = +84°$ is shown in Fig. 5. This survey alone should detect nearly 2×10^6 sources above ≈ 2.5 mJy and be sensitive to the low-brightness ($T_{\rm B} \approx 1$ K) disk emission of normal spiral galaxies. The rms position uncertainties are $\sigma_p < 0''.4$ for strong ($S \gg 25\,\mathrm{mJy}$) sources, increasing to $\approx 1''$ at $S = 10\,\mathrm{mJy}$ and $\approx 5''$ at the survey limit.

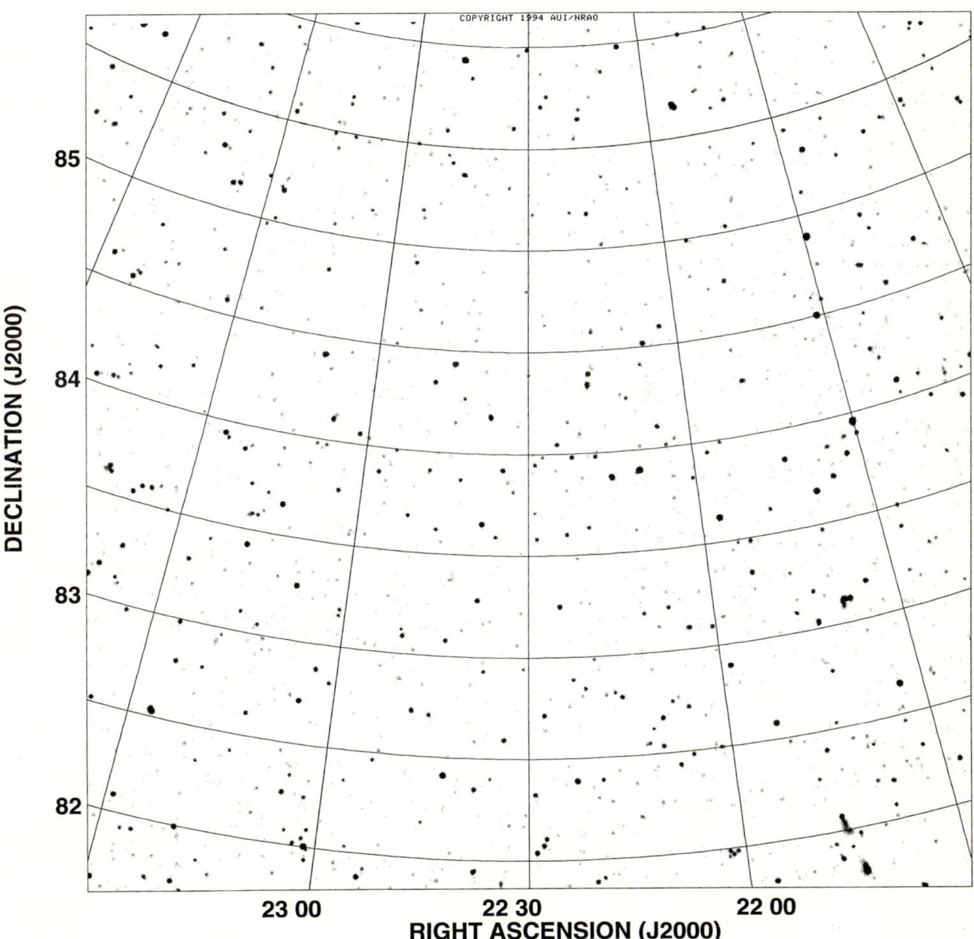

Fig. 5. Gray-scale representation of the $4° \times 4°$ I-polarization NVSS image centered on J2000 $\alpha = 22^{\mathrm{h}}\,30^{\mathrm{m}}$, $\delta = +84°$.

4 Radio Surveys and the Early Universe

The new radio surveys have many applications to studying the early universe. Some examples are:

(1) The first quasars appeared at $z > 5$. Absorption in the Lyα forest and shortward of the Lyman continuum cutoff reddens high-redshift quasars, making color surveys an effective method to find them. However, at very high redshifts quasars are not well separated in color from red galactic stars. Radio emission can be used to choose the best color-selected quasar candidates for follow-up optical spectroscopy (McMahon 1991).
(2) The strongest flat-spectrum radio sources are usually produced by optically luminous quasars, and many of the known quasars with $z > 3$ are the optical identifications of flat-spectrum radio sources. Accurate radio positions from the new surveys will also aid the discovery of quasars detected in other wavebands. For example, the majority of high-redshift X-ray sources found by ROSAT are flat-spectrum radio quasars (Brinkmann *et al.* 1995).
(3) The first galaxies were formed at redshifts $z > 4$. At any flux-density level, the most distant galaxies have the highest radio luminosities and, generally, the steepest radio spectra ($\alpha > 1$). Identifying steep-spectrum radio sources is still the best way to find galaxies with very high redshifts (cf. Lacy *et al.* 1994; Röttgering 1995).
(4) "Protogalaxies" are thought to undergo tremendous bursts of star formation which are quickly obscured by interstellar dust. The *IRAS* source F10214+4724 at $z \approx 2.3$ (Rowan-Robinson *et al.* 1991) may be the first example of such a galaxy. It obeys the FIR/radio correlation, so the 1.4 GHz VLA surveys should be able to detect others, if they exist, and provide the more accurate positions needed to identify FIR/radio sources with optically obscured galaxies. The new Green Bank 100 m Telescope (GBT) will have a very broadband spectrometer and be a powerful "redshift machine" for measuring redshifts of obscured starburst galaxies in the CO $J = 1 \rightarrow 0$ line.
(5) Surveys such as the NVSS that detect $\gg 10^5$ sources over most of the sky may be able to measure the motion of the Earth relative to the background of radio galaxies and quasars. The Earth's motion with velocity v induces a Doppler dipole anisotropy in the source counts with amplitude $[2 + (\gamma - 1)(1 + \alpha)](v/c)$, where $\gamma \approx 2$ is the differential source-count exponent and $\alpha \approx 0.7$ is the mean source spectral index (Ellis & Baldwin 1984).
(6) Fluctuations in the sky density of radio sources reflect the large-scale clustering of galaxies at redshifts $z \approx 0.8$. Since radio sources are spread over a Hubble distance along the line-of-sight, only the largest clusters would be detectable against statistical fluctuations in the numbers of foreground and background sources. However, the new surveys will find enough sources to measure clustering in the young universe or set interesting upper limits to it (Benn & Wall 1995).
(7) Gravitational lenses measure the distribution of mass along the lines-of-sight to distant quasars. High-resolution radio searches for gravitational lensing in samples of flat-spectrum sources are very effective at finding objects with galaxian masses and can be used to constrain the mass distribution of galaxies in the young universe (cf. Patnaik *et al.* 1993).

5 NVSS Astrophysical Results

At the time of this meeting, the published astrophysical results from the NVSS can be summarized in the box below.

The box is empty because the the first NVSS maps have just been produced. There is still plent of room in it for results from *your* work. I encourage you to take advantage of this opportunity. Detailed information about the NVSS is available on the WWW (via the NRAO master home page, URL address `http://www.nrao.edu`) or by anonymous ftp. Your first ftp session might be:

```
          ftp gibbon.cv.nrao.edu
   login: anonymous
password: your e-mail address
          cd pub/nvss
          get README
          quit
```

If you have any further questions about the survey, contact me at Internet address `jcondon@nrao.edu`.

References

Baldwin, J.E., Boysen, R.C., Hales, S.E. G., Jennings, J.E., Waggett, P.C., Warner, P.J., & Wilson, D.M.A. 1985, *Mon. Not. R. Astr. Soc.* **217**, 717

Bash, F.N., & Douglas, J.N. 1974, *Astron. Astrophys. Suppl.* **15**, 457

Becker, R.H., White, R.L., & Edwards, A.L. 1991, *Astrophys. J. Suppl.* **75**, 1

Becker, R.H., White, R.L., & Helfand, D.J. 1994, in *Astronomical Data Analysis Software and Systems III*, ASP Conference Series **61**, 165

Benn, C.R., & Wall, J.V. 1995, *Mon. Not. R. Astr. Soc.* **272**, 678

Brinkmann, W., Siebert, J., Reich, W., Fürst, E., Reich, P., Voges, W., Trümper, J., & Wielebinski, R. 1995, *Astron. Astrophys. Suppl.* **109**, 147

Condon, J.J. 1984, *Astrophys. J.* **287**, 461

Condon, J.J., & Broderick, J.J. 1985, *Astron. J.* **90**, 2540

Condon, J.J., & Broderick, J.J. 1986, *Astron. J.* **91**, 1051

Condon, J.J., Broderick, J.J., & Seielstad, G.A. 1989, *Astron. J.* **97**, 1064

Condon, J.J., Broderick, J.J., Seielstad, G.A., Douglas, K., & Gregory, P.C. 1994a, *Astron. J.* **107**, 1829

Condon, J.J., Cotton, W.D., Greisen, E.W., Yin, Q.F., Perley, R.A., & Broderick, J.J. 1994b, in Astronomical Data Analysis Software and Systems III, ASP Conference Series **61**, 155

Condon, J.J., Griffith, M.R., & Wright, A.E. 1993, *Astron. J.* **106**, 1095
de Bruyn, G., Miley, G., Tang, Y., Bremer, M., Brouw, W., Rengelink, R., Bremer, M., & Röttgering, H. 1994, NFRA Newsletter, extra issue, 4
Ellis, G.F.R., & Baldwin, J.E. 1984, *Mon. Not. R. Astr. Soc.* **206**, 377
Ficarra, A., Grueff, G., & Tomassetti, G. 1985, *Astron. Astrophys. Suppl.* **59**, 255
Gregory, P.C., & Condon, J.J. 1991, *Astrophys. J. Suppl.* **75**, 1011
Gregory, P.C., Vavasour, J.D., Scott, W.K., & Condon, J.J. 1994, *Astrophys. J. Suppl.* **90**, 173
Griffith, M.R., Wright, A.E., Burke, B.F., & Ekers, R.D. 1994, *Astrophys. J. Suppl.* **90**, 179
Gunn, J.E., & Knapp, G.R. 1993, in *Sky Surveys: Protostars to Protogalaxies*, ASP Conference Series, **43**, 267
Hales, S.E.G., Baldwin, J.E., & Warner, P.J . 1988, *Mon. Not. R. Astr. Soc.* **234**, 919
Hales, S.E.G., Masson, C.R., Warner, P.J., & Baldwin, J.E. 1990, *Mon. Not. R. Astr. Soc.* **246**, 256
Hales, S.E.G., Masson, C.R., Warner, P.J., Baldwin, J.E., & Green, D.A. 1993, *Mon. Not. R. Astr. Soc.* **262**, 1057
Hales, S.E.G., Mayer, C.J., Warner, P.J., & Baldwin, J.E. 1991, *Mon. Not. R. Astr. Soc.* **251**, 46
Lacy, M., *et al.* 1994, *Mon. Not. R. Astr. Soc.* **271**, 504
Large, M.I., Campbell-Wilson, D., Cram, L.E., Davison, R.G., & Robertson, J.G. 1994, PASAu, **11**, 44
McGilchrist, M.M., Baldwin, J.E., Riley, J.M., Titterington, D.J., Waldram, E.M., & Warner, P.J. 1990, *Mon. Not. R. Astr. Soc.* **246**, 110
McMahon, R. 1991, in *The Space Distribution of Quasars*, ASP Conference Series, **21**, 129
Moshir, M., *et al.* 1992, Explanatory Supplement to the *IRAS Faint Source Survey*, Version 2, JPL D-10015 8/92 (Jet Propulsion Laboratory: Pasadena)
Patnaik, A.R., Browne, I.W.A., King, L.J., Muxlow, T.W.B., Walsh, D., & Wilkinson, P.N. 1993, *Mon. Not. R. Astr. Soc.* **261**, 435
Rees, N. 1990, *Mon. Not. R. Astr. Soc.* **244**, 233
Röttgering, H. 1995, this volume p.97
Rowan-Robinson, M., *et al.* 1991, Nature, **351**, 719
Tasker, N.J., Condon, J.J., Wright, A.E., & Griffith, M.R. 1994, AJ, **107**, 2115
Visser, A.E., Riley, J.M., Röttgering, H.J.A., & Waldram, E.M. 1995, *Astron. Astrophys. Suppl.* **110**, 419
White, R.L., & Becker, R.H. 1992, *Astrophys. J. Suppl.* **79**, 331
Windhorst, R.A., Mathis, D.F., & Neuschaefer, L.W. 1990, in *Evolution of the Universe of Galaxies*, ASP Conference Series, **10**, 389
Wright, A.E., Griffith, M.R., Burke, B.F., & Ekers, R.D. 1994, *Astrophys. J. Suppl.* **91**, 111

High Redshift Galaxies: Problems and Prospects

George Efstathiou

Department of Physics, University of Oxford, Keble Road, Oxford OX1 3RH, England

1 Introduction

The many new results presented at this meeting show clearly how our knowledge of the high redshift Universe has improved over the last few years. It is interesting to contrast the present status of the field to that in 1988, as summarized in the NATO workshop conference *The Epoch of Galaxy Formation* (Frenk *et al.* 1989). At that meeting, there was much debate about the epoch of galaxy formation and a fair amount of skepticism concerning the low formation redshifts predicted by some theorists. The faint blue galaxies at $B \gtrsim 26$ had only just been discovered and only a few preliminary results from the first multiobject deep redshift surveys were mentioned. A few tentative results on the evolution of the quasar luminosity function at $z > 3$ were presented at the meeting, and new observations showing that damped Lyα systems contain a mass in neutral hydrogen comparable to that in stars at the present epoch seemed problematic to many participants. Evidently, tremendous advances have been made since 1988 and, as the impressive results from Keck have shown, there is every reason to expect this progess to continue as more 8-10m class telescopes become operational.

In this summary, I review the observations from a theoretical perspective. As a narrow minded theorist, I have no doubt misinterpreted the data to fit my favourite theoretical model, particularly in Section 4. I hope that readers will be tolerant of my speculations.

2 Galaxy Formation in Hierarchical Clustering Theories

Figure 1 shows an updated version of a diagram from Efstathiou & Rees (1988). It shows the comoving number density of dark matter haloes with mean overdensity $\Delta \approx 200$ and mass $> M$ plotted as a function of redshift. These curves were derived using the Press-Schechter (1974) theory assuming the power spectrum of the standard cold dark matter (CDM) model (*i.e.* scale invariant fluctuations,

$\Omega = 1$ and $h = 0.5)^1$. The present *rms* amplitude of the mass fluctuations in spheres of radius $8h^{-1}$Mpc has been chosen to be $\sigma_8 = 0.59$, corresponding to bias factor of $b = 1/\sigma_8 = 1.7$. The normalization in Figure 1 is higher than that adopted by Efstathiou & Rees, in line with current fashion (see *e.g.* Strauss & Willick 1995) but the calculation is identical apart from this change.

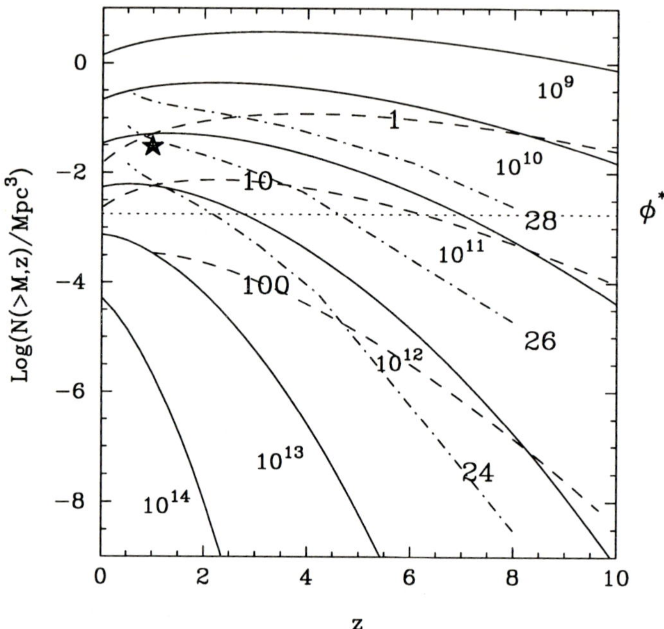

Fig. 1. The solid curves show the comoving number density of haloes with mass $> M$ plotted against redshift for the standard CDM model normalized as described in the text. The mass limits (in $M_\odot$) are listed next to each curve. The dashed lines show locii at which the fiducial star formation rate given by equation (1) is equal to 100, 10 and $1 M_\odot$/yr. The dotted lines show locii at which haloes undergoing star formation at the fiducial rate have blue magnitudes of 24, 26 and 28. The dotted line labelled ϕ^* shows the local space density of L^* galaxies, and the star shows the approximate comoving space density of $B \sim 26$ galaxies if they are located at $z \sim 1$.

Figure 1 can be used to illustrate some key features of galaxy formation in hierarchical theories, albeit in a simplified form. We make the following remarks:

[1] The specific CDM model in Figure 1 fails to match the amplitude of the microwave background anisotropies and observations of large-scale structure in

[1] Where Ω is the ratio of the mean mass density divided by the critical density of the Einstein de-Sitter model and h is Hubble's constant in units of 100 km s^{-1}Mpc^{-1}. Unless otherwise stated all densities and masses in this paper are computed assuming $h = 0.5$.

the Universe (see *e.g.* Efstathiou, Bond & White 1992). However, the *general* features of Figure 1 will apply to any CDM-like model, *e.g.* mixed (cold+hot) dark matter models (Klypin *et al.* 1993). The reader should be aware that certain numbers, *e.g.* the space densities of of high mass haloes, are extremely sensitive to the normalization and the shape of the power spectrum of density irregularities on small scales.

[2] The dotted line labelled ϕ^* shows the space density of luminous (L^*) galaxies measured in nearby redshift surveys. Adopting a baryonic density of $\Omega_b \approx 0.05$, as suggested by primordial nucleosynthesis if $h \approx 0.5$ (Walker *et al.* 1991), it seems reasonable to associate each L^* galaxy with a halo of mass $M_H \sim 2.5 \times 10^{12} M_\odot$. Figure 1, therefore implies that most bright galaxies formed at relatively low redshifts $z \sim 3$. The low redshift of galaxy formation is a generic feature of CDM-like models in which the power spectrum assymptotically approaches $P(k) \propto k^{-3}$ on small scales, though the characteristic formation redshift is sensitive to the parameters of the model (see *e.g.* Heyl *et al.* 1995). Observations of galaxy formation and evolution are therefore capable of setting strong constraints on theoretical models.

[3] The comoving space densities of low mass haloes, $M_H \lesssim 10^{11} M_\odot$ are predicted to be much higher than the present day space density of L^* galaxies. If the power spectrum of the mass fluctuations is approximated by a power law, $P(k) \propto k^n$, then according to the Press-Schechter model $N(\geq M_H) \propto M_H^{-1-(3+n)/6}$ at low masses. Thus, in any hierarchical theory of galaxy formation, the cumulative number density of low mass haloes is predicted to increase steeply with decreasing mass. In contrast, the cumulative number density of low luminosity galaxies in local redshift surveys is observed to increase logarithmically with decreasing luminosity ($N(> L) = \int \phi(L) dL$, $\phi(L) \propto L^\alpha$ at $L \ll L^*$, with $\alpha \approx -1$, *e.g.* Loveday *et al.* 1992; but see Marzke *et al.* 1995, and Section 3.2 below).

[4] CDM-like models predict such a large resevoir of low mass haloes that only a small fraction need be the sites of star formation to dominate galaxy counts at faint magnitudes. To see this, it is useful to define a fiducial star formation rate, $\dot{M}_f$, given by the mass in baryons within a halo of mass M_H divided by the age of the Universe at redshift z,

$$\dot{M}_f = 1.0 \left(\frac{\Omega_b}{0.05}\right) \left(\frac{M_H}{10^{11} M_\odot}\right) \left[\frac{(1+z)}{2}\right]^{3/2} M_\odot \mathrm{yr}^{-1} \,. \tag{1}$$

If the star formation rate is much lower than $\dot{M}_f$, then there must be a mechanism that either prevents gas from collapsing to a high density in the first place (*e.g.* inability of the gas to cool) or inhibits star formation once the gas has collapsed. A star formation rate much larger than $\dot{M}_f$ implies that star formation occurs in bursts. The three dashed lines in Figure 1 show the locii at which the fiducial star formation rate is equal to 1, 10 and $100 \dot{M}_\odot/\mathrm{yr}$.

The B-band magnitude of a galaxy at $z \gtrsim 1$ is dominated by the light from massive, short lived, stars and so is fixed approximately by the instantaneous star formation rate. For a normal stellar initial mass faction,

$$B \approx 28.2 - 2.5\log_{10}\dot{M} + 5\log_{10}[(1+z)^{1/2} - 1] \quad (2)$$

(White & Frenk 1991). The dot-dashed lines in Figure 1 show the locii at which haloes undergoing star formation at the fiducial rate (1) have B magnitudes of approximately $B = 24$, 26 and 28. The faint galaxy counts of Tyson (1988), Cowie *et al.* (1988), Metcalfe *et al.* (1995) and others, give a surface density of $\mathcal{N} \approx 4.3 \times 10^8$ galaxies per steradian at $B = 26$. The star in Figure 1 shows the implied space density of these galaxies in an $\Omega = 1$ universe if their typical redshift is $z \sim 1$ (as suggested by the gravitational lensing of faint arclets, *e.g.* Smail *et al.* 1994). We can see from this figure that the faint counts can be explained if $\sim 10^{11} M_\odot$ haloes are undergoing star formation at the fiducial rate (1), or if a proportion of lower mass haloes are undergoing bursts of star formation. Evidently, large numbers of faint galaxies seem quite plausible in CDM-like models, because of the high space densities of low mass haloes (see *e.g.* White & Frenk 1991, Lacey *et al.* 1993).

Figure 1 raises a number of important problems which I discuss in the rest of this summary.

- The predicted space densities of low mass ($M \lesssim 10^{11} M_\odot$) haloes are high even at redshifts $z \sim 10$. Are there any observational consequences?
- We have a rough idea of the redshifts of the faint blue galaxies. What more can we learn? Can we test the idea that they are associated with low mass haloes?
- What has happened to the faint blue galaxies?
- Figure 1 implies a large amount of evolution at low redshifts, especially in the space density of galaxy clusters. How can we test this?
- What are the most promising ways of finding 'normal' galaxies at high redshifts?

3 Observations

3.1 Low Mass Haloes at High Redshifts

Can we detect the large numbers of haloes at high redshifts predicted in hierarchical clustering theories? Their typical radii and circular speeds are

$$r_h \approx 120 M_{11}^{1/3}(1+z)^{-1} h_{50}^{-2/3} \left(\frac{\Delta}{200}\right)^{-1/3} \text{kpc} \quad (3a)$$

$$v_c \approx 60 M_{11}^{1/3}(1+z)^{1/2} h_{50}^{1/3} \left(\frac{\Delta}{200}\right)^{1/6} \text{km/s} \quad (3b)$$

where M_{11} is the mass of a halo in units of $10^{11} M_\odot$ and h_{50} is H_0 in units of 50km s^{-1}Mpc^{-1}. Evidently according to Figure 1, the first generation of objects will collapse at high redshifts ($z > 10$) and will have low masses and circular speeds. Star formation and nuclear activity can then photoionize the intergalactic medium (IGM), raising the temperature to $T \sim 10^4 K$ (Couchman

& Rees 1986, Shapiro, Giroux & Babul 1994). The sound speed of the IGM is $c_s \approx 12(T/10^4K)^{1/2}$km/s, comparable to the circular speeds of low mass haloes. Photoionization of the IGM therefore provides a feedback mechanism to suppress the collapse of low mass haloes. However, the highly ionized diffuse gas in these haloes will be visible as Lyα absorption lines in quasar spectra (Rees 1986). In fact, the typical sizes and abundances of low mass haloes are of about the right order of magnitude to explain the observed numbers of lines (Mo *et al.* 1993). Hydrodynamical simulations support this general picture (Cen *et al.* 1994), and show that typical Lyα clouds are elongated, non-equilibrium, structures (see also Figure 2). The detailed numerical modelling of Lyα clouds is likely to lead to strong constraints on the intensity and perhaps spectrum of the photoionizing background at high redshifts. On the observational side there remain many interesting questions capable of testing this picture. For example:

• Is there significant differential evolution in the numbers of Lyα lines as a function of equivalent width?

• Are the Lyα clouds clustered? Are there any trends of the clustering properties with equivalent width?

• How well are the lines fitted by Voigt profiles?

• Are there 'proximity' effects in the line widths as well as the line numbers (Miralda-Escudé & Rees 1994)?

• Are the statistics of coincidences between clouds seen along neighbouring sight lines in gravitationally lensed quasars consistent with highly elongated structures?

On the theoretical side, it should be possible to improve the hydrodynamic modelling of Lyα clouds to the point where astrophysical uncertainties are likely to limit the reliability of the predictions. For example, when and how is the IGM heated? What are the effects of local heating, *i.e.* energy injection from star formation or nuclear activity in the cores of clouds?

The high density core regions of Lyα clouds are likely to be the sites of star formation, as are haloes with circular speeds $v_c \gg c_s$. What factors determine the characteristic redshifts and luminosities of the faint blue galaxies seen in deep CCD images? One explanation might be the steep decline in the intensity of the photoinizing background radiation (Babul & Rees 1992). The proximity effect indicates a value of the photoionizing flux at the Lyman limit of $J \sim 10^{-21}$ergcm^{-2}s^{-1}Hz^{-1} at $z > 2$ and this could fall by as much as a factor of 100 between $z = 2$ and the present epoch as the number density of quasars and other photoionizing sources declines sharply. Radiative cooling will become more effective in low mass haloes as the photoionizing flux declines (Efstathiou 1992) and this could lead to the collapse of gas and to increased rates of star formation. The position of the star in Figure 1 shows that a modest rate of star formation $\sim 1M_\odot$/yr in $10^{11}M_\odot$ haloes at $z \sim 1$ could acount for the observed surface density of faint galaxies.

Another, possibly related problem, concerns the density of neutral hydrogen in damped Lyα systems. Figure 3 shows Ω_b in damped Lyα systems as a function of redshift (Lanzetta *et al.* 1993). As noted by a number of authors, these observations provide important constraints on CDM-like models (Mo & Miralda-

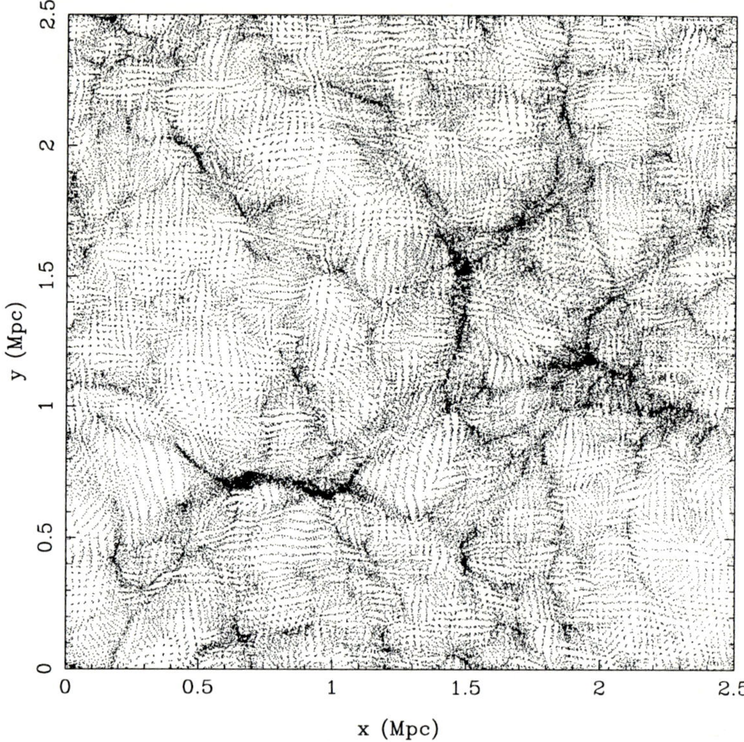

Fig. 2. A slice 125kpc deep through a CDM universe at a redshift $z = 3$. The CDM model is normalized as described in the text. Each particle has a mass of $3.4 \times 10^7 M_\odot$ and the axes give physical dimensions (*i.e* proper distances). Notice the highly elongated filamentary structure.

Escudé 1994, Kauffmann & Charlot 1994). The three lines in Figure 3 show the fraction of the baryonic material associated with virialized haloes with circular speeds $v_c > 100$, 200 and 300km/s (computed as in Figure 1 and assigning a baryonic mass $M_b = \Omega_b M_H$ to each halo). The observations at $z \geq 3$ are most easily explained if the damped Lyα systems are associated with small galaxies, rather than the discs of normal ($v_c \gtrsim 200$km/s) spiral galaxies. This conclusion applies even more forcefully, if dust obscuration in damped Lyα systems introduces a significant bias against finding damped Lyα systems in the spectra of redenned quasars (Fall & Pei 1995, and see the dashed line in Figure 3). Recent observations suggest that the damped Lyα systems have low metallicities (Pettini *et al.* 1994) but there are few constraints on their sizes and circular speeds (see Section 3.4).

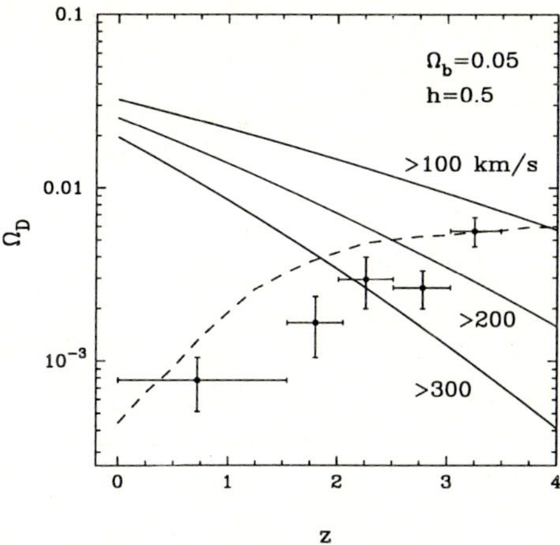

Fig. 3. The points show the baryon density in damped Lyman α systems inferred from the observations of Lanzetta *et al.* (1993) plotted as a function of redshift. The three lines show the baryon density associated with haloes of circular speeds > 100, 200 and 300km/s for the standard CDM model normalized to $\sigma_8 = 0.59$. The dashed line shows an estimate of Ω_{HI} from a model of Fall and Pei (1995) that corrects for biases in the selection of damped Lyα systems caused by dust.

3.2 The Nature of the Faint Galaxies

The previous paragraph described one model for the faint galaxies but many others have been proposed, ranging from the idea that they are subunits that merge to make up ordinary luminous galaxies (Broadhurst, Ellis & Glazebrook 1992) to the suggestion by Gronwall & Koo (1995) that the degree of evolution to $B \sim 26$ has been greatly overestimated because of uncertainties in the local luminosity function and cosmological model. It is useful to summarize some of the recent observations:

• The B-band counts continue to rise to $B \sim 27.5$, though there is tentative evidence for a flattening in the counts at $B \gtrsim 26$ (Metcalfe *et al.* 1995).

• Redshift surveys to $B \approx 24$ show that the median redshift is low ($z \approx 0.46$) implying an increase in the number density of $\lesssim L^*$ galaxies (Glazebrook *et al.* 1995a). In fact, new results from the I-band limited CFRS redshift survey (Lilly *et al.* 1995) and the $B \leq 24$ surveys of Ellis, Colless and collaborators (Colless, 1995) show an increase in the number density of blue galaxies over the redshift range $z \sim 0.2$–1, with some indications of a steepening of the faint end slope of the luminosity function. In contrast, the luminosity function of red galaxies (redder than present day Sbc galaxies) hardly changes over this redshift range

• The mean [O II] equivalent width in B-selected redshift surveys increases with redshift, indicating higher rates of star formation in the past (Colless *et al.* 1993).

• K-band counts to K=24 are consistent with modest evolution if $q_0 = 0.5$ and are only slightly higher than the predictions of no evolution models if $q_0 = 0$ (Djorgovski *et al.* 1995). K-band limited redshift surveys show that there is little evolution in the shape of the K-band luminosity function to redshifts $z \approx 0.5$

(Glazebrook *et al.* 1995b, Cowie *et al.* 1994, though these authors find different normalizations).

• The optical morphologies of $I \leq 22$ galaxies observed with HST show that the excess blue galaxies have the morphologies of late-type/irregular or peculiar galaxies (Glazebrook *et al.* 1995c).

• Weak gravitational lensing of faint galaxies by foreground clusters suggests that most of the galaxies at $B \sim 26$ are at redshifts $z \sim 1$ (Smail, Ellis & Fitchett 1994). This is broadly in accord with spectroscopic redshifts measured for a small number of faint arclets (Soucail 1992).

• The faint blue galaxies seem to be more weakly clustered than present day L^* galaxies (*e.g.* Efstathiou *et al.* 1991, Neuschaefer & Windhorst 1991, 1995).

The detection of a change in the shape of the luminosity function with redshift provides direct evidence that evolution is taking place. The redshift surveys suggest that most normal $\sim L^*$ galaxies are in place by $z \sim 1$ and evolve modestly by the present day. The evolving galaxy population (at least to $I \sim 22$) has blue colours, late-type morphologies and spectral characteristics consistent with the idea that they are sub-L^* galaxies undergoing enhanced star formation rates. These observations seem difficult to reconcile with the 'no evolution' models advocated by Gronwall & Koo (1995).

The deep redshift surveys have revealed another interesting problem. Even at $z \sim 0.2$, the normalizations of the luminosity functions determined from the CFRS and Colless *et al.* redshift surveys are about twice the value measured from nearby surveys (*e.g.* Loveday *et al.* 1992). A similar high normalization is determined from the K-band redshift survey of Glazebrook *et al.* (1995b) and from the luminosity function of Mg II absorber-selected galaxies (Steidel, Dickinson & Persson, 1995). Furthermore a high normalization is required to explain the counts of 'normal' galaxies in the HST images of Glazebrook *et al.* (1995c). This discrepancy is not understood at present, but it seems unlikely that evolution is responsible since this would require a doubling in the comoving space densities of galaxies of all luminosities, colours and morphological types from $z = 0$ and $z = 0.2$, which seems implausible. It is more likely that the discrepancy is the result of a surface brightness selection effect caused by the high isophotal threshold used to identify nearby galaxies on photographic plates (see *e.g.* Ferguson & McGaugh 1995).

We make two further points here. Firstly, the faint redshift and HST surveys have succeeded in finding a rapidly evolving $z \lesssim 1$ blue population at $I \lesssim 22$, but it may be dangerous to extrapolate these trends to fainter magnitudes. For example, Figure 1 shows that a large proportion of the faint galaxies at $B \sim 28$ could, conceivably, lie at redshifts $z \sim 4$. This could be tested by searching for a population of ultra-faint objects with Lyman continuum breaks in the B-band. Secondly, the fate of these faint blue galaxies is still unknown. It seems unlikely on dynamical grounds that they merge with high efficiency into present day L^* galaxies (and the weak clustering of these objects argues against this). It is more plausible that they fade and become low luminosity, low surface brightness galaxies. Recently, Marzke *et al.* (1995) have estimated the luminosity function of Sm/Im galaxies in the CfA redshift surveys and find a steep faint end slope

($\alpha \approx -1.87$). The space density of these galaxies exceeds that of the rest of the galaxy population at luminosities $L \lesssim 0.05L^*$ and it is conceivable that they are related to the faint blue galaxies seen at higher redshifts.

3.3 The Evolution of Clusters of Galaxies

One of the most striking predictions of CDM-like models is the rapid evolution of rich clusters of galaxies (*e.g.* Frenk *et al.* 1990). For example, in Figure 1 the comoving space density of $> 10^{14}M_\odot$ virialized systems declines by about two orders of magnitude between $z = 0$ and $z \sim 1.5$. Nevertheless, this effect is difficult to detect because it requires complete samples of high redshift clusters with well defined selection criteria (preferably based on X-ray temperature or velocity dispersion). Strong evidence for a decline in the space density of rich clusters to redshifts $z \sim 0.3$ comes from X-ray flux limited surveys (Gioia *et al.* 1990, Henry *et al.* 1992). However, it is difficult to predict the evolution of the X-ray luminosity function theoretically because the emission is sensitive to the central gas density and hence on the baryon density and thermodynamic state of the gas.

Optical surveys (*e.g.* Gunn *et al.* 1986, Couch *et al.* 1991) have revealed some examples of rich clusters with redshifts $z \gtrsim 0.5$, but it is not straightfoward to compare these observations with theoretical predictions because of the possibility of contamination by groups and clusters seen in projection, and because the masses are unknown. In an interesting investigation, Castander *et al.* (1994) have made deep ROSAT observations of 5 optically selected clusters with $z \sim 0.7$–0.9. They succeeded in detecting two clusters, but their X-ray luminosities are much lower than expected for present day clusters of similar optical richnesses. This is consistent with a decline in the comoving space density of X-ray luminous clusters beyond $z = 0.3$.

There have been several claims of clusters and superclusters of galaxies at high redshift. Two recent examples are the detection of a $z \approx 1$ cluster of galaxies in the CFRS survey (possibly associated with a QSO) with velocity dispersion of 955km/s (Le Fèvre *et al.* 1994) and the possible detection of a cluster of galaxies at $z = 3.4$ associated with a damped Lyα system (Giavalisco *et al.* 1994).

The construction of complete samples of rich clusters at high redshift remains a formidable challenge. One promising technique is to use K-band imaging, since distant clusters should appear at higher contrasts with respect to the background in the near-IR. Dickinson *et al.* (1995) find several density enhancements in K-band images of 3CR galaxies that may be rich clusters at $z > 1$. With the availability of large format near-IR detectors it may even become possible to survey a large enough random area of sky to identify distant clusters in useful numbers.

3.4 Galaxies at High Redshift

High redshift radio galaxies: Almost all galaxies with redshifts $z > 2$ have been discovered because of their extreme luminosities at long radio wavelengths.

In fact, the highest redshift galaxy known, at $z = 4.25$, is associated with the radio source 8C 1435+635 (Lacy *et al.* 1994). At first, it was thought that these high redshift radio galaxies might be 'passively' evolving giant elliptical galaxies, similar to their low redshift counterparts (*e.g.* Lilly & Longair 1984). However, it is now clear that these objects are much more complex. Many $z > 1$ radio galaxies display complex elongated structures that are aligned with the radio structure. The origin of these alignments remains unclear. Jet induced star formation is one possibility, though the detection of significant polarization in some objects suggests that electron or dust scattering of AGN light is involved at some level (see McCarthy 1993, and references therein). HST images may help unravel the causes of the aligned optical and radio morphologies (*e.g.* Miley *et al.* 1992). Recently, Eales and Rawlings (1993) have shown that emission lines powered by the AGN make substantial contributions to the K-band fluxes of many $z > 2$ systems. A dust-redenned quasar nucleus can also mimic the colours of an old stellar population, as in 3C22 (Rawlings *et al.* 1995). Correcting for the emission line contribution, Eales and Rawlings find flat spectral energy distributions for several galaxies, as expected for protogalaxies. Is there evidence for an evolved stellar population in high redshift radio galaxies? Evidently, broad band colours can be misleading and high signal to noise spectra, from which the emission lines can be subtracted, are required to reveal an underlying old stellar continuum. For example, Lacy *et al.* (1995) and Stockton *et al.* (1995) find evidence for a 4000Å break in the spectrum of 3C65 at $z = 1.175$, suggesting the presence of an old stellar population of age $\sim 3 - 4$Gyr. Similar observations of higher redshift galaxies are clearly required to establish the fraction of light attributable to old stars.

Emission line surveys: Lyα emission line surveys of blank fields of sky have so far proved unsuccessful (*e.g.* Djorgovski *et al.* 1993). These surveys reach sensitivity levels of $\sim 10^{-17}$ergs^{-1}cm^{-2}arcsec^{-2}, which is about the Lyα luminosity expected at $z \sim 3$ of a dust free Milky Way sized system making stars at rate of $\sim 1 M_\odot$/yr. The most plausible explanation for the failure of these surveys is that the Lyα emission from star forming galaxies is severly attenuated by dust extinction; even a gas to dust ratio of a few percent of that measured in the Milky Way can suppress the Lyα flux by between one and two orders of magnitude (Charlot & Fall 1991).

Narrow band Hα (6563Å) surveys in the near-IR seem much more promising. Hα lies in the K-band in the redshift range $2.08 < z < 2.66$. It is much less affected by dust than Lyα and provides a more direct estimate of the star formation rate (Kennicutt 1993). Two preliminary projects have been undertaken so far. Thompson *et al.* (1994) sample 0.75 arcmin^2 to a narrow band flux limit of $\sim 10^{-15}$ergcm^{2}s^{-1}, but find no candidate protogalaxies. Bunker *et al.* (1995) have conducted a deeper survey in the field of the quasar PHL957. They succeeded in detecting a galaxy[2] at $z = 2.313$ that lies near to a damped Lyα

[2] One of a very few galaxies detected in Lyα emission line surveys of damped Lyα systems, Lowenthal *et al.* 1991

system at $z = 2.309$ and infer a star formation rate of $\approx 18h^{-2}M_\odot/\mathrm{yr}$. These studies have demonstrated that narrow band imaging in the near-IR is capable of reaching interesting sensitivity levels. The crude arguments of Section 2 suggest that more sensitive Hα surveys may well find copious numbers of galaxies at $z > 2$.

Damped Lyα and Mg II absorber galaxies: We do not yet have a clear picture of the nature of damped Lyα systems and in my view the idea that they are large spiral discs at an early stage of evolution rests on a shaky foundation. Unfortunately, very few of these systems have been seen in Lyα emission. There is evidence that some of these systems have sizes of $\sim 10h^{-1}$kpc. In the best studied case, Briggs *et al.* (1989) deduce a size of $\gtrsim 8h^{-1}$ for the $z = 2.04$ system in PKS0458-020 by comparing the 21cm absorption line profiles from the extended and compact radio structure. The narrow line profiles suggest that we may be seeing a large disc galaxy face on, but it is possible that similar structure could arise from a sheet of neutral gas in a low velocity dispersion halo. Steidel & Hamilton (1992) suggest that a faint galaxy detected within 2.8 arcsec of Q0000-263 may be responsible for the damped Lyα system at $z = 3.390$, implying a size of $\sim 10h^{-1}$kpc. At lower redshifts, Steidel *et al.* (1994, 1995b) identify a damped Lyα system at $z = 0.868$ with a low luminosity galaxy ($\approx 0.2L^*$) and another at $z = 0.692$ with a luminous ($\approx 0.8L^*$) but low surface brightness galaxy. Both objects thus differ from 'normal' disc galaxies. One particularly exciting result is the discovery of Lyα emission from three galaxies associated with the $z = 2.811$ damped Lyα system seen in the spectrum of PKS0528-250 (Møller & Warren 1993, Warren & Møller 1995). The redshifts of all three systems have been confirmed spectroscopically and they are separated by $\sim 100h^{-1}$kpc, with the nearest lying just 1.22arcsec from the quasar, implying a size of $\sim 4h^{-1}$kpc and an H I mass of $\sim 10^9M_\odot$ if it is responsible for the damped Lyα emission. ¿From the continuum fluxes they deduce star formation rates of $\sim 1h^{-2}M_\odot/yr$ for the two more distant blobs, consistent with the star formation rates required to produce the Lyα emission line fluxes, indicating a low gas-to-dust ratio. Furthermore, the velocity difference between the emission line and absorber redshifts indicates a relatively low dynamical mass of $\sim 4 \times 10^{10}h^{-1}M_\odot$ for the damped Lyα galaxy. Warren & Møller conclude that this system resembles a small group of sub-L^* galaxies rather than a large disc galaxy. They also draw attention to the alignment of the three galaxies and to the similarity with the filamentary structure seen in N-body simulations (*cf* Figure 2).

The possibility of studying the chemical evolution of protogalaxies to $z > 3$ is another exciting prospect. In an important paper, Pettini *et al.* (1994) have measured metallicities for 17 damped Lyα systems in the redshift range 1.78-3.03. They find typical metallicities of between 0.1 and 0.01 of solar abundance and infer dust-to-gas ratio of about 1/10th of that in the Milky Way, consistent with estimates based on the the reddening of quasars behind damped Lyα systems (Pei, Fall & Bechtold 1991). Pettini *et al.* find some evidence of a trend between metallicity and redshift, with the higher metallicity systems lying at the low-end of the redshift range, but they also find a large scatter in the metallicities suggesting that we are seeing a very non-uniform class of objects. These observations indicate a potential problem with the idea that the damped Lyα systems are the progenitors of normal spiral discs. If this were the case, it is natural to interpret the decline in the H I density in damped systems with decreasing redshift as evidence that gas is being turned into stars. However, simple 'closed box' models of chemical evolution predict metallicities of $\sim 0.35 Z_\odot$ when half the gas is consumed, much higher than Pettini *et al.* observe (Lanzetta, Wolfe & Turnshek 1995). Another way of phrasing this problem is by comparing the metallicities of damped Lyα systems with the metallicities of old disk stars in our Galaxy (Pettini *et al.* 1995) which rarely have metallicities less than $0.1 Z_\odot$. Evidently, the picture of chemical evolution inferred from the damped Lyα systems is very different from that of disc stars in our Galaxy. An attractive solution to this problem has been proposed by Fall and Pei (1995), who show that dust obscuration leads to significant biases against finding high-metallicity damped Lyα systems in optically selected quasar samples. Fall and Pei construct consistent models of the chemical evolution of these systems which require a higher density of neutral hydrogen than inferred directly from the observations (see the dashed line in Figure 3), thus exacerbating the discrepancy with the HI content in normal spiral discs. In summary, it seems to me plausible that the damped Lyα systems span a wide range of galaxy types and masses and that we are catching most of them at a special stage in their evolution, before they have converted a significant fraction of their gas into stars.

Perhaps surprisingly, damped Lyα systems have been detected in CO emission (Brown & Vanden Bout 1993; Frayer, Brown & Vanden Bout 1994). The conversion of CO line luminosities to H_2 gas mass is extremely uncertain, but if the clouds in the damped Lyα systems are similar to those in our Galaxy, the inferred gas masses are high $\sim 10^{12} h^{-2} M_\odot$. The large beams are an additional complication and it is possible that the gas is shared amongst a number of galaxies within the field. Nevertheless, these results are intriguing and, if the inferred gas masses are correct, suggest large galaxies in which molecular gas is a major constituent.

Finally in this section, we mention the important survey of Mg II absorber galaxies by Steidel, Dickinson & Persson (1995a). which spans the redshift range 0.2-1. They find no evidence for any change in the characteristic B or K band absolute magnitude with redshift and a conspicuous lack of low luminosity galaxies in the B-band. Evidently, the Mg II selection biases against the faint blue galaxies seen in field surveys, but picks up normal galaxies of all spectroscopic types.

4 Speculations

What are we to make of these results? The deep redshift and Mg II absorber surveys suggest that most normal L^* galaxies were already in place at $z \sim 1$. The red galaxies seem to be consistent with passive or mild evolution, implying a much earlier formation epoch. Normal disc systems appear compatible with a roughly constant star formation rate since $z \sim 1$. However, in addition to ordinary spirals and ellipticals, there is a rapidly evolving blue population that may well be confusing our attempts to unravel the formation histories of normal galaxies. I have argued that modest rates of star formation in low mass haloes might explain these galaxies. Such a model can account for the the high space density and weak clustering of the blue galaxy population (Efstathiou 1995) and seems an inevitable consequence of hierarchical clustering. For similar reasons, I suspect that most of the damped Lyα systems at high redshift will turn out to be dwarf galaxies, caught before they have expelled most of their gas via supernova driven winds. The discs of normal $\sim L^*$ spirals must be present amongst the damped Lyα systems, and may perhaps be found by imaging the more metal rich systems at lower redshifts.

Although the faint blue galaxy population may contribute a significant fraction of the total metallicity production in the Universe[3] (Cowie, 1988) my hunch is that they are largely irrelevant to the formation of normal $\sim L^*$ galaxies. We therefore seem to be in an interesting situation where we have irrefutable evidence for evolution at redshifts $z \lesssim 1$, but have yet to identify normal galaxies that are obviously in the process of formation. Quasar evolution provides indirect evidence that galaxy formation peaks at a redshift $z \sim 3$, though there are so many uncertainties in estimating the efficiency of massive black hole formation that it is difficult to turn this into a quantitative argument (see Haehnelt & Rees 1993, for a comprehensive discussion).

Galaxy formation may well be a gentle process (Baron & White 1987, *cf* Figure 1), in which case protogalaxies at $z \gtrsim 2$ will be hard to find. Perhaps the most promising ways forward are to extend the absorption line selected surveys to higher redshifts and to search for protogalaxies at far-IR and sub-mm wavelengths (see *e.g.* Franceschini *et al.* 1994).

Acknowledgements: I thank PPARC for the award of a Senior Fellowship, Mark Lacy for comments on some parts of the manuscript and various LT projects from which I have learned a lot.

[3] This apparent coincidence may be related to that fact that haloes in CDM-like models contribute about the same mass density in each logarithmic mass interval.

References

Babul, A., Rees, M.J. 1992, *Mon. Not. R. Astr. Soc.* **255**, 346

Baron, E. & White, S.D.M. 1987, *Astrophys. J.* **322**, 585

Briggs, F.H., Wolfe, A.M., Liszt, H.S., Davis, M.M., Turner, K.L. 1989, *Astrophys. J.* **341**, 650

Broadhurst, T.J., Ellis, R.S., Glazebrook, K. 1992, *Nature* **355**, 55

Brown, R.L., Vanden Bout, P.A. 1993, *Astrophys. J. Lett.* **412**, L21

Bunker, A.J., Warren, S.J., Hewett, P.C., Clements, D.L. 1995, *Mon. Not. R. Astr. Soc.* **273**, 513

Castander, F.F., Ellis, R.S., Frenk, C.S., Dressler, A., Gunn, J.E. 1994, *Astrophys. J. Lett.* **424**, L79

Cen, R., Miralda-Escudé, J., Ostriker, J.P., Rauch, M. 1994, *Astrophys. J. Lett.* **437**, L9

Charlot, S., Fall, S.M. 1991, *Astrophys. J.* **378**, 471

Colless, M.M. 1995, in *Wide Field Spectroscopy and the Distant Universe*, eds S.J. Maddox and A. Alfonso-Aragon, World Scientific Publishing Co., in press

Colless, M.M., Ellis, R.S., Broadhurst, T.J., Taylor, K. & Peterson, B.A. 1993, *Mon. Not. R. Astr. Soc.* **261**, 19

Couch, W.J., Ellis, R.S., Malin, D.F., MacLaren, I. 1990, *Mon. Not. R. Astr. Soc.* **249**, 606

Couchman, H.M.P., Rees, M.J. 1986, *Mon. Not. R. Astr. Soc.* **221**, 53

Cowie, L.L. 1988, in *The Post Recombination Universe*, eds N. Kaiser, A. Lasenby, Kluwer, Dordrecht, p.1

Cowie, L.L., Lilly, S.J., Gardner, J.P., McLean, I.S. 1988, *Astrophys. J. Lett.*, **332** L29

Cowie, L.L., Songalia, A., Hu, E.M. 1995, *Astrophys. J.*, in press

Dickinson, M., *et al.* 1995, in preparation

Djorgovski, S., Thompson, D., Smith, J.D. 1993, in *First Light in the Universe*, eds B. Rocca-Volmerange, M. Dennefeld, B. Guiderdoni and J. Tran Thanh Van, Editions Frontières, Gif-sur-Yvette, p.67

Djorgovski, G. *et al.* 1995, *Astrophys. J. Lett.* **438**, L13

Efstathiou, G. 1992, *Mon. Not. R. Astr. Soc.* **256**, 43p

Efstathiou, G. 1995, *Mon. Not. R. Astr. Soc.* **272**, L25

Efstathiou, G.,Rees M.J. 1988, *Mon. Not. R. Astr. Soc.* **230**, 5p

Efstathiou, G., Bernstein, G., Katz, N., Tyson, J.A., Guhathakurta, P. 1991, *Astrophys. J. Lett.*, **380**, L47

Efstathiou, G., Bond, J.R., White, S.D.M. 1992, *Mon. Not. R. Astr. Soc.* **258**, 1p

Fall, S.M., Pei, Y.C., in *QSO Absorption Lines*, ed G. Meylan, Springer-Verlag, in press

Ferguson, H.C., McGaugh, S.S. 1995, *Astrophys. J.* **440**, 470

Franceschini, A., Mazzei, P., de Zotti, G., Danese, L. 1994, *Astrophys. J.* **427**, 140

Frenk, C.S., Ellis, R.S., Shanks, T., Heavens, A.F., Peacock, J.A. 1989, *The Epoch of Galaxy Formation*, Kluwer, Dordrecht

Frenk, C.S., White, S.D.M., Efstathiou, G., Davis, M. 1990, *Astrophys. J.* **351**, 10

Freyer, D.T., Brown, R.L., Vanden Bout, P.A. 1994, *Astrophys. J. Lett.* **433**, L5

Giavalisco, M., Steidel, C.C., Szalay, A.S. 1994, *Astrophys. J. Lett.* **425**, L5

Gioia, I.M., Henry, J.P., Maccacaro, T., Morris, S.L., Stocke J.T, Wolter A. 1990, *Astrophys. J. Lett.* 356, L35

Glazebrook, K., Ellis, R., Colless, M., Broadhurst, T., Allington-Smith,J., Tanvir, N. 1995a, *Mon. Not. R. Astr. Soc.* **273**, 157

Glazebrook, K., Peacock, J.A., Miller, L., Collins, C.A. 1995b, *Mon. Not. R. Astr. Soc.*, in press

Glazebrook, K, Ellis, R., Santiago, B., Griffiths, R. 1995c, *Mon. Not. R. Astr. Soc.*, in press

Gronwall, C., Koo, D.C. 1995, *Astrophys. J. Lett.* **440**, L1

Gunn J.E., Hoessel J., Oke J.B. 1986, *Astrophys. J.* **306**, 30

Haehnelt M.G., Rees M.J. 1993, *Mon. Not. R. Astr. Soc.* **263**, 168

Henry, J.P., Gioia, I.M., Maccacaro, T., Morris, S.L., Stocke, J.T., Wolter A. 1992, *Astrophys. J.* **386**, 408

Heyl, J.S., Cole, S., Frenk, C.S., Navarro, J.F. 1995, *Mon. Not. R. Astr. Soc.* **274**, 755

Kauffmann, G., Charlot, S. 1994, *Astrophys. J. Lett.* **430**, L97

Kennicutt, R.C. 1983, *Astrophys. J.* **272**, 54

Klypin, A, Holtzmann, J., Primack, J., Regos, E. 1993, *Astrophys. J.* **416**, 1

Lacey, C., Guiderdoni, B., Rocca-Volmerange, B., Silk, J. 1993, *Astrophys. J.* **402**, 15

Lacy, M. *et al.* 1994, *Mon. Not. R. Astr. Soc.* **271**, 504

Lacy, M., Rawlings S., Eales S., Dunlop J.S. 1995, *Mon. Not. R. Astr. Soc.* **273**, 821

Lanzetta, K.M., Turnshek, D.A., Wolfe, A.M. 1993, *Astrophys. J. Suppl.* **84**, 1

Lanzetta, K.M., Wolfe, A., Turnshek, D.A. 1995, *Astrophys. J.* **440**, 435

Le Fèvre, O., Crampton, D., Hammer, F., Lilly, S.J., Tresse, L. 1994, *Astrophys. J. Lett.* **423**, L89

Lilly, S.J., Longair, M. 1984, *Mon. Not. R. Astr. Soc.* **211**, 833

Lilly, S.J., Tresse, L., Hammer, F., Crampton, D., Le Fevre, O. 1995, preprint

Loveday, J., Peterson, B.A., Efstathiou, G., Maddox, S.J. 1992, *Astrophys. J.* **390**, 338

Lowenthal, J.D., Hogan, C.J., Green, R.F., Caulet, A., Woodgate, B.E., Brown, L., Foltz, C.B. 1991, *Astrophys. J. Lett.* **377**, L73

Marzke, R.O., Geller, M.J., Huchra, J.P., Corwin, H.G. 1995, *Astron. J.*, in press

McCarthy, P.J. 1993, *Ann. Rev. Astr. Astrophys.* **31**, 639

Metcalfe, N., Shanks, T., Fong, R., Roche, N. 1995, *Mon. Not. R. Astr. Soc.* **273**, 257

Miley, G.K., Chambers, K.C., van Breugel, W.J.M., Macchetto, F. 1992, *Astrophys. J. Lett.* **401**, L69

Miralda-Escudé, J., Rees, M.J. 1994, *Mon. Not. R. Astr. Soc.* **266** 343

Mo, H.J., Miralda-Escudé, J, Rees, M.J. 1993, *Mon. Not. R. Astr. Soc.* **264**, 705

Mo, H.J., Miralda-Escudé, J. 1994, *Astrophys. J. Lett.* **430**, L25

Møller, P., Warren, S.J. 1993, *Astron. Astrophys* **270**, 43

Neuschaefer, L., Windhorst, R.A. 1995, *Astrophys. J.*, in press

Neuschaefer, L., Windhorst, R.A., Dressler A. 1991, *Astrophys. J.* **382**, 32

Pei, Y.C., Fall, S.M., Bechtold J. 1991, *Astrophys. J.* **378**, 6

Pettini, M., King, D.L., Smith L.J., Hunstead, R. 1995, preprint

Press, W.H., Schechter, P. 1974, *Astrophys. J.* **187**, 425

Rees, M.J. 1986, *Mon. Not. R. Astr. Soc.* **218**, 25p

Rawlings, S., Lacy, M., Sivia, D.S., Eales, S.A. 1995, *Mon. Not. R. Astr. Soc.* **274**, 428

Shapiro, P.R., Giroux, M.L., Babul, A. 1994, *Astrophys. J.* **427**, 25

Smail I., Ellis, R.S., Fitchett, M.J. 1994, *Mon. Not. R. Astr. Soc.* **270**, 245

Soucail, G. 1992, in *Clusters and Superclusters of Galaxies*, ed Fabian A.C., Kluwer, Dordrecht, p.199

Steidel, C.C., Hamilton, D. 1992, *Astron. J.* **104**, 3

Steidel, C.C., Pettini, M., Dickinson, M., Persson, S.E. 1994, *Astron. J.* **108** 2046

Steidel, C.C., Dickinson, M., Persson, S.E. 1995a, *Astrophys. J.*, in press

Steidel, C.C., Bowen, D.V., Blades, J.C., Dickinson, M. 1995b, *Astrophys. J. Lett.* **440**, L45

Stockton, A., Kellog, M., Ridgway, S.E. 1995, *Astrophys. J. Lett.* **443**, L69

Strauss, M.A., Willick, J.A. 1995, *Physics Reports*, in press

Thompson, D., Djorgovski, S., Beckwith, S.V.W. 1994, *Astron. J.* **107**, 1

Tyson, J.A. 1988, *Astron. J.* **96**, 1

Warren, S.J., Møller, P. 1995, *Astron. Astrophys*, submitted

Walker, T., Steigman, G., Schramm, D.N., Olive, K., Kang, H.S. 1991, *Astrophys. J.* **376**, 51

White, S.D.M., Frenk, C.S. 1991, *Astrophys. J.* **379**, 25

Springer-Verlag and the Environment

We at Springer-Verlag firmly believe that an international science publisher has a special obligation to the environment, and our corporate policies consistently reflect this conviction.

We also expect our business partners – paper mills, printers, packaging manufacturers, etc. – to commit themselves to using environmentally friendly materials and production processes.

The paper in this book is made from low- or no-chlorine pulp and is acid free, in conformance with international standards for paper permanency.

Lecture Notes in Physics

For information about Vols. 1–425
please contact your bookseller or Springer-Verlag

Vol. 426: L. Mathelitsch, W. Plessas (Eds.), Substructures of Matter as Revealed with Electroweak Probes. Proceedings, 1993. XIV, 441 pages. 1994

Vol. 427: H. V. von Geramb (Ed.), Quantum Inversion Theory and Applications. Proceedings, 1993. VIII, 481 pages. 1994.

Vol. 428: U. G. Jørgensen (Ed.), Molecules in the Stellar Environment. Proceedings, 1993. VIII, 440 pages. 1994.

Vol. 429: J. L. Sanz, E. Martínez-González, L. Cayón (Eds.), Present and Future of the Cosmic Microwave Background. Proceedings, 1993. VIII, 233 pages. 1994.

Vol. 430: V. G. Gurzadyan, D. Pfenniger (Eds.), Ergodic Concepts in Stellar Dynamics. Proceedings, 1993. XVI, 302 pages. 1994.

Vol. 431: T. P. Ray, S. Beckwith (Eds.), Star Formation and Techniques in Infrared and mm-Wave Astronomy. Proceedings, 1992. XIV, 314 pages. 1994.

Vol. 432: G. Belvedere, M. Rodonò, G. M. Simnett (Eds.), Advances in Solar Physics. Proceedings, 1993. XVII, 335 pages. 1994.

Vol. 433: G. Contopoulos, N. Spyrou, L. Vlahos (Eds.), Galactic Dynamics and N-Body Simulations. Proceedings, 1993. XIV, 417 pages. 1994.

Vol. 434: J. Ehlers, H. Friedrich (Eds.), Canonical Gravity: From Classical to Quantum. Proceedings, 1993. X, 267 pages. 1994.

Vol. 435: E. Maruyama, H. Watanabe (Eds.), Physics and Industry. Proceedings, 1993. VII, 108 pages. 1994.

Vol. 436: A. Alekseev, A. Hietamäki, K. Huitu, A. Morozov, A. Niemi (Eds.), Integrable Models and Strings. Proceedings, 1993. VII, 280 pages. 1994.

Vol. 437: K. K. Bardhan, B. K. Chakrabarti, A. Hansen (Eds.), Non-Linearity and Breakdown in Soft Condensed Matter. Proceedings, 1993. XI, 340 pages. 1994.

Vol. 438: A. Pękalski (Ed.), Diffusion Processes: Experiment, Theory, Simulations. Proceedings, 1994. VIII, 312 pages. 1994.

Vol. 439: T. L. Wilson, K. J. Johnston (Eds.), The Structure and Content of Molecular Clouds. 25 Years of Molecular Radioastronomy. Proceedings, 1993. XIII, 308 pages. 1994.

Vol. 440: H. Latal, W. Schweiger (Eds.), Matter Under Extreme Conditions. Proceedings, 1994. IX, 243 pages. 1994.

Vol. 441: J. M. Arias, M. I. Gallardo, M. Lozano (Eds.), Response of the Nuclear System to External Forces. Proceedings, 1994, VIII. 293 pages. 1995.

Vol. 442: P. A. Bois, E. Dériat, R. Gatignol, A. Rigolot (Eds.), Asymptotic Modelling in Fluid Mechanics. Proceedings, 1994. XII, 307 pages. 1995.

Vol. 443: D. Koester, K. Werner (Eds.), White Dwarfs. Proceedings, 1994. XII, 348 pages. 1995.

Vol. 444: A. O. Benz, A. Krüger (Eds.), Coronal Magnetic Energy Releases. Proceedings, 1994. X, 293 pages. 1995.

Vol. 445: J. Brey, J. Marro, J. M. Rubí, M. San Miguel (Eds.), 25 Years of Non-Equilibrium Statistical Mechanics. Proceedings, 1994. XVII, 387 pages. 1995.

Vol. 446: V. Rivasseau (Ed.), Constructive Physics. Results in Field Theory, Statistical Mechanics and Condensed Matter Physics. Proceedings, 1994. X, 337 pages. 1995.

Vol. 447: G. Aktaş, C. Saçlıoğlu, M. Serdaroğlu (Eds.), Strings and Symmetries. Proceedings, 1994. XIV, 389 pages. 1995.

Vol. 448: P. L. Garrido, J. Marro (Eds.), Third Granada Lectures in Computational Physics. Proceedings, 1994. XIV, 346 pages. 1995.

Vol. 449: J. Buckmaster, T. Takeno (Eds.), Modeling in Combustion Science. Proceedings, 1994. X, 369 pages. 1995.

Vol. 450: M. F. Shlesinger, G. M. Zaslavsky, U. Frisch (Eds.), Lévy Flights and Related Topics in Physics. Proceedigs, 1994. XIV, 347 pages. 1995.

Vol. 451: P. Krée, W. Wedig (Eds.), Probabilistic Methods in Applied Physics. IX, 393 pages. 1995.

Vol. 452: A. M. Bernstein, B. R. Holstein (Eds.), Chiral Dynamics: Theory and Experiment. Proceedings, 1994. VIII, 351 pages. 1995.

Vol. 453: S. M. Deshpande, S. S. Desai, R. Narasimha (Eds.), Fourteenth International Conference on Numerical Methods in Fluid Dynamics. Proceedings, 1994. XIII, 589 pages. 1995.

Vol. 454: J. Greiner, H. W. Duerbeck, R. E. Gershberg (Eds.), Flares and Flashes, Germany 1994. XXII, 477 pages. 1995.

Vol. 455: F. Occhionero (Ed.), Birth of the Universe and Fundamental Physics. Proceedings, 1994. XV, 387 pages. 1995.

Vol. 456: H. B. Geyer (Ed.), Field Theory, Topology and Condensed Matter Physics. Proceedings, 1994. XII, 206 pages. 1995.

Vol. 457: P. Garbaczewski, M. Wolf, A. Weron (Eds.), Chaos – The Interplay Between Stochastic and Deterministic Behaviour. Proceedings, 1995. XII, 573 pages. 1995.

Vol. 458: I. W. Roxburgh, J.-L. Masnou (Eds.), Physical Processes in Astrophysics. Proceedings, 1993. XII, 249 pages. 1995.

Vol. 459: G. Winnewisser, G. C. Pelz (Eds.), The Physics and Chemistry of Interstellar Molecular Clouds. Proceedings, 1993. XV, 393 pages. 1996.

Vol. 461: R. López-Peña, R. Capovilla, R. García-Pelayo, H. Waelbroeck, F. Zertuche, (Eds.), Complex Systems and Binary Networks. Lectures, México 1995. X, 223 pages. 1995.

Vol. 463: H. Hippelein, K. Meisenheimer, H.-J. Röser (Eds.), Galaxies in the Young Universe. Proceedings, 1994. XV, 314 pages. 1995.

New Series m: Monographs

Vol. m 1: H. Hora, Plasmas at High Temperature and Density. VIII, 442 pages. 1991.

Vol. m 2: P. Busch, P. J. Lahti, P. Mittelstaedt, The Quantum Theory of Measurement. XIII, 165 pages. 1991.

Vol. m 3: A. Heck, J. M. Perdang (Eds.), Applying Fractals in Astronomy. IX, 210 pages. 1991.

Vol. m 4: R. K. Zeytounian, Mécanique des fluides fondamentale. XV, 615 pages, 1991.

Vol. m 5: R. K. Zeytounian, Meteorological Fluid Dynamics. XI, 346 pages. 1991.

Vol. m 6: N. M. J. Woodhouse, Special Relativity. VIII, 86 pages. 1992.

Vol. m 7: G. Morandi, The Role of Topology in Classical and Quantum Physics. XIII, 239 pages. 1992.

Vol. m 8: D. Funaro, Polynomial Approximation of Differential Equations. X, 305 pages. 1992.

Vol. m 9: M. Namiki, Stochastic Quantization. X, 217 pages. 1992.

Vol. m 10: J. Hoppe, Lectures on Integrable Systems. VII, 111 pages. 1992.

Vol. m 11: A. D. Yaghjian, Relativistic Dynamics of a Charged Sphere. XII, 115 pages. 1992.

Vol. m 12: G. Esposito, Quantum Gravity, Quantum Cosmology and Lorentzian Geometries. Second Corrected and Enlarged Edition. XVIII, 349 pages. 1994.

Vol. m 13: M. Klein, A. Knauf, Classical Planar Scattering by Coulombic Potentials. V, 142 pages. 1992.

Vol. m 14: A. Lerda, Anyons. XI, 138 pages. 1992.

Vol. m 15: N. Peters, B. Rogg (Eds.), Reduced Kinetic Mechanisms for Applications in Combustion Systems. X, 360 pages. 1993.

Vol. m 16: P. Christe, M. Henkel, Introduction to Conformal Invariance and Its Applications to Critical Phenomena. XV, 260 pages. 1993.

Vol. m 17: M. Schoen, Computer Simulation of Condensed Phases in Complex Geometries. X, 136 pages. 1993.

Vol. m 18: H. Carmichael, An Open Systems Approach to Quantum Optics. X, 179 pages. 1993.

Vol. m 19: S. D. Bogan, M. K. Hinders, Interface Effects in Elastic Wave Scattering. XII, 182 pages. 1994.

Vol. m 20: E. Abdalla, M. C. B. Abdalla, D. Dalmazi, A. Zadra, 2D-Gravity in Non-Critical Strings. IX, 319 pages. 1994.

Vol. m 21: G. P. Berman, E. N. Bulgakov, D. D. Holm, Crossover-Time in Quantum Boson and Spin Systems. XI, 268 pages. 1994.

Vol. m 22: M.-O. Hongler, Chaotic and Stochastic Behaviour in Automatic Production Lines. V, 85 pages. 1994.

Vol. m 23: V. S. Viswanath, G. Müller, The Recursion Method. X, 259 pages. 1994.

Vol. m 24: A. Ern, V. Giovangigli, Multicomponent Transport Algorithms. XIV, 427 pages. 1994.

Vol. m 25: A. V. Bogdanov, G. V. Dubrovskiy, M. P. Krutikov, D. V. Kulginov, V. M. Strelchenya, Interaction of Gases with Surfaces. XIV, 132 pages. 1995.

Vol. m 26: M. Dineykhan, G. V. Efimov, G. Ganbold, S. N. Nedelko, Oscillator Representation in Quantum Physics. IX, 279 pages. 1995.

Vol. m 27: J. T. Ottesen, Infinite Dimensional Groups and Algebras in Quantum Physics. IX, 218 pages. 1995.

Vol. m 28: O. Piguet, S. P. Sorella, Algebraic Renormalization. IX, 134 pages. 1995.

Vol. m 29: C. Bendjaballah, Introduction to Photon Communication. VII, 193 pages. 1995.

Vol. m 30: A. J. Greer, W. J. Kossler, Low Magnetic Fields in Anisotropic Superconductors. VII, 161 pages. 1995.

Vol. m 31: P. Busch, M. Grabowski, P. J. Lahti, Operational Quantum Physics. XI, 230 pages. 1995.

Vol. m 32: L. de Broglie, Diverses questions de mécanique et de thermodynamique classiques et relativistes. XII, 198 pages. 1995.

Vol. m 33: R. Alkofer, H. Reinhardt, Chiral Quark Dynamics. VIII, 115 pages. 1995.

Vol. m 34: R. Jost, Das Märchen vom Elfenbeinernen Turm. VIII, 286 pages. 1995.

Vol. m 35: E. Elizalde, Ten Physical Applications of Spectral Zeta Functions. XIV, 228 pages. 1995.

Vol. m 36: G. Dunne, Self-Dual Chern-Simons Theories. X, 217 pages. 1995.

Vol. m 37: S. Childress, A.D. Gilbert, Stretch, Twist, Fold: The Fast Dynamo. XI, 410 pages. 1995.